Polymers as Aids
in Organic Chemistry

Polymers as Aids in Organic Chemistry

N. K. MATHUR
C. K. ÑARANG
Department of Chemistry
University of Jodhpur
Jodhpur, India

R. E. WILLIAMS
Division of Biological Sciences
National Research Council of Canada
Ottawa, Ontario, Canada

1980

ACADEMIC PRESS
A Subsidiary of Harcourt Brace Jovanovich, Publishers
New York   London   Toronto   Sydney   San Francisco

COPYRIGHT © 1980, BY ACADEMIC PRESS, INC.
ALL RIGHTS RESERVED.
NO PART OF THIS PUBLICATION MAY BE REPRODUCED OR
TRANSMITTED IN ANY FORM OR BY ANY MEANS, ELECTRONIC
OR MECHANICAL, INCLUDING PHOTOCOPY, RECORDING, OR ANY
INFORMATION STORAGE AND RETRIEVAL SYSTEM, WITHOUT
PERMISSION IN WRITING FROM THE PUBLISHER.

ACADEMIC PRESS, INC.
111 Fifth Avenue, New York, New York 10003

United Kingdom Edition published by
ACADEMIC PRESS, INC. (LONDON) LTD.
24/28 Oval Road, London NW1 7DX

Library of Congress Cataloging in Publication Data

Mathur, N. K.
  Polymers as aids in organic chemistry.

  Includes bibliographies and index.
  1. Polymers and polymerization. 2. Chemical tests
and reagents. 3. Chemistry, Organic. I. Narang,
C. K., joint author. II. Williams, R. E., joint author.
III. Title.
QD381.8.M37   547'.1'39   79–52789
ISBN 0–12–479850–0

PRINTED IN THE UNITED STATES OF AMERICA

80 81 82 83   9 8 7 6 5 4 3 2 1

Contents

Preface   xi

1. Introduction
I.	History	1
II.	Development of Polymer Science and Technology	2
III.	Definition and Classification of Polymers	2
IV.	Preparation of Synthetic Polymers	4
V.	Properties of Polymers	5
VI.	Synthesis of Functionalized Polymers	6
VII.	Types of Functionalized Polymers	7
VIII.	General Chemical Reactions of Polymers	7
IX.	Polymers as Aids in Organic Synthesis	8
X.	Kinetics of Polymer-Analogous Reactions	9
XI.	Literature on Solid-Phase Synthesis	12
	References	12

2. Polymeric Support Materials
I.	Introduction	14
II.	Styrene-Based Polymers	15
III.	Functionalization of Styrene-Based Polymers via Chloromethylation and Other Methods	18
IV.	Miscellaneous Polymer Supports	25
	References	32

3. Determination of Functionalization in Polymer Supports
I.	Introduction	37
II.	Chemical Methods for Functional Group Analysis of Polymers	38
III.	Physical and Physicochemical Methods of Determining Functional Groups in Polymers	41
IV.	Physical and Chemical Nature of Immobilization of Reactive Sites on Polymers	46

V.	Use of Radiolabeled Reagents to Follow the Changes in Resin Functionalities	49
VI.	Reporting of Results	50
	References	51

4. Polypeptide Synthesis on Polymer Supports

I.	Introduction and History	53
II.	Basic Principles of Merrifield's Solid-Phase Peptide Synthesis	55
III.	Supports for Solid-Phase Peptide Synthesis	56
IV.	Linkage of the First Amino Acid to the Polymer	62
V.	Protecting Groups Used in Solid-Phase Peptide Synthesis	63
VI.	Coupling of Successive Amino Acids to Resin-Bound Amino Acids	64
VII.	Cleavage of the Resin–Peptide Bond	65
VIII.	Monitoring of Solid-Phase Peptide Synthesis	67
IX.	Automation in Solid-Phase Peptide Synthesis	69
X.	Racemization Problems in Solid-Phase Peptide Products	70
XI.	Purification of Solid-Phase Peptide Products	70
XII.	Problems in Solid-Phase Synthesis	71
XIII.	Solid-Phase Coupling of N-Carboxylanhydrides (NCA)	73
XIV.	Solid-Phase Synthesis Using Side-Chain Functionalities for Attachment to Polymers and Bidirectional Extension of Peptide Chains	73
XV.	Fragment Condensation Strategy in Solid-Phase Peptide Synthesis	74
	References	75

5. Oligonucleotide Synthesis on Polymer Supports

I.	Introduction	81
II.	General Principles of Solid-Phase Oligonucleotide Synthesis	84
III.	Polymer Supports	85
IV.	Functionalization of Polymer Supports	95
V.	Strategies Used for Oligonucleotide Synthesis on Polymer Supports	95
VI.	Protection of Reactive Groups	96
VII.	Cleavage of the Protecting Groups	97
VIII.	Attachment of the Polymeric Carrier to the Nucleotide or Nucleoside	97
IX.	Elongation of the Nucleotide Chain on the Polymer Support	98
X.	Cleavage of the Polymer–Nucleoside/Nucleotide Bond	98
XI.	Monitoring in Polymer-Supported Oligonucleotide Synthesis	100
XII.	Purification of Synthetic Oligonucleotides	100
XIII.	Synthesis of Oligoribonucleotides on Polymer Supports	100
XIV.	Miscellaneous Application of Polymers in Polynucleotide Synthesis	101
XV.	Advantages and Limitations	101
	References	102

6. Oligosaccharide Synthesis on Polymer Supports

I.	Introduction	105
II.	Basic Principles of Polymer-Supported Oligosaccharide Synthesis	107
III.	Polymer Supports for the Synthesis of Oligosaccharides	108
IV.	Linkage of the First Sugar Molecule to the Polymer Support and Product Removal	110
V.	Protecting Groups in Oligosaccharide Synthesis Employing Polymer Supports	111
VI.	Mechanism and Steric Control in Successive Coupling of Monosaccharide Residues on a Polymer Support	112
VII.	Miscellaneous Applications of Polymers in the Carbohydrate Field	113
VIII.	Monitoring of Solid-Phase Oligosaccharide Synthesis	114
IX.	Advantages, Limitations, and Future Scope of the Use of Polymer Supports in Polysaccharide Synthesis	114
	References	116

7. Peptide Synthesis Using Polymeric Active Esters

I.	Introduction	117
II.	Principles of Peptide Synthesis Using Polymeric Active Esters	117
III.	Polymeric Active Esters Used for Peptide Synthesis	118
IV.	Synthesis of Cyclic Peptides Using Polymeric Active Esters	119
V.	Scope and Limitations of the Polymeric Active Ester Method for Peptide Synthesis	122
	References	123

8. Solid-Phase Sequencing of Peptides and Proteins

I.	Introduction	125
II.	Solid-Phase Edman Degradation Using Polymeric Reagents	126
III.	Solid-Phase Degradation Employing Polymer-Bound Peptides	127
IV.	Other Polymer Supports for Solid-Phase Sequencing	129
V.	Attachment of the Peptide to the Polymer Support	129
VI.	Automation in Solid-Phase Sequencing	133
VII.	Solid-Phase Sequencing of Peptides from the Carboxyl Terminus	133
VIII.	Scope and Limitations of Solid-Phase Sequencing Methods	134
	References	136

9. Polymeric Supports in General Organic Chemistry

I.	Introduction	138
II.	Alkylation and Acylation of Esters Using Functionalized Carriers	139
III.	Dieckmann Cyclization of Polymer-Bound Esters	143
IV.	Cyclization of Large Ring Compounds on Polymeric Supports	145
V.	Monofunctionalization of Polymer-Bound Compounds	146
VI.	Synthesis of Threaded Macrocyclic Systems (Hooplanes)	152
VII.	Photochemical Applications	153
	References	153

10. Polymer-Supported Asymmetric Synthesis and Resolution of Racemates Using Asymmetric Polymeric Materials

    I. Introduction   155
    II. Asymmetric Syntheses of Polymeric Supports   156
    III. Resolution of Racemates Using Polymeric Materials   158
    References   162

11. Application of Polymeric Supports in Identifying Reaction Intermediates

    I. Introduction   165
    II. General Strategy Used for Trapping Reaction Intermediates Employing Polymeric Supports   166
    References   172

12. Polymer-Bound Reagents

    I. Introduction   174
    II. Polymeric Oxidizing Reagents   175
    III. Polymeric Oxidation–Reduction Reagents   180
    IV. Polymeric Reducing Reagents   181
    V. Polymeric Group Transfer Reagents   183
    VI. Polymeric Coupling Agents   191
    VII. Miscellaneous Reagents   194
    References   195

13. Polymer-Bound Catalysts (I)

    I. Introduction   198
    II. Ion-Exchange Resins as General Acid–Base Catalysts   199
    III. Polystyrene–Aluminum Chloride as a Lewis Acid Catalyst   205
    IV. Polymer-Based "Super Acid" Catalysts   205
    V. Polymeric Esterolytic Catalysts   206
    VI. Polymer-Supported Phase-Transfer Catalysts   209
    VII. Polymeric Triphase Catalysts   212
    VIII. Polymer-Based Photosensitizers   213
    References   216

14. Polymer-Bound Catalysts (II) Transition Metal Complexes Bound to Polymers

    I. Polymer-Supported Transition Metal Catalysts   220
    II. Principles of Homogeneous Transition Metal Complex Catalysts   221
    III. Preparation of Polymer-Bound Transition Metal Complexes   222
    IV. Structure of Polymeric Catalysts   228

V.	Types of Reactions Catalyzed by Polymer-Anchored Catalysts	229
VI.	Asymmetric Organic Synthesis via Transition Metal Catalysts Bound to Polymeric Chiral Ligands	233
	References	236

15. Polymers as Aids in Related Areas of Chemistry

I.	Introduction	239
II.	Applications in Analytical Chemistry	239
III.	Polymer-Bound Agriculturally and Pharmacologically Active Agents	242
IV.	Applications in Biochemistry	242
	References	252

Index   255

Preface

Many areas of scientific endeavor have felt the effect of the utilization of polymeric materials. Organic chemistry is no exception. Originally used as catalysts, organic and inorganic polymeric materials are now used to support molecules during their transformation and to support reagents that must be easily separated from the final product. The impetus to research in the latter two areas is provided by the ease with which the products or reagent molecules may be recovered after reaction.

During the past 15 years a rapid increase in the knowledge pertaining to the use of polymeric materials in organic chemistry has been accompanied, as usual, by a rapidly increasing, vast and quite extensive literature. Many areas of organic chemistry have been touched in the process. It is the purpose of this volume to indicate the wide-ranging influence the use of polymeric materials has had. In order to do this we have had to classify the uses of polymeric materials under the general headings: supports, reagents, and catalysts. To illustrate the uses to which the polymeric materials have been put in each category we have used a limited number of examples from the literature. The reader wishing more information has been referred to pertinent reviews that cover many of the aspects in greater depth than is possible here. Where it was felt to be necessary, i.e., where adequate reviews did not exist or where a large number of more recent examples have appeared since the last review of the area, the literature has been covered more extensively and attempts have been made to bring coverage up to date. In this regard the pertinent literature was searched until mid-1979. Even though we have tried to cover the literature fully we may have inadvertently neglected to include some references. For these omissions and any other errors we apologize.

This volume has been set up to reflect the broad classifications mentioned earlier. A brief introduction to polymer chemistry is followed by a tabulation of the various types of polymers that have been used and the methods for their characterization. Thereafter, sections follow that touch on the use of polymers as supports. Examples are given where polymers have been used as supports in peptide chemistry, in oligonucleotide chemistry, less extensively in oligosaccharide chemistry, in peptide sequencing, in the preparation of monofunctionalized difunctional compounds, as aids in asymmetric syntheses, and as trapping agents in the determination of reaction intermediates. In the next section the use of polymers either to support reagents or be reagents themselves is considered. Many areas of chemistry are touched and include peptide chemistry, oxidation and reduction reactions, and nucleophilic displacement reactions. In the subsequent sections the use of polymers as catalysts is described. In most instances the polymer has been derivatized to carry the catalytic functionality. One of the most extensive areas in this regard has been that in which transition metals have been immobilized in the polymer matrix and used as catalysts. Finally, the last chapter deals with an often neglected area of organic chemistry. Polymer-immobilized compounds, enzymes, and whole cells have been used to carry out a large number of reactions, most of which impinge on the area of organic chemistry.

Help with the preparation of this volume was welcomed and the contributions of the following are gratefully acknowledged: Drs. K. K. Banerji and C. R. Menon for reading some parts of the book; Drs. A. Patchornik and K. Brunfeldt for supplying us with preprints and reprints of their articles; Dr. K. E. Norris for his help in compiling the chapter on oligonucleotide synthesis; Drs. M. K. Sahni and K. C. Gupta for providing some of the drawings and schemes; the drafting staff of the National Research Council (Miss C. Clyde and Mrs. D. H. Ladouceur) for their efforts in preparing figures and schemes in their final form; Mrs. S. L. Khatri, Mrs. H. Letaif, Mrs. M. Nadon, and Miss M. Manson for typing the manuscript; and finally Ms. S. Kielland for her efforts in reading and checking the completed manuscript.

N. K. Mathur
C. K. Narang
R. E. Williams

1 Introduction

I.	History	1
II.	Development of Polymer Science and Technology	2
III.	Definition and Classification of Polymers	2
IV.	Preparation of Synthetic Polymers	4
V.	Properties of Polymers	5
	A. Bonding Forces in Polymers	5
	B. Crystallinity	5
	C. Steric Configuration	5
	D. First- and Second-Order Transition Temperatures	6
	E. Miscibility and Solubility	6
	F. Solution Properties of Polymers	6
VI.	Synthesis of Functionalized Polymers	6
VII.	Types of Functionalized Polymers	7
VIII.	General Chemical Reactions of Polymers	7
IX.	Polymers as Aids in Organic Synthesis	8
X.	Kinetics of Polymer-Analogous Reactions	9
XI.	Literature on Solid-Phase Synthesis	12
	References	12

I. HISTORY

High-molecular-weight compounds are among the most common naturally occuring substances. Indeed, some of them form the very basis of animate nature. Naturally occurring polymers have been utilized throughout the ages. Commercial utilization of modified natural polymers began quite early in the last century. For example, several derivatives of natural polymers were discovered (e.g., cellulose nitrate in 1838 and cellulose acetate in 1870) and used much earlier than the beginning of the systematic development of purely synthetic polymeric materials. Although styrene was polymerized as early as 1839, isoprene in 1879, and methacrylic acid in 1880, and although the oldest of the purely synthetic plastics, phenol–formaldehyde resin (Bakeland's Bakelite), was produced on a small commercial scale as early as 1907, it was not until the 1930s that the science of high polymers truly began.

II. DEVELOPMENT OF POLYMER SCIENCE AND TECHNOLOGY

About a century ago, when the unique properties of natural polymers were recognized, the term "colloid" was proposed to distinguish them from materials that could be obtained in crystalline form. It was soon recognized that certain crystalline substances could be transformed into colloids and the concept of a "colloidal state of matter" was developed. "Collodial particles" were considered to be built up of a large number of small molecules, by physical association. This concept, which was extended to cover the naturally occurring polymers, was to a very great extent responsible for the delay in the development of a polymer science.

The acceptance of macromolecular theory came in 1920, largely due to the research of Herman Staudinger. Even then, the existence of macromolecules was questioned by contemporary chemists who doubted the presence of end groups in such molecules. Since the chemical methods of those days were not able to detect the end groups, Staudinger suggested that no end groups were needed to saturate the terminal valencies of the long chains and they were considered to be "unreactive" simply because of the large size of the molecules. As an alternative explanation, the concept of large ring structures was also put forward. It became clear later, as the chemical methods for end group determination were studied and developed, that the ends of long-chain molecules consist of normal, valence-satisfied structures.

Early industrial developments in the field of polymer science and technology were concerned with the modification and utilization of natural polymers. The commercial production of purely synthetic polymers was started in the early 1900s, when some commercially important polymers were prepared. It was the late 1930s and the beginning of the Second World War that saw the development of all but a handful of the wide variety of synthetic polymers now in commercial use.

Subsequent developments in polymer science are so diverse as to be beyond the scope of this book and are accessible through several monographs and edited works (Mark *et al.*, 1940; Mark *et al.*, 1964; Flory, 1953; Huggins, 1958; Fettes, 1964; Miller, 1966; Ravve, 1967; Billmeyer, 1971).

III. DEFINITION AND CLASSIFICATION OF POLYMERS

A polymer is a giant molecule built up of relatively small, chemically bonded, repeating units. The molecular weight of such molecules may run from very low values into the millions, and an ordinary polymer generally

III. Definition and Classification of Polymers

consists of a mixture of molecules of different molecular weights. Thus, the molecular weight of a polymer refers to a weight-average.

The size of a polymer molecule is expressed in terms of the average number of repeat units in the molecule and is called the degree of polymerization (DP). From the known DP and the known molecular weight of the monomer (repeat unit), the average molecular weight of a polymer is easily computed:

Average molecular weight =
$$\text{DP} \times \text{molecular weight of monomer}$$

The constitution of a polymer is generally described in terms of the structural units. When only one type of monomer unit is present in a polymer, it is called a homopolymer; a polymer having two or more structural units is referred to as a copolymer. In "linear polymers," the monomer units are joined together in a straight, open-chain fashion, whereas "cross-linked polymers" have a three-dimensional network.

The repeat-unit in a polymer molecule is generally equivalent to the monomer—the starting material from which the polymer is formed. The polymer is generally named by adding the prefix "poly" to the name of the monomer. Thus poly(vinyl chloride) molecules contain the repeat unit $= CH_2CHCl =$ and its monomer is vinyl chloride ($CH_2 = CHCl$). Copolymers are generally classified according to the arrangements of the monomer units in their molecules. A copolymer may have these features.

1. An ordered sequence of two or more monomers (a sequential polymer) such as a co(polyethylenemaleic anhydride):

$$-(CH_2-CH_2-CH-CH-)_n$$
$$\quad\quad\quad\quad\quad\quad\quad \diagdown\;\diagup$$
$$\quad\quad\quad\quad\quad O\;\;\;O\;\;\;O$$

2. A random sequence of the monomers in which the distribution of each monomer is random, e.g.,

$$A-A-B-A-B-B-B-A-A \ldots$$

3. The monomers in blocks of individual monomers, e.g., a block copolymer of styrene and isoprene may be represented as

$$(-CH_2-\underset{\phi}{CH}-CH_2-\underset{\phi}{CH}-CH_2-CH)_n-(CH_2-\underset{CH_3}{C}=CH-CH_2-CH-\underset{CH_3}{C}=CH-CH_2-)_m$$

4. Polymer chains of a different monomer grafted onto the main polymer backbone, e.g.,

```
—A—A—A—A—A—A—A—A—A—A—
     |                   |
     B                   B
     |                   |
     B                   B
     |                   |
     B                   B
```

IV. PREPARATION OF SYNTHETIC POLYMERS

Synthetic polymers are formed by the polymerization of monomers. Polymerization processes are basically of two types: addition polymerization or polycondensation. The resulting polymers are classified by their mode of formation as either "addition polymers" or "condensation polymers." This classification of polymerization processes and, hence of the resulting polymers, leads to an incongruous situation. For example, polyethylene, which is normally produced by addition polymerization of the monomer, ethylene, can also be produced from diazomethane by polycondensation, e.g.,

$$n\text{CH}_2 = \text{CH}_2 \rightarrow -(\text{CH}_2-\text{CH}_2)_n$$

$$n\,\text{CH}_2\text{N}_2 \rightarrow -(\text{CH}_2-\text{CH}_2)_{n/2} + n\,\text{N}_2$$

On the other hand, nylon-6, normally considered to be a condensation polymer, is actually produced by the addition polymerization of caprolactam:

$$n\,(\text{CH}_2)_5-\underset{\underset{\text{NH}}{|}}{\text{C}}=\text{O} \longrightarrow -[(\text{CH}_2)_5-\text{CO}-\text{NH}]_n-$$

Addition polymers are generally based on olefinic monomers and can thus be distinguished from condensation polymers which are generally formed by reaction of two different functional groups involving the elimination of some simpler molecules. Condensation polymerizations have also been called "step-reaction polymerizations."

Addition polymerizations proceed either by free-radical or by ionic mechanisms and can be carried out either in bulk solution, i.e., on the neat monomer, or in suspension or emulsion. Each method has its own advantages and disadvantages. The choice of method of polymerization also depends to a very great extent both on the nature of the monomer and on the product desired. Polycondensations or step-reactions proceed according to the mechanism demanded by the reactive functional groups. Some common step-reactions are esterification, amidification, and urethane formation, as well as ring-opening or transesterification.

V. PROPERTIES OF POLYMERS

Although polymeric substances (natural and synthetic—inorganic or organic) are easily recognized by their physical appearance and certain specific properties (such as low or negligible solubility in common solvents, mechanical strength, elasticity, fiber-forming properties, and dimensional stability), they may still differ considerably in their physical properties. Polymers may be in the form of readily soluble liquids or low-melting, waxy, or even very hard and brittle, solids.

Many properties of polymers appear anomalous when compared to those of low-molecular-weight compounds. However, the presumably anomalous properties of polymers can be interpreted as normal for such materials when molecular size and stabilizing forces are taken into consideration.

A. Bonding Forces in Polymers

Primary chemical bonds along polymer chains are generally completely satisfied. Secondary bond forces, e.g., van der Waals forces, various types of dipole interactions, and hydrogen bonding are, however, also present. Whereas these secondary bond forces play only a relatively minor role in influencing the properties of smaller molecules, in polymers they assume an extremely important role. The high molecular weight of the polymer permits these forces to build up sufficient strength to impart to it the observed excellent mechanical strength and rigidity. These intermolecular forces also influence other properties of the polymers, e.g., swelling, gelation, miscibility, and solubility in certain solvents.

B. Crystallinity

When polymer molecules possess symmetry, they will also have an accompanying tendency to form crystalline regions. Unlike small molecules, polymers may be amorphous and yet have regions of crystallinity. The crystalline regions of the polymers have increased mechanical strength and differ in other properties from the amorphous regions in the polymers.

C. Steric Configuration

Depending on the method of polymerization, polymers can be made that are either isotactic, i.e., substituents around the polymer backbone are in an ordered configuration, or atactic, i.e., substituents have a random distribution. Crystallinity and other physical properties of a polymer are dependent upon the substituent's configuration.

D. First- and Second-Order Transition Temperatures

These refer either to the melting temperatures of crystal regions in crystalline polymers or to the softening temperature in amorphous regions, respectively. These temperatures are not as "sharp" as those of low-molecular-weight solids. The softening of polymers results from the increased kinetic energy of the molecules as it becomes large enough to overcome secondary bond forces.

E. Miscibility and Solubility

These properties are determined by intermolecular forces. Compared to the dissolution of low-molecular-weight substances, the dissolving of a polymer is a slow process and takes place in two stages. First, solvent molecules slowly diffuse into the polymer, resulting in swelling and gelation. This may be all that happens if strong polymer–polymer intermolecular forces are present because of cross-linking, crystallinity, and strong hydrogen bonding. In the case of linear polymers, the first stage is followed by a second stage in which a truly homogeneous solution results from diffusion of solvated polymer molecules into the solvent. For polymers that are to be used as insoluble reagents, swelling rather than solubility is the required property.

F. Solution Properties of Polymers

Dilute solutions of completely soluble polymers exhibit the usual colligative properties of solutions. These properties have frequently been used to determine polymeric molecular weights; e.g., viscosity and light-scattering measurements are frequently made on polymer solutions for molecular weight determinations.

VI. SYNTHESIS OF FUNCTIONALIZED POLYMERS

There is very little new synthetic organic chemistry involved in the synthesis and transformation of polymers. The organic chemistry involved is usually the application of known, solution-phase organic reactions to polymeric chemistry.

In organic chemistry, hydrocarbons are considered parent compounds, while other organic compounds are considered derivatives. This analogy can be extended to organic polymers as well, where polymeric hydrocarbons can be considered parent polymers from which other functionalized polymers are derived. The exceptions, of course, are the heterochain polymers, where small carbon chains are linked through heteroatoms. Polymethylenes are the simplest of the organic polymers, but other hydrocarbon polymers such as polyalkanes, polycycloalkanes, polyalkenes, and polyarenes are also known. The polyarenes are "purely" aromatic polymers such as polyphenylene. The more important commercial arene polymers are, however, derived from those containing arylsubstituents on an alkane chain, e.g., polystyrene.

From the synthetic point of view, there are two possible methods of preparing a functionalized polymer. The first method involves starting with a properly functionalized monomer and then polymerizing it. The chief advantage of this method is that the resulting polymer is truly homogeneous, and the degree of functionalization in such polymers is also fixed and high. Monomer instability and incompatible polymerization conditions tend to limit preparations by this route to relatively simple polymers. The second, and more frequently used, method involves first forming a polymeric carrier and subsequently introducing functional groups into the preformed polymer structure. The degree of functionalization is easily controlled in this case, but the distribution of the groups on the polymer matrix may not be uniform.

VII. TYPES OF FUNCTIONALIZED POLYMERS

Polymers can be prepared that contain practically any organic functional group found in low-molecular-weight compounds. Polymers containing halogens, hydroxyls, ethers, aldehydes, carboxylic acid groups and their derivatives (such as esters, acid chlorides, amides), sulfonic acid groups, thio, nitro, amino (primary, secondary, or tertiary), and quaternary ammonium groups are well known. Certain heterocyclic systems (pyridine, quinoline, etc.) as well as such less common groups as triaryl phosphine, N-haloimides, and peroxy acids have also been incorporated into polymers.

VIII. GENERAL CHEMICAL REACTIONS OF POLYMERS

A polymer possessing a number of diverse properties may be required to perform a particular task. The available, simple homopolymers may not possess the required properties, and hence it becomes necessary to transform them into new polymers by carrying out chemical reactions on them.

The chemical reactions of polymers can be classified as follows.

1. Those affecting degree of polymerization (DP). These involve further polymerization (including cross-linking) of already formed polymers and the synthesis of a graft or a block copolymer, as well as degradation reactions. Such reactions have been classified as "macromolecular."
2. Those not affecting DP. These involve reactions of functional groups already contained in the polymer molecules: Some of these reactions are reversible and are referred to as polymer-analogous transformations. These reactions have been used in polymer-mediated organic syntheses and will be the subject of the bulk of this book.

Some overlap between categories can occur. For example, the sequential synthesis of a polypeptide on a polymer support is equivalent to grafting the polypeptide onto the original polymer support, whereas each step of the solid-phase peptide synthesis can be regarded as a polymer-analogous transformation.

Polymer-analogous reactions can be carried out to modify the properties of commercial polymers. A well-known sequence of polymer-analogous reactions is the conversion of poly(vinyl acetate) to poly(vinyl acetal) via poly(vinyl alcohol).

In general, polymers undergo chemical reactions much in the same way as do low-molecular-weight compounds, providing of course the site of reaction is accessible. For example, carboxylic polymers readily undergo esterification, amidification, peracid formation, and anhydride formation. The carbonyl group in polymers can also undergo its usual reactions, e.g., oximation and reduction. Benzene rings in styrene polymers can undergo such reactions as nitration, sulfonation, halogenation, alkylation, chloromethylation, and acylation. Many of these reactions have been used for preparing functionalized polystyrene-based reagents.

Wittig reactions have been carried out on polymer-containing carbonyl groups while an alkene synthesis with low-molecular-weight aldehydes or ketones has been carried out with a polymeric phosphorous ylide (Wittig's reagent). Similarly, polymers can undergo expoxidation with organic peroxides, while polymers containing peroxy groups can oxidize small alkene substrates. In general, there is an almost unlimited choice in the use of polymeric reagent and low-molecular-weight substrates, or vice versa.

IX. POLYMERS AS AIDS IN ORGANIC SYNTHESIS

Prior to 1963, the reactions of polymers were mainly carried out with the object of improving or modifying their structural properties and of making them suitable for specific purposes. Excluding, of course, the use of ion-exchangers, there are only a few scattered references to the use of polymers as chemical reagents for synthesis. The credit for the systematic introduction of polymers as reagents for organic synthesis goes both to Merrifield (1963) and to Letsinger and Kornet (1963).

When a reagent (or a substrate) is covalently bound to a polymer, it acquires the physical properties of the latter. Consequently the functionalized polymer (if reasonably cross-linked) remains insoluble in common organic solvents. If the polymer is porous and swells in a suitable solvent, the functional groups anchored on it are easily available for chemical transformation. Covalent attachment of the functional groups to a polymer helps in "keeping track" of the transformed product in a chemical synthesis, resulting in simplication of the processes. A classification of polymer-mediated reactions has been suggested by Patchornik and Kraus (1976b), but basically the reactions fall into three main categories (Mathur and Williams, 1976).

1. The first type includes reactions in which the polymer acts as a carrier for the substrate. The product remains attached to the support while the by-products, excess of reagents, and solvents all remain in solution and can be removed by filtration. The synthesis may involve a single step (such as acylation of an enolizable polymeric ester), or it may be a sequential synthesis of biopolymers, where the successive addition of monomers is carried out as a graft on the basic polymer chain. The last stage in such a synthesis involves cleavage of the product from the polymer backbone.

2. The second type includes reactions in which a polymer incorporating a conventional synthetic reagent, e.g., a peracid, N-bromoimide, metal hydride, is reacted with a low-molecular-weight substrate which is transformed into the product. The excess of the polymeric reagent and the spent polymer remain insoluble whereas the product goes into solution.

3. The third type includes reactions of polymeric reagents carrying catalytic groups. These reactions are not basically different from the reactions classified under 2. In this case, however, the by-product polymer is the same as the functionalized polymer.

X. KINETICS OF POLYMER-ANALOGOUS REACTIONS

Although it is more than two and a half decades since polymer-mediated syntheses were put into practice, few systematic kinetic studies of such reactions have been made. Even detailed kinetic studies of earlier known polymer-analogous transformations, such as the esterification of vinyl alcohols or the hydrolysis of the corresponding esters, have been lacking. Kinetic studies of certain step-reactions in linear polymerizations have been made, however, and the concept of functional group reactivity being independent of molecular weight has been developed. Such step-reactions were carried in solution, whereas syntheses with polymeric reagents are carried in the heterogenous phase. In other words, we do not have enough data or parallel examples to predict with certainty the factors that influence reaction rates of polymeric reagents used in synthesis.

When we consider step-reactions of linear polymers in solution, it is necessary to make some simplifying assumptions, without which the analysis of kinetic data would be hopelessly difficult. The following assumptions are made.

1. The rate constants for monofunctional and bifunctional reagents are identical when sufficiently long chains separate the reactive groups in bifunctional compounds.

2. The chemical reaction between reactive groups results after a period of many collisions and before the reactants can diffuse away. Although long polymer chains diffuse slowly in solution, the mobility of the terminal functional groups of the chain is much greater than that of the entire chain. Such groups can diffuse readily, over a considerable region through rearrangements of the conformation of nearby chain segments.

3. Even though a lower diffusion rate prolongs the time before the two reactive groups diffuse into the same region, it also prolongs the time during which they are close and colliding.

These simplifying assumptions have been made for a reaction involving a linear polymer having two functional groups, one at each end of the polymer chains, and another bifunctional small molecule. It is also assumed that the rate of reaction of a group must be independent of the size of the molecule to which it is attached. The assumption is amply justified by experimental evidence involving the rate constants of condensation reactions in a homologous series. The measured rate constants reach asymptotic values, independent of chain length, and show no tendency to drop off with increasing molecular size. However, the following conditions must be met during the reaction.

1. The reactions must take place in a homogeneous phase, e.g., in a liquid medium. All reactants and products must thus be soluble.

2. Only one polymer-attached functional group participates in each elementary step of the reaction. The remaining species must be small and mobile.

3. The low-molecular-weight homolog must be chosen with sufficient

X. Kinetics of Polymer-Analogous Reactions

care, so that all steric factors occurring in the immediate vicinity of the chain are taken into consideration.

In the case of polyfunctional step-reactions resulting in the formation of three-dimensional polymers, the situation is further complicated by gelation. The onset of gel formation is marked by division of the mixture into an insoluble rigid gel and the surrounding solution. The functional groups on the gel are not free to move and the low-molecular-weight substances must diffuse to these fixed reactive sites in the rigid-gel structure.

In the case of an insoluble polymeric reagent participating in a synthetic reaction, the situation is somewhat similar to the polyfunctional step-reaction polymerization, beyond the state of gelation. Some deviation takes place from linear step-reaction polymerization kinetics, and although the assumption of the chemical reactivity of the functional group being independent of the size of the molecule is still made, it still amounts to an oversimplification.

In the case of the chemical reactions of polyfunctionalized insoluble polymers, the situation is further complicated by the fact that the reactive groups on the polymer backbone are randomly distributed throughout the entire length of the chain and not confined to the ends of the polymer chain. Folding of the polymer chain and proximate groups are bound to affect its reactivity.

In contrast to the simplifying assumptions made for the stepwise polymerization of linear molecules, let us now consider those necessary to study the ion-exchange processes (Helfferich, 1962): (1) Reactive groups are randomly distributed throughout each particle of the ion-exchanger; (2) swollen particles possess a gel-like structure in which the solvent and low-molecular-weight substances can diffuse freely, but in which the reactive exchange groups are rigidly fixed in the gel structure; (3) the overall exchange reaction involves the following steps: (a) diffusion of ions through the solution to the surface of the exchange particle, (b) diffusion of these ions through the gel particles, (c) exchange of these ions with those already in the exchanger, (d) diffusion of the displaced ions to the surface of the exchanger, and, (e) diffusion of the displaced ions through the solution.

The overall reaction rate could depend either on the two-step diffusion rate or on the actual exchange rate at the reactive (exchange) site. In the case of the strongly acidic sulfonic acid resins, exchange rates are very fast and the rate of exchange is mainly governed by the diffusion rates. On the other hand, in weakly acidic carboxylic acid resins, the exchange rates are slow and can be the rate-determining factor.

Although there is a lack of experimental data, it is reasonable to assume that reaction rates between covalent molecules and functional groups on swollen, rigid polymer beads will be governed by the reaction rate of the functional group. Since the rate of reaction of the functional group depends upon its nature and can only be changed by employing catalysts and elevated temperature, the observed reaction rate would also depend on several factors such as: (1) concentration of the low-molecular-weight species in solution in contact with the resin, (2) stirring or mixing rate, (3) diameter of resin particles, (4) the diffusion rate of the low-molecular-weight species (this, in turn, will depend on the degree of cross-linking in the resin and the solvent employed), and (5) the temperature of the solution.

It is hoped that, in the near future, more experimental work in this field will be carried out.

XI. LITERATURE ON SOLID-PHASE SYNTHESIS

Literature (books and reviews) on solid-phase peptide synthesis, immobilized catalysts and enzymes, and affinity chromatography will be referred to in the respective chapters. Literature covering other aspects of solid-phase synthesis includes reviews and books by Frankhauser and Brenner (1973), Patchornik *et al.* (1973), Ledwith and Sherrington (1974), Leznoff (1974), Overberger and Sannes (1974), Blossey and Neckers (1975), Patchornik and Kraus (1976a,b), Crosby (1976), Crowley and Rapoport (1976), Mathur and Williams (1976), Weinshenker and Crosby (1976), Heitz (1977), Leznoff (1978), Hodge (1978), Neckers (1978), Patchornik (1978), and Manecke and Storck (1978).

REFERENCES

Billmeyer Jr., F. W. (1971). "Text Book of Polymer Science," 2nd ed. Wiley, New York.
Blossey, E. C., and Neckers, D. C., eds. (1975). "Solid-Phase Synthesis." Dowden Hutchinson and Ross, Stroudsburg, Pennsylvania.
Crosby, G. A. (1976). *Aldrichimica Acta* **9**, 15.
Crowley, J. L., and Rapoport, H. (1976). *Acc. Chem. Res.* **9**, 135.
Fettes, E. M., ed. (1964). "Chemical Reactions on Polymers." Wiley (Interscience), New York.
Flory, P. J. (1953). "Principles of Polymer Chemistry." Cornell Univ. Press, Ithaca, New York.
Fankhauser, P., and Brenner, M. (1973). *In* "The Chemistry of Polypeptides" (P. G. Katsoyannis, ed.), pp. 389–411. Plenum, New York.
Heitz, W. (1977). *Adv. Poly. Sci.* **23**, 2.

References

Helferrich, F. G. (1962). "Ion Exchange." McGraw-Hill, New York.
Hodge, P. (1978). *Chem. in Britain* **237.**
Huggins, M. L. (1958). "Physical Chemistry of High Polymers." Wiley, New York.
Ledwith, A., and Sherrington, D. C. (1974). *In* "Molecular Behaviour and the Development of Polymeric Materials" (A. Ledwith and A. M. North, eds.). Chapman & Hall, London.
Letsinger, R. L., and Kornet, M. J. (1963). *J. Am. Chem. Soc.* **85,** 3045.
Leznoff, C. C. (1974). *Chem. Soc. Rev.* **3,** 65.
Leznoff, C. C. (1978). *Accts. Chem. Res.* **11,** 327.
Manecke, G., and Storck, W. (1978). *Angew Chem. Int. ed.* **17,** 657.
Mark, H. F., *et al.,* eds. (1940). "High Polymers." Wiley (Interscience), New York.
Mark, H. F., *et al.,* eds. (1964). "Encyclopaedia of Polymer Science and Technology." Vol. 1. Wiley (Interscience), New York.
Mathur, N. K., and Williams, R. E. (1976). *J. Macromol. Sci. Rev. Macromol. Chem.* C(15), 117.
Merrifield, R. B. (1963). *J. Am. Chem. Soc.* **85,** 2149.
Miller, M. L. (1966). "The Structure of Polymers." Van Nostrand-Reinhold, New York.
Neckers, D. C. (1978). *Chem. Tech.* 108.
Overberger, C. G., and Sannes, N. K. (1974). *Angew. Chem. Int. Ed.* **13,** 99.
Patchornik, A. (1978). *Israel J. Chem.* No. 4, **17.**
Patchornik, A., and Kraus, M. A. (1976a). *Pure Appl. Chem.* **46,** 183.
Patchornik, A., and Kraus, M. A. (1976b). "Encyclopedia of Polymer Science and Technology," Supplement No. 1, p. 468. Wiley (Interscience), New York.
Patchornik, A., Fridkin, M., and Katchalski, E. (1973). *In* "The Chemistry of Polypeptides" (P. G. Katsoyannis, ed.), p. 315. Plenum, New York.
Ravve, A. (1967). "Organic Chemistry of Macromolecules." Dekker, New York.
Weinshenker, N. M., and Crosby, G. A. (1976). *Annu. Rep. Med. Chem.* **11,** 281.

2 Polymeric Support Materials

I.	Introduction	14
II.	Styrene-Based Polymers	15
III.	Functionalization of Styrene-Based Polymers via Chloromethylation and Other Methods	18
IV.	Miscellaneous Polymer Supports	25
	References	32

The first and foremost requirement for a synthesis using a polymeric reagent is the polymer itself. Polymers are prepared by polymerization of the appropriate monomers, but in some cases natural or modified natural polymers have also been used.

I. INTRODUCTION

Polymers have been designed to play three main roles in organic synthesis. They have been used to immobilize substances on which reactions are being done, to serve as reagents in reactions, and, finally, to catalyze reactions. In order to be effective in each of these roles, the polymer should have the following properties.

1. Where a solid-phase reaction is desired, the support should be totally insoluble in common solvents.
2. The polymer should be either of the relativity rigid ("nonswellable") or of the quite flexible ("swelling") type;
3. It should be capable of functionalization to a high degree, and the functional groups should be uniformly distributed in the polymer. Suitable analytical methods for determination of functionalization should also be available.
4. The functional groups in the polymer should be easily accessible, either in the rigid or in the swelled form, to the reagents and solvents. Improvements in accessibility are sometimes achieved by grafting of the

reactive functional groups to the polymer backbone by a "long handle" or "spacer-arm."

5. The functionalized polymer should undergo straightforward reaction with the reagents and be free of any side reactions.

6. The functionalized polymer should be compatible with the solvents and reagents used. The compatibility of polymers can be increased by the incorporation of certain solvent- or reagent-indifferent functionalities.

7. The polymer should be easy to handle and should not undergo mechanical fracturing during synthetic operations.

8. As far as possible, the by-product polymer should be capable of being regenerated by a simple, low-cost, high-yield reaction.

The proper choice of the polymer is an important factor for success in polymer-mediated synthesis. A wide range of polymers are available, including both aliphatic and aromatic monomer-based organic and inorganic polymers. The polymers are usually prepared by polymerization of the appropriate monomers; however, in some cases natural or modified natural polymers have been used.

Polystyrene has been the most widely used of polymers, for various reasons that will be discussed in detail subsequently. A random survey of nearly 100 syntheses employing polymeric reagents reveals that about 80% of the polymers were based on styrene. Limitations on its use have been found in the synthesis of oligosaccharides and oligonucleotides whose polarity is incompatible with the hydrophobic and nonpolar nature of polystyrene. Other polymers used include poly(vinyl alcohol), polymethacrylate, poly(ethylene glycol), polyethylenimine, polyacrylamide, poly(amino acid)s, poly(vinyl chloride), co[poly(allyl chloride)–divinylbenzene], co(polyethylene–maleic anhydride), poly(4-vinylpyridine), and various phenol–formaldehyde resins. In addition to these synthetic polymers, natural polymers based upon cellulose, dextrans (e.g., Sephadexes), and agar (e.g., Sepharoses) have been used. Inorganic polymer matrices (e.g., silica or porous glass with organic groups on its surface) have also been used.

II. STYRENE-BASED POLYMERS

Styrene–DVB (divinylbenzene) polymers with different cross-linkings and bead size are commercially available (Table 2-1). These polymers are produced by heterogeneous (suspension) polymerization. The size of the polymer bead depends on the extent of dispersion in solution, the amount of agitation, the temperature, and the initiator used during polymeriza-

TABLE 2-1

Product Specifications of Some Commercially Available Sytrene–DVB Bio-Beads

Product[a]	Mesh size	Molecular-weight exclusion limit
Bio-Beads S-X_1	200–400	600–14,000
Bio-Beads S-X_2	200–400	100–2,700
Bio-Beads S-X_3	200–400	up to 2,000
Bio-Beads S-X_4	200–400	up to 1,400
Bio-Beads S-X_8	200–400	up to 1,000
Bio-Beads S-X_{12}	200–400	up to 400
Bio-Beads SM-2	20–50	600–14,000

[a] Bio-Beads S are swellable and microporous whereas Bio-Beads SM are macroporous and very nearly nonswellable. The approximate percentage of cross-linking (n) is represented by the subscript in S-X_n. All products are available from Bio-Rad Laboratories, Richmond, California. Similar products are available from Rohm and Haas, Philadelphia, Pennsylvania (e.g., the microporous Amberlites XAD-1, XAD-2, XAD-4, and swellable macroporous XE-305).

tion. When free-radical polymerization is initiated, tough, insoluble, and almost completely spherical cross-linked beads of the polymer precipitate out. The polymers can be easily synthesized in the laboratory from monomers, but the commericial products are more uniform in size and cross-linking. Experimental details for the preparation of popcorn polymers have been described (Amos *et al.*, 1952; Letsinger and Hamilton, 1959). In addition to these styrene–DVB polymers, their chloromethylated derivatives are also commercially available. These are commonly referred to as Merrifield resins because of their widespread use in the polypeptide synthesis process initiated by R. B. Merrifield (Merrifield, 1963).

Polystyrene–DVB polymers have been extensively used in peptide synthesis and in a great variety of other syntheses. Styrene-based polymers have many advantages over other resins. (1) Aromatic ring functionalization is achieved easily to give reactive, yet selective styrene-based reagents. (2) The type and degree of cross-linking can easily be controlled. Since the degree of cross-linking in the polymer influences its swelling nature, polymer beads of both a swelling and nonswelling type can be made. (3) Being hydrocarbon-like, these polymers are compatible with organic solvents so that functional groups are easily accessible to the reagents and solvents. (4) The polymers are not

II. Styrene-Based Polymers

degraded by most chemical reagents under ordinary conditions and can withstand the chemical treatments and physical handling required in sequential synthesis.

Pore dimension within polystyrene polymers can be controlled during manufacture by regulating the concentration of DVB. To a certain extent, however, the degree of cross-linking may further change during functionalization reactions such as chloromethylation. Pore dimensions are also influenced by the solvent employed, being maximal in relatively nonpolar solvents. When maximally swollen, molecular-weight exclusion limits for commercially available polymers (e.g., Bio-Rad Bio-Beads SX_n) range from 400 to 14,000 (as determined by gel permeation). These limits are certainly altered during functionalization and loading of the polymer. Thus, for the synthesis of organic molecules of widely different sizes, a wide choice of pore dimensions is available.

The relatively rigid, macroporous gel is another type of styrene polymer that, once solvated, does not appreciably change dimensions as a function of solvent polarity. Chemical transformation of swellable polymers will only take place inside the polymer if conducted under conditions in which it is swollen. In the case of macroporous polymers, however, internal regions are highly solvated and readily open to reaction. Even so, rigid sections within the hydrocarbon network remain totally unsolvated and may be totally inaccessible to chemical transformation. Several groups of workers (Blackburn *et al.*, 1969; Letsinger *et al.*, 1964; Fyles and Leznoff, 1976) have made a comparative study of "swelling" (microporous) and "nonswelling" (macroporous or macroreticular) resins. It has been concluded that the macroreticular polymers may be used in almost any solvent since much less swelling of the polymer matrix is required prior to reaction. For the swellable polymers, it becomes essential to use solvents with good swelling properties, such as dioxane, tetrahydrofuran, chloroform, methylene chloride, or benzene.

Swellable resins were found to offer distinct advantages over the nonswellable ones: (1) They are less fragile and require less care in handling (Stewart and Young, 1969); (2) higher reaction rates can be achieved during the reactions of polymer functionalization; (3) their loading capacity is higher.

The macroreticular resins, however, have the advantages of (1) ease of filtration from the reaction medium after reaction; (2) more accessible reactive groups, and (3) large pore sizes which offer less hindrance to the diffusion of the reactants.

Linear or "soluble" polystyrene polymers (MW 50,000–300,000) have been used for a number of syntheses (Hayatsu and Khorana, 1966, 1967; Cramer *et al.*, 1966; Kabachink *et al.*, 1970). Soluble polystyrene-supported reactions have been shown to give yields comparable to those in syntheses in the homogeneous phase. After the synthetic operations, the separation of the polymer-bound product can be achieved by ultrafiltration, dialysis, and gel filtration using Sephadex LH20 (Potapov et al., 1972). The recovery of material by these methods is good, but time-consuming. To improve the recovery rate, precipitation methods have been used; but they are not quantitative and involve loss of material. Precipitation methods of recovering material have also prompted the use of isotactic polystyrene. Since it has a crystalline nature, it is nearly insoluble in organic solvents. It can, however, be recovered completely by washing with polar solvents such as water, methanol, or ethanol. This differential solubility has been exploited in certain sequential syntheses (Tsou and Yip, 1973; Potapov et al., 1971).

III. FUNCTIONALIZATION OF STYRENE-BASED POLYMERS VIA CHLOROMETHYLATION AND OTHER METHODS

Functionalization of sytrene polymers involves electrophilic substitution on the aromatic ring. Chloromethylation has been the most widely used reaction (Merrifield, 1963). Chloromethylation of styrene polymers is carried out using a Lewis acid catalyst and chloromethyl methyl ether as the solvent [Eq. (1)]. Carbon disulfide or chloroform have also been employed as cosolvents.

$$\text{\textcircled{P}}-\text{C}_6\text{H}_4 + \text{ClCH}_2\text{OCH}_3 \xrightarrow{\text{catalyst}} \text{\textcircled{P}}-\text{C}_6\text{H}_4-\text{CH}_2\text{Cl} \quad (1)$$

The more effective Friedel–Crafts catalyst, anhydrous aluminum chloride, is not desirable since it is incorporated into the polymer as a complex, cannot be washed away completely with common solvents, and resists even hydrolysis (Neckers et al., 1972). In addition to anhydrous $SnCl_4$ (Merrifield, 1963; Stewart and Young, 1969), improved procedures employing BF_3 (Sparrow, 1975) and anhydrous $ZnCl_4$ (Feinberg and Merrifield, 1974) have been described.

The degree of chloromethylation in the resin is easily assayed by determining the chlorine content. Merrifield, in his original synthesis of a tetrapeptide (Merrifield, 1963), employed a resin in which 22% of the benzene rings were substituted. In such a moderately functionalized polymer, the anchored substrates were said to be easily accessible and no extensive cross-linking was observed during chloromethylation.

When anchoring a substrate on a polymer, the covalent bonding of the

III. Functionalization of Styrene-Based Polymers

substrate also serves to block one of the groups whose participation in the reaction is undesirable. In this respect, resins containing benzyl chloride groups have a distinct advantage. When substrate–support linkages via carboxylic groups are desired, the benzyl esters are easily formed (often in quantitative yield) by reacting the carboxyl group in the presence of a base such as triethylamine. Benzyl esters have the additional advantage of undergoing acid-catalyzed (HBr–AcOH) cleavage while remaining intact during many base-catalyzed reactions. For these reasons, benzyl esters have been extensively used in peptide synthesis.

In addition to their direct use, chloromethyl groups are readily modified into other functional groups. The more important functional groups that have been introduced via chloromethyl groups are shown in Table 2-2. Modification reactions may also be phase-transfer catalyzed (Frechet *et al.*, 1979). Other functional groups may be directly introduced into the styrene support polymer by well-known reaction sequences (Table 2-3) (Patterson, 1971; Frechet and Farrall, 1977).

Among the many methods of functionalization of styrene polymers, halogenation followed by metallation and quenching with appropriate reactants appears to be the most important. One-step, direct metallation (nucleophilic substitution) using tetramethylethylenediamine and butyllithium has been reported to be less satisfactory than the two-step bromination–lithiation process (Farrall and Frechet, 1976).

Another method of preparing functionalized polymers involves copolymerization of substituted styrene monomers plus styrene and/or DVB to give the functionalized polymer directly (Table 2-4). Introduction of functional groups into styrene polymers by copolymerization of suitably substituted styrene monomers is reported to give polymers of more uniform functionalization. In addition, they are not contaminated by small proportions of other functional groups remaining from incomplete prior chemical transformation.

Often it has been possible to prepare the same functionalized polymer by two different methods. For example, the polymer may be prepared by functionalization of a suitably cross-linked styrene polymer, or by copolymerization of preformed (substituted or functionalized) vinyl monomers in the presence of DVB. Typical examples are benzoic acid-group-bearing polymers and triarylphosphine-group-bearing polymers that have been synthesized either by functionalization of styrene polymers or by copolymerization of the respective functionalized monomers (Schemes 2-1 and 2-2). In one case (Guthrie *et al.*, 1971), the monomer was loaded with the reactant, which was supposed to undergo subsequent reaction on the polymer support, and then the resulting preloaded monomer was polymerized. (For more details, see Chapter 6.)

TABLE 2-2

Functional Groups That Can Be Introduced into Copolystyrene–DVB via Chloromethylation

Functional group	Conditions of functionalization	References
—CH_2I	NaI in acetone	Snyder et al., 1972
—CH_2Br	HBr/AcOH acting on the polymer benzyl ester; this is a by-product in many reactions involving cleavage of polymer ester on Merrifield resin	Tilak, 1968
—$CH_2S^+Me_2Cl^-$	Me_2S	Snyder et al., 1972; Tilak, 1968
—CH_2OH	By hydrolysis of the above polymers	Letsinger et al., 1964
—CH_2OCOCl	Phosgenation of the hydroxymethyl polymer	Letsinger et al., 1964
—$CH_2OCOOCOPh$	Reaction of sodium benzoate with the above chloroformate polymer	Shambhu and Digenis, 1974
—CH_2NH_2	Gabriel synthesis, from the chloromethyl polymer	Weinshenker and Shen, 1972; Mitchell et al., 1976
—$CH_2N^+R_3Cl^-$	Quaternization of the chloromethyl polymer	Helfferich, 1962; Regen and Lee, 1974
—CH_2NHR	Amination with the appropriate amine	Laursen, 1971; Collman and Reed, 1973
⟨NO_2(Br)⟩—CH_2Cl	Bromination or nitration of the chloromethyl polymer	Merrifield, 1963
—CHO	Me_2SO oxidation of the chloromethyl polymer	Frechet and Pelle, 1975
—$CO_2COC_6H_5$	Reaction of PhCOOONa with the acid chloride polymer	Shambhu and Digenis, 1973
—CH_2PPh_2	Reaction of Ph_2PLi with the chloromethyl polymer	Issleib and Tzschach, 1959; Grubbs and Kroll, 1971; Capka et al., 1971
—$CH_2OCH_2CHOHCH_2OH$	Reaction of 2,2-dimethyl-1,3-dioxolane-4-methanol (Na salt) with the chloromethyl polymer, followed by hydrolysis	Leznoff and Wong, 1973

III. Functionalization of Styrene-Based Polymers

TABLE 2-2 (*Continued*)

Functional group	Conditions of functionalization	References
—CH$_2$COOH	Reaction of KCN with the chloromethyl polymer followed by hydrolysis	Kusama and Hyatsu, 1970
—CH$_2$COCl	Reaction of PCl$_5$ with the above polymer	Kusama and Hayatsu, 1970
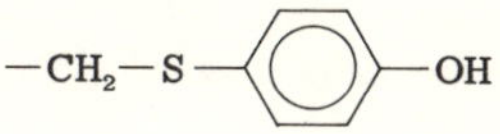 —CH$_2$—S—⟨C$_6$H$_4$⟩—OH	Reaction of sodium *p*-mercaptophenol with the chloromethyl polymer	Flanigan and Marshall, 1970
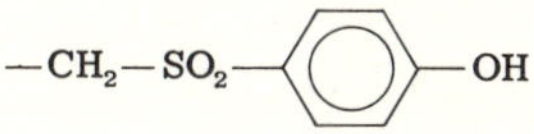 —CH$_2$—SO$_2$—⟨C$_6$H$_4$⟩—OH	Peracid oxidation of the above polymer	Marshall and Liener, 1970
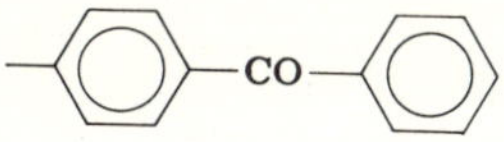 —CH$_2$O$_2$C—⟨C$_6$H$_4$⟩—CO—⟨C$_6$H$_5$⟩	Reaction of the corresponding acid with the chloromethyl polymer in presence of a base	Blossey and Neckers, 1974; Blossey *et al.*, 1973; Panse and Laufer, 1970; Harrison and Harrison, 1967
—CH$_2$O—Rose Bengal or other dyes		
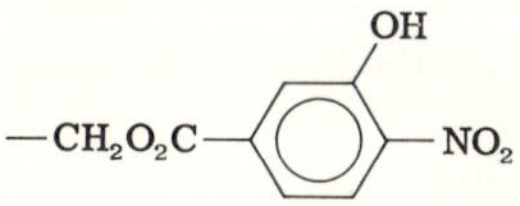 —CH$_2$O$_2$C—⟨C$_6$H$_3$(OH)(NO$_2$)⟩		
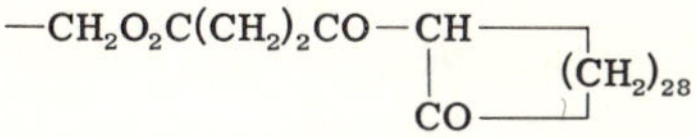 —CH$_2$O$_2$C(CH$_2$)$_2$CO—CH—(CH$_2$)$_{28}$—CO		

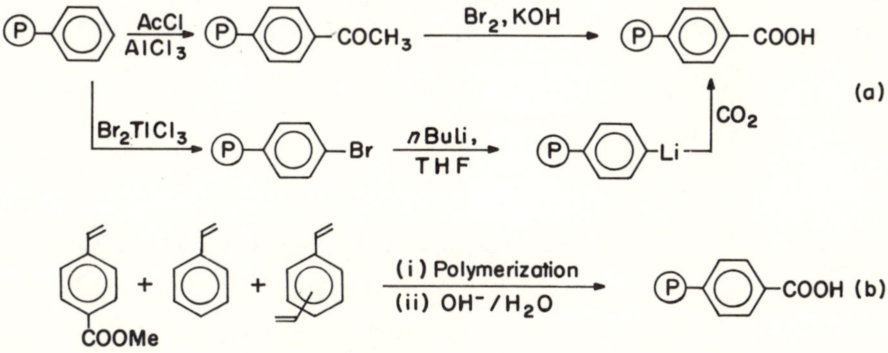

Scheme 2-1

(a) P—⟨C$_6$H$_5$⟩ $\xrightarrow{\text{AcCl}/\text{AlCl}_3}$ P—⟨C$_6$H$_4$⟩—COCH$_3$ $\xrightarrow{\text{Br}_2, \text{KOH}}$ P—⟨C$_6$H$_4$⟩—COOH

$\downarrow$ Br$_2$TlCl$_3$

P—⟨C$_6$H$_4$⟩—Br $\xrightarrow{n\text{BuLi}/\text{THF}}$ P—⟨C$_6$H$_4$⟩—Li $\xrightarrow{\text{CO}_2}$

(b) CH$_2$=CH—⟨C$_6$H$_4$⟩—COOMe + CH$_2$=CH—⟨C$_6$H$_5$⟩ + CH$_2$=CH—⟨C$_6$H$_4$⟩—CH=CH$_2$ $\xrightarrow[\text{(ii) OH}^-/\text{H}_2\text{O}]{\text{(i) Polymerization}}$ P—⟨C$_6$H$_4$⟩—COOH

TABLE 2-3

Functional Groups That Can Be Introduced in Copolystyrene–DVB Polymer via Routes Other Than Chloromethylation

Functional group	Conditions of functionalization	References
—COCH$_3$	AcCl/AlCl$_3$	Blackburn et al., 1969; Letsinger et al., 1964
—COCH$_2$Br	Bromination of the acetyl polymer	Wegand, 1968
—CO$_2$H	NaBrO oxidations of the acetyl polymer	Blackburn et al., 1969; Letsinger et al., 1964
—CONPh$_2$	Ph$_2$NCOCl/AlCl$_3$	Letsinger et al., 1964
—COOH	Hydrolysis of the above carboxyamide polymer	Letsinger et al., 1964
—CO$_2$OH	H$_2$O$_2$–MeSO$_3$H oxidation of the carboxyl polymer	Helfferich and Luten, 1964; Takagi, 1967; Harrison and Hodge, 1974; Frechet and Haque, 1975
—COCl	Reaction of SOCl$_2$ with the carboxyl polymer	Letsinger and Mahadevan, 1966
—CH$_2$OH	Reduction of the carboxyl polymer with LiAlH$_4$	Blackburn et al., 1969
—CH$_2$OCOCl	Phosgenation of the hydroxymethyl polymer	Letsinger et al., 1964; Felix and Merrifield, 1970
—SO$_3$H	Sulfonation of polystyrene	Helfferich, 1962
—SO$_2$Cl	Chlorosulfonation of polystyrene or reaction of SOCl$_2$/PCl$_5$ with the sulfonic acid polymer	Helfferich, 1962
—SO$_2$N$_3$	Reaction of NaN$_3$ with the chlorosulfonic polymer	Rousch et al., 1974
—SO$_2$NH$_2$	Reaction of NH$_3$ with the chlorosulfonic polymer	Kenner et al., 1971
—SO$_2$NH–(phenanthroline)	Reaction of 5-aminophenanthroline with chlorosulfonic polymer	Rebek and Gavina, 1974
—NO$_2$	Nitration of polystyrene	Dowling and Stark, 1969
—NH$_2$	SnCl$_2$ reduction of the nitro polymer	Dowling and Stark, 1969
—NCS	Reaction of CSCl$_2$ with the amino polymer	Dowling and Stark, 1969

III. Functionalization of Styrene-Based Polymers

TABLE 2-3 (*Continued*)

Functional group	Conditions of functionalization	References
—Cl, —Br	Halogenation [bromination, Tl(OAc)$_3$- or FeCl$_3$-catalyzed]	Farrall and Frechet, 1976; Camps *et al.*, 1971a; Heitz and Michels, 1971
—Li	(a) Lithiation of the brominated polystyrene with *n*BuLi in THF	Camps *et al.*, 1971a; Farrall and Frechet, 1976
	(b) Reaction of the 1:1 complex of *n*BuLi and *N,N,N,N*-tetramethylethylenediamine of polystyrene	Chalk, 1968; Evans *et al.*, 1976; Braun, 1959
Polymers prepared via lithiation		Farrall and Frechet, 1976
(i) —CH$_2$CH$_2$OH	$\begin{matrix} H_2C \\ \diagdown \\ O \\ \diagup \\ H_2C \end{matrix}$	
(ii) —COOH	CO$_2$	
(iii) —SH	S$_8$	
(iv) —SCH$_3$	CH$_3$SSCH$_3$	
(v) —B(OH)$_2$	B(OCH$_3$)$_3$	
(vi) —CONHC$_6$H$_5$	C$_6$H$_5$NCO	
(vii) —Si(CH$_3$)$_2$Cl	Si(CH$_3$)$_2$Cl$_2$	
(viii) —P(C$_6$H$_5$)$_2$	ClP(C$_6$H$_5$)$_2$	
(ix) —CH$_2$CH$_2$Br	Br(CH$_2$)$_2$Br	
(x) —CHO	(CH$_3$)$_2$NCHO	
(xi) —C(C$_6$H$_5$)$_2$OH	C$_6$H$_5$COC$_6$H$_5$	
Polymer–Wittig reagent —PPh$_2$=CHR	Reaction of RCH$_2$X with triphenyl phosphine resin (viii), followed by treatment with a base	Camps *et al.*, 1971b; McKinley and Rakshys, 1972; Heitz and Michels, 1972
—P(Ph$_2$)ML$_{n-x}$	Reaction of the metal complex (ML$_n$) with the triphenyl polymer (viii)	Michalska and Webster, 1975
Polymeric organotin dihydride reagent —SnH$_2$—*n*Bu	From the lithiated polymer, by the reaction sequence: (1) MgBr$_2$–etherate (2) SnCl$_3$–*n*Bu (3) LiAlH$_4$	Weinshenker *et al.*, 1975
Polymeric 1-hydroxybenzotriazole —H$_2$C—(benzotriazole ring with N—N–OH)	From polystyrene by the reaction sequence: (1) 3-nitro-4-chlorobenzyl bromide/AlCl$_3$ (2) hydrazine (3) HCl	Kalir *et al.*, 1975

(*Continued*)

TABLE 2-3 (*Continued*)

Functional group	Conditions of functionalization	References
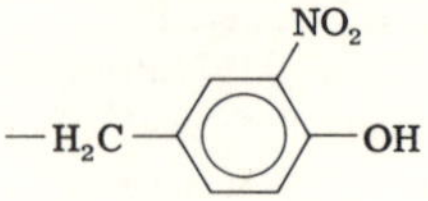 —H₂C—C₆H₃(NO₂)—OH	Friedel–Crafts alkylation of polystyrene with 3-nitro-4-hydroxybenzyl chloride	Kalir et al., 1974; Warshawsky et al., 1978
Trityl or substituted trityl chloride —CPh₂Cl	From polystyrene, by the reaction sequence: (1) Friedel–Crafts benzoylation (2) C₆H₅MgBr/H₂O (3) AcCl	Hayatsu and Khorana 1966; 1967; Cramer et al., 1966; Melby and Strobach, 1967
—C(=N₂⁺=N⁻)—C₆H₅	From polystyrene, by the reaction sequence: (1) Friedel–Crafts benzoylation (2) hydrazine (3) HNO₂	Southard et al., 1969; Chapman and Walker, 1975
—AlCl₃	AlCl₃ addition to polystrene in CS₂	Neckers et al., 1972
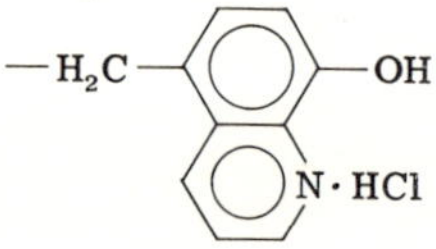 —H₂C—(quinolinol)—OH, N·HCl	Friedel–Crafts alkylation of polystyrene with 5-chloromethyl-8-hydroxyquinoline	Warshawsky et al., 1978
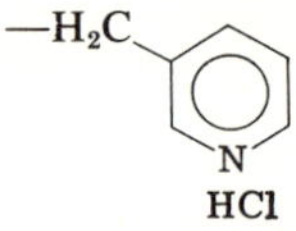 —H₂C—(pyridyl), HCl	Friedel–Crafts alkylation of polystyrene with picolyl chlorides	Warshawsky et al., 1978

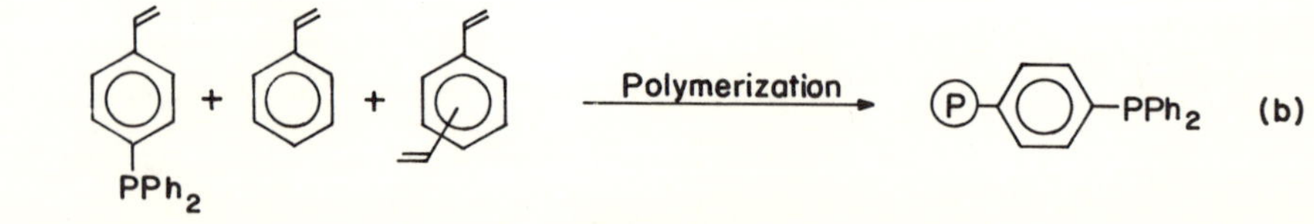

$$\text{(P)}-\text{C}_6\text{H}_5 \xrightarrow[\text{(ii)}n\text{BuLi}]{\text{(i)}\text{Br}_2,\text{TlCl}_3} \text{(P)}-\text{C}_6\text{H}_4-\text{Li} \xrightarrow{\text{Cl PPh}_2} \text{(P)}-\text{C}_6\text{H}_4-\text{PPh}_2 \quad \text{(a)}$$

$$\text{CH}_2=\text{CH}-\text{C}_6\text{H}_4-\text{PPh}_2 + \text{CH}_2=\text{CH}-\text{C}_6\text{H}_5 + \text{CH}_2=\text{CH}-\text{C}_6\text{H}_4 \xrightarrow{\text{Polymerization}} \text{(P)}-\text{C}_6\text{H}_4-\text{PPh}_2 \quad \text{(b)}$$

Scheme 2-2

IV. Miscellaneous Polymer Supports

TABLE 2-4

Vinyl Monomers Used for the Preparation of Substituted Copolystyrene Resins

Monomer	References
4-Vinylbenzyl alcohol (acetate)	Letsinger et al., 1964
2-(4-Vinylphenyl)ethanol (acetate)	Leebrick and Ramsden, 1958
4-Vinylbenzoic acid (methyl ester)	Letsinger et al., 1964
1,3-Diethyl-4-vinylbenzene	Rubinstein and Patchornik, 1972, 1975
2-Nitro-4-vinylthiophenol	Skylarov et al., 1966
Methylthio(4-vinylphenyl)	Tanimoto et al., 1967
2-Nitro-4-vinylphenol	Fridkin et al., 1965, 1966
4-Chloromethylstyrene	Arshady et al., 1976
4-Methoxystyrene	Arshady et al., 1976
Diphenyl(4-vinylphenyl)carbinol	Cramer and Koster, 1968
4-Vinylbromo(iodo)benzene	Glaser et al., 1973; Braun and Selig, 1964
Sugar–(4-vinylbenzoate)	Guthrie et al., 1971
Sugar–(4-vinylbenzenesulfonate)	Guthrie et al., 1973
2-p-Styryl-4,5-bistosyloxymethyl-1,4-dioxane 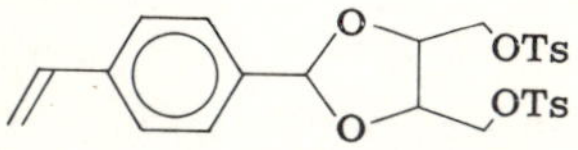	Takaishi et al., 1976
4-Vinylmonobenzo-15-crown-5 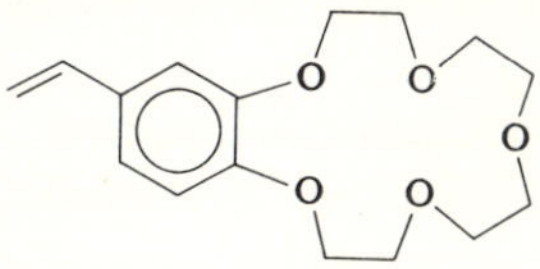	Kopolow et al., 1973
9,10-Di-p-styrylanthracene	Rosenthal and Acher, 1974

IV. MISCELLANEOUS POLYMER SUPPORTS

Vinyl group polymerization has been used to prepare polymers incorporating heterocyclic systems such as pyridine, quinoline, and imidazole rings. They have been prepared by polymerization of the respective monomers either alone or as copolymers with DVB (Table 2-5). These

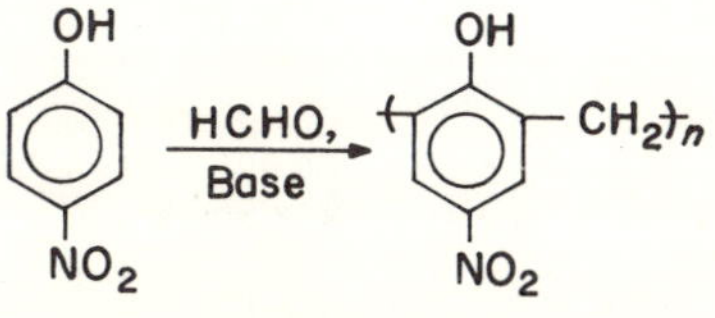

Scheme 2-3

TABLE 2-5

Vinyl Monomers Containing Heterocycles Used for the Preparation of Polymeric Reagents

Monomer	References
2- or 4-Vinylpyridine	Kurimura et al., 1971; Biedermann et al., 1973; Hallensleben, 1974
6-Isopropenylquinoline	Williams et al., 1972; Brown and Williams, 1971
N-Vinyl-2-pyrrolidinone	Bayer and Geckeler, 1974
2-Methyl-5-vinyl-8-hydroxyquinoline	Manecke and Haake, 1968
N-Methylvinylimidazole	Overberger and Salomone, 1969a,b; Letsinger and Klaus, 1964

heterocyclic-group-bearing polymers have been functionalized to give polymeric reagents for synthesis. Phenol–formaldehyde resins have also been used for making functionalized polymers (Table 2-6). For example, a polymer for preparing amino acid active esters was synthesized from *p*-nitrophenol and formaldehyde. In these phenoplast resins, *p*-nitrophenol acts as a bifunctional group and yields linear, soluble polymers (Scheme 2-3). In one additional example, salicylic acid was copolymerized with anisole and formaldehyde. This particular polymer owes its reactivity to neighboring groups participating in the transacylation reactions (Sahni et al., 1977b).

Vinyl polymers incorporating acrylamide, ethylene–maleic anhydride, styrene–maleic anhydride, maleimide, etc. have also been prepared and functionalized to serve as polymeric supports in organic synthesis (Table 2-7).

Poly(ethylene glycol) (MW 20,000) has been used as a support for N-protected amino acids in peptide synthesis (Mutter et al., 1971) and in oligonucleotide synthesis (Koster, 1972b). This polymer support was functionalized to incorporate the trityl group. The resulting hydrophilic

TABLE 2-6

Phenol–Formaldehyde Polymeric Reagents

Polymeric reagent	References
p-Nitrophenol–formaldehyde	Skylarov et al., 1966
4-(Methylthiophenol)–formaldehyde	Flanigan and Marshall, 1970
Bis(*p*-hydroxyphenyl)sulfone–formaldehyde	Wieland and Birr, 1966; Wieland et al., 1972
Salicylic acid–formaldehyde	Sahni et al., 1977b

IV. Miscellaneous Polymer Supports

polymer is soluble, permitting homogeneous-phase synthesis. After the reaction, the low-molecular-weight reactants and by-products are separated by ultrafiltration (membrane filtration) or dialysis. It has been claimed that a better control of the coupling reaction is obtained. However, the method does not appear to have gained the wide acceptance of Merrifield's fully solid-phase method.

Polyamide supports such as poly-L-lysine (MW 80,000) (Chapman and Kleid, 1973), polydimethylacrylamide (Gait and Sheppard, 1976, 1977), and polyacrylmorpholide (Narang et al., 1977) have been reported for the synthesis of oligonucleotides. Polyamide resins, being more polar, are claimed to be more compatible with oligonucleotide synthesis than are their styrene-based counterparts. Polypeptide synthesis has also been reported on a polydimethylacrylamide-based support (Atherton et al., 1975).

A water soluble polyethylenimine support has been developed for peptide synthesis using N-carboxylic anhydrides in a sequential method of peptide synthesis. The final polymer–peptide cleavage was effected by tryptic digestion (Pfaender et al., 1975).

Peracids based on poly(methacrylic acid) have been prepared and used in epoxidation reactions (Takagi, 1967). The resulting polymer peracids were unstable and tended to detonate. Polymer peracids incorporated in polystyrene polymer were relatively more stable and did not explode on impact (Harrison and Hodge, 1974).

Reagent polymers prepared from poly(maleic anhydride) and its copolymers were functionalized to contain N-bromo- and N-hydroxymide groups (Yaroslavsky et al., 1970; Yaroslavsky and Katchalski, 1972; Fridkin et al., 1972). These reagents are polymer analogs of N-bromosuccinimide and N-hydroxysuccinimide, respectively.

Polysaccharide-based hydrophilic polymer supports, prepared from cellulose, agarose, Sepharose, and Sephadex, are well known as gel-filtration media in chromatographic processes for the fractionation of substances of different molecular weights. Some of these gels have been functionalized and used in affinity chromatography. These are compatible with polar substances such as oligonucleotides and sugars and have been used as supports for nucleotide and sugar solid-phase synthesis. Typical examples of functionalization of Sephadex incorporating trityl, dehydrolipoic acid, and carboxymethyl groups are given in Table 2-7.

Inorganic matrix supports have been developed for anchoring reactive groups to glass surfaces or to silica gel. (Table 2-8). It is claimed that such groups are more accessible to the reagents in solution compared with those embedded in the polymer bead. This is to be expected because on such supports the reactive groups are mainly confined to the surface. The inorganic matrix used should have a large surface area. Porous glass

TABLE 2-7

Miscellaneous Organic Polymers

Polymer/functional group	Reaction use	References
Poly(ethylene glycol)/ tritylmethyl group	Polymer support for oligonucleotide synthesis (via trityl methyl ether)	Koster, 1972b
Poly(ethylene glycol)/NH_2 group	Polymer support for oligonucleotide synthesis (via phosphoramidate linkage)	Brandstetter et al., 1973
Poly(ethylene glycol)/OH group	Polymer support for peptide synthesis	Mutter et al., 1971; Mutter and Bayer, 1974; Bayer et al., 1974
Copolymer of 2-vinylpyrrolidinone and vinyl alcohol/OH group	Polymer support for peptide synthesis (via amino acid ester)	Bayer and Geckeler, 1974
Copolymer of 2-vinylpyrrolidinone and vinyl alcohol/OH group	Polymer support for oligonucleotide synthesis (via mixed carbonate ester)	Seliger and Aumann, 1973
Poly(vinyl alcohol)/OH group	Polymer support for oligonucleotide synthesis (via 5-phosphate ester)	Schott et al., 1973
Hydroxypropylated dextran (Sephadex LH20)/OH group	Polymer support for peptide synthesis (via amino acid ester)	Vlasov et al., 1973; Bilibin et al., 1973; Bilibin and Vlasov, 1973
Hydroxypropylated dextran (Sephadex LH20)/OH group	Polymer support for oligonucleotide synthesis (via 5'-phosphate ester)	Koster and Heyns, 1972
Carboxymethylated dextran/ ϵ-NH_2-lys	Polymer support for peptide synthesis (via —NH_2-lys amide)	Livshits and Vasil'ev, 1973
Lipoic acid incorporated in Sephadex, Sepharose, and cellulose/ $-\underset{\underset{O}{\|}}{C}-(CH_2)_4-\underset{\underset{S}{\|}}{CH}-CH_2-\underset{\underset{S}{\|}}{CH_2}$	Polymeric reagent for reduction of disulfides	Gorecki and Patchornik, 1973
ϵ-N-(3-Nitrotyrosylated) poly-DL-lys/ —lys— $\|$ $NHCOCH(NH_2)-CH_2-\langle\rangle-OH(COR)$ $\underset{NO_2}{}$	Polymeric active ester support for peptide synthesis	Fridkin et al., 1965

IV. Miscellaneous Polymer Supports

TABLE 2-7 (*Continued*)

Polymer/functional group	Reaction use	References
ε-N-(p-Aminobenzoylated) poly-L-lysine/ 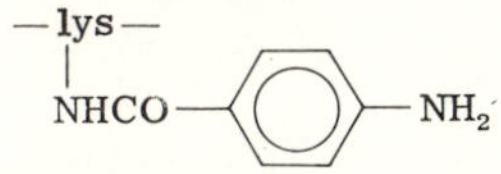 —lys— | NHCO—⟨◯⟩—NH$_2$	Polymer support for oligonucleotide synthesis (via phosphoramidate linkage)	Chapman and Kleid, 1973
Copolymer of dimethylacrylamide, N,N'-bisacrolylethylenediamine and β-alanyl-N-acryloylhexamethylenediamine/ —COCH$_2$CH$_2$NH$_2$	Polymeric support for peptide synthesis (via amino acid amide)	Atherton *et al.*, 1975
The above copolymer incorporating the following spacer-arm and "safety catch" —CO—(CH$_2$)$_2$— —S(CH$_2$)$_2$OH	Polymer support for oligonucleotide synthesis (via phosphate ester)	Gait and Sheppard, 1976, 1977
Aminoethyl groups incorporated in polyacrylmorpholide/ —CH$_2$CH$_2$NH$_2$	Polymer support for oligonucleotide synthesis (via amide bond formation) with 5'-O-(p-carboxymethyloxytrityl)-thymidine	Narang *et al.*, 1977
Peracid incorporated in polymethylacrylic acid/ ⓟ—CO$_3$H	Polymeric peracid reagent for epoxidation	Takagi, 1967

N-Bromo(chloro)polymaleimide/ 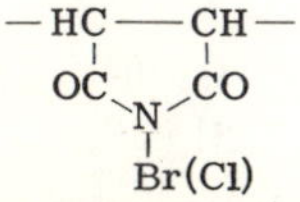
 —HC——CH—
 | |
 OC CO
  \\N/
   |
  Br(Cl)
	Polymeric halogenating reagent	Yaroslavsky *et al.*, 1970; Yaroslavsky and Katchalski, 1972

N-Bromocopolyethylene maleimide/ 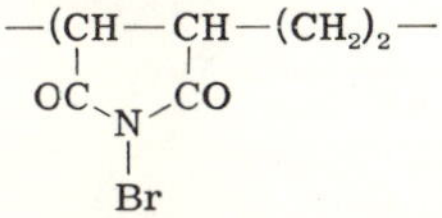
 —(CH—CH—(CH$_2$)$_2$—
 | |
 OC CO
  \\N/
   |
   Br
	Polymeric halogenating reagent	Sahni, 1977

N-Hydroxycopolyethylene maleimide/ 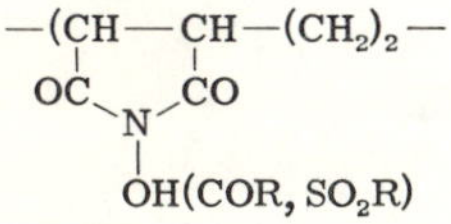
 —(CH—CH—(CH$_2$)$_2$—
 | |
 OC CO
  \\N/
   |
  OH(COR, SO$_2$R)
	Polymeric active ester for acylation and peptide synthesis	Laufer *et al.*, 1968; Fridkin *et al.*, 1972; Sahni *et al.*, 1977b

(*Continued*)

TABLE 2-7 (*Continued*)

Polymer/functional group	Reaction use	References
N-Hydroxyco(polystyrene maleimide)/ 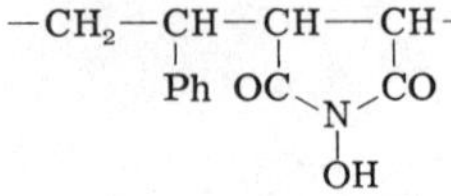 $-CH_2-CH-CH-CH-$ $PhOC\diagdown_N\diagup CO$ $\vert$ OH	Polymeric active ester for acylation and peptide synthesis	Rogozhin et al., 1973
Carbodiimide group incorporated in polyethylene/ 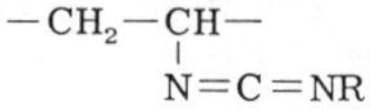 $-CH_2-CH-$ $\vert$ $N=C=NR$	Polymeric condensing agent for peptide and anhydride synthesis	Fridkin et al., 1971
Polyhexamethylenecarbodiimide/ $-(CH_2)_6N=C=N-(CH_2)_6-$	Polymeric condensing agent for peptide and anhydride synthesis	Wolman et al., 1967
Polyethylenimine/ 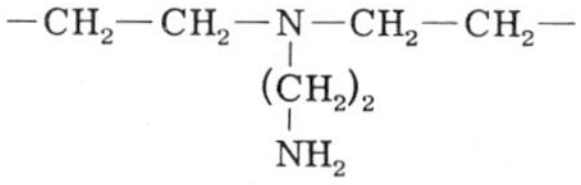 $-CH_2-CH_2-N-CH_2-CH_2-$ $(CH_2)_2$ $\vert$ NH_2	Polymer support for peptide synthesis by controlled NCA coupling (via amino acid amide bond)	Pfaender et al., 1975

TABLE 2-8

Inorganic Polymers

Polymer/support	Reaction use	References
Merrifield-type silanized glass support/ 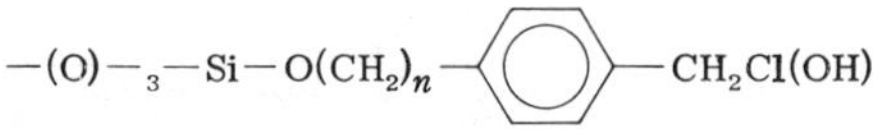 $-(O)-_3-Si-O(CH_2)_n-\bigcirc-CH_2Cl(OH)$	Solid support for peptide synthesis (via amino acid benzyl ester)	Bayer et al., 1972; Parr and Grohamm, 1972; Parr et al., 1974
Corning 3-aminopropyl glass (APG)/$\equiv$SiO(CH$_2$)$_3$NH$_2$	Solid support for peptide sequencing	Wachter et al., 1975
Corning β-N-aminoethyl-(2-aminopropyl) glass (β-APG)/$\equiv$SiO(CH$_2$)$_3$NH(CH$_2$)$_2$NH$_2$	Same as above	Wachter et al., 1975
Trityl chloride incorporated into silica gel/$\equiv$Si—TrCl	Polymer support for oligonucleotide synthesis (via the trityl ether bond)	Koster, 1972a

IV. Miscellaneous Polymer Supports

TABLE 2-8 (*Continued*)

Polymer/support	Reaction use	References
Glass-bound dye/ —(O)—$_3$Si(CH$_2$)$_3$NH-indicator	Acid–base indicator for pH measurement	Harper, 1975
Silica gel—dry	Inorganic support for FeCl$_3$; MeO$^-$Na$^+$; KMnO$_4$	Keinan and Mazur, 1977a, 1978; Regen and Koteel, 1977
Silica–alumina and alumina—dry	Inorganic support for oxidations and reductions of sulfides and alcohols and oxidative rearrangements of ketones and olefins	Taylor *et al.*, 1976; San Filippo and Chern, 1977; Keinan and Mazur, 1977b; Posner *et al.*, 1977; Liu and Tong, 1978; Posner, 1978
Clays and celite	Inorganic support for oxidative rearrangements of ketones and olefins; oxidations of alcohols to ketones by KMnO$_4$ and Collins reagent	Andersen and Uh, 1973; Kakis *et al.*, 1974 Fetizon and Mourges, 1974; Taylor *et al.*, 1976 Regen and Koteel, 1977
Graphite–potassium	Support for reduction of ketones, alkylhalides, alkyl sulfonate esters; formation of enolate anions; reduction of transition metals	Bergbreiter and Killough, 1978 and references therein
Graphite–bromine®	Support for bromination of alkenes and ketones	Page-Lecuyer *et al.*, 1973
Graphite–antimony pentachloride	Support for halogen displacement	Luche *et al.*, 1974
Graphite–chromium trioxide	Support for oxidizing alcohols to aldehydes	Lalancette *et al.*, 1972
Graphite–aluminum trichloride	Support for alkylation of aromatic systems	Lalancette *et al.*, 1974
Graphite–bisulfate	Support for esterification and ketal formation	Bertin *et al.*, 1974

beads used for such supports are manufactured by Corning Glass and marketed by Pierce Chemical Co., Rockford, Ill. under the name Corning Biochemical Supports® ("Pierce Handbook and General Catalogue," 1977, **78**, p. 274). Silane coupling reagents are used to activate the porous glass, and in most cases the reactive organic molecules are linked to the support via spacer-arms.

Such supports have been frequently used in preparing supported catalysts. Transition metal complexes are easily bound to phosphenated silica. Benzyl halide groups bound to glass surfaces have been used for peptide synthesis. One of the most recent and interesting applications of porous glass supports has been in binding of acid–base indicators to produce re-usable glass-bound pH indicators. In addition to their use in batch reactions, the porous glass supports have been used in column-based applications where dimensional stability of the support is required.

Besides being modified by silation, silica gel alone (Table 2-8) has recently been reported as a carrier for ferric chloride, sodium methoxide, and potassium permanganate (Keinan and Mazur 1977a, 1978; Regen and Koteel, 1977). Silica–alumina and alumina (Table 2-8) have both been used as carriers and alone to promote reactions (for leading references see Taylor et al., 1976; San Filippo and Chern, 1977; Keinan and Mazur, 1977b; Posner et al., 1977; Liu and Tong, 1978; Posner, 1978).

In addition to the silica and alumina supports mentioned previously, clays (Taylor et al., 1976; Regen and Koteel, 1977) and Celite (Kakis et al., 1974; Fetizon and Mourges, 1974; Andersen and Uh, 1973) have been used for thallium, silver, and chromium salts.

Finally, some mention must be made of graphite insertion compounds. These compounds, in which reactive compounds such as potassium metal are intercalated between the planes of carbon atoms in the graphite, have been used as catalysts (Boersma, 1974) and as reagents (Kagan, 1976a,b) in organic chemistry. They have many of the characteristics of insolubilized organic and inorganic polymers. Some of the various insertion compounds that have been prepared are found in Table 2-8. Many of these materials are available from the Alfa Division, Ventron Corporation under the trade name Graphimets®. Their pyrophoric nature and the fact that many of the reactions in which they may be used can be done by alternate methods have tended to limit their use in organic chemistry.

REFERENCES

Amos, J. L., Coulter, K. E., and Tennant, F. M. (1952). "Styrene" (R. H. Boundy, and R. F. Boyer, eds.), p. 729. Van Nostrand-Reinhold, New York.
Anderson, N. H., and Uh, H. (1973). *Synth. Commun.* **3**, 115.
Arshady, R., Kenner, G. W., and Ledwith, A. (1976). *Makromol. Chem.* **177**, 2911.
Atherton, E., Clive, D. L. J., and Sheppard, R. C. (1975). *J. Am. Chem. Soc.* **97**, 6584.
Bayer, E., and Geckeler, K. (1974). *Annalen,* 1671.
Bayer, E., Breitmaier, E., Jung, G., and Parr, W. (1972). 2nd Am. Pept. Symp. Cleveland 1970, "Progress in Peptide Research," Vol. II, (S. Lande, ed.), Gordon & Breach, New York.
Bayer, E., Mutter, M., Uhmann, R., Polster, J. and Mauser, H. (1974). *J. Am. Chem. Soc.* **96**, 7333.
Bergbreiter, D. E., and Killough, J. M. (1978). *J. Am. Chem. Soc.* **100**, 2126.
Bertin, J., Kagan, H. B., Luche, J. L., and Setton, R. (1974). *J. Am. Chem. Soc.* **96**, 8113.
Biedermann, H. G., Griessl, E., and Wichmann, K. (1973). *Makromol. Chem.* **172**, 49.
Bilibin, A. Yu., and Vlasov, G. P. (1973). *J. Gen. Chem. USSR* **43**, 1828.

References

Bilibin, A. Yu., Kozhevnikova, N. Yu., Vlasov, G. P. (1973). *Zh. Obshch. Khim.* 43, 2046 [C.A. 1974, **80,** 15176].
Blackburn, G. M., Brown, M. J., and Harris, M. P. (1969). *J. Chem. Soc.* **C,** 676.
Blossey, E. C., and Neckers, D. C. (1974). *Tetrahedron Lett.,* 323.
Blossey, E. C., Neckers, D. C., Thayer, A. L., and Schaap, A. P. (1973). *J. Am. Chem. Soc.* **95,** 5820.
Boersma, M. A. M. (1974). *Catal. Rev. Sci. Eng.* **10,** 243.
Brandstetter, F., Schott, H., and Bayer, E. (1973). *Tetrahedron Lett.,* 2997.
Braun, D. (1959). *Makromol. Chem.* **30,** 85.
Braun, D., and Selig, E. (1964). *Chem. Ber.* **97,** 3098.
Brown, J., and Williams, R. E. (1971). *Can. J. Chem.* **49,** 3764.
Camps, F., Castells, J., Ferrando, M. J., and Font, J. (1971a). *Tetrahedron Lett.,* 1713.
Camps, F., Castells, J., Font, J., and Vela, F. (1971b). *Tetrahedron Lett.,* 1715.
Capka, M., Svoboda, P., Cerny, M., and Hetflej, J. (1971). *Tetrahedron Lett.,* 4787.
Chalk, A. J. (1968). *J. Polym. Sci. Part B.* **6,** 649.
Chapman, T. M., and Kleid, D. G. (1973). *J. Chem. Soc. Chem. Commun.,* 193.
Chapman, P. H., and Walker, D. (1975). *J. Chem. Soc. Chem. Commun.,* 690.
Collmann, J. P., and Reed, C. A. (1973). *J. Am. Chem. Soc.* **95,** 2048.
Cramer, F., and Koster, H. (1968). *Angew. Chem. Int. Ed.* **7,** 473.
Cramer, F., Helbig, R., Hettler, H., Scheit, K. H., Seliger, H. (1966). *Angew. Chem. Int. Ed.* **12,** 640.
Dowling, L. M., and Stark, G. R. (1969). *Biochemistry* **8,** 4728.
Evans, D. C., Phillips, L. Barrie, J. A., and George, M. H. (1974) *J. Polym. Sci. Polym. Chem. Ed.* **12,** 199.
Farrall, M. J., and Frechet, J. M. J. (1976). *J. Org. Chem.* **41,** 3877.
Feinberg, R. S., and Merrifield, R. B. (1974). *Tetrahedron Lett.,* 3204.
Felix, A. M., and Merrifield, R. B. (1970). *J. Am. Chem. Soc.* **92,** 1385.
Fetizon, M., and Mourges, P. (1974). *Tetrahedron* **30,** 327.
Flanigan, E., and Marshall, G. R. (1970). *Tetrahedron Lett.,* 2403.
Frechet, J. M. J., and Farrall, J. (1977). "Chemistry and Properties of Crosslinked Polymers" (S. S. Labana, ed.), p. 59. Academic Press, New York.
Frechet, J. M. J., and Haque, K. E. (1975). *Marcomolecules* **8,** 130.
Frechet, J. M. J., and Pelle, G. (1975). *J. Chem. Soc. Chem. Commun.,* 225.
Frechet, J. M. J., de Smet, M. D., and Farrall, M. J. (1979). *J. Org. Chem.* **44,** 1774.
Fridkin, M., Patchornik, A., and Katchalski, E. (1965). *J. Am. Chem. Soc.* **87,** 4646.
Fridkin, M., Patchornik, A., and Katchalski, E. (1966). *J. Am. Chem. Soc.* **88,** 3164.
Fridkin, M., Patchornik, A., and Katchalski, E. (1971). *Peptides 1969, Proc. Eur. Peptide Symp., 10th,* p. 164. (E. Scoffone, ed.), North-Holland Publ., Amsterdam.
Fridkin, M., Patchornik, A., and Katchalski, E. (1972). *Biochemistry* **11,** 466.
Fyles, T. M., and Leznoff, C. C. (1976). *Can. J. Chem.* **54,** 935.
Gait, M., and Sheppard, R. (1976). *J. Am. Chem. Soc.* **98,** 8514.
Gait, M. J., and Sheppard, R. C. (1977). *Nucl. Acids Res.* **4,** 1135, 4391.
Glaser, R., Sequin, U., and Tamm, C. (1973). *Helv. Chim. Acta.* **56,** 654.
Gorecki, M., and Patchornik, A. (1973). *Biochim. Biophys. Acta* **303,** 36.
Grubbs, R. H., and Kroll, L. C. (1971). *J. Am. Chem. Soc.* **93,** 3062.
Guthrie, R. D., Jenkins, A. D., and Stehlicek, J. (1971). *J. Chem. Soc.* **C** 2690.
Guthrie, R. D., Jenkins, A. D., and Roberts, G. (1973). *J. Chem. Soc. Perkin 1,* 2414.
Hallensleben, M. L. (1974). *J. Polym. Sci. Polym. Symp.* **47,** 1.
Harper, G. B. (1975). *Anal. Chem.* **47,** 348.
Harrison, I. T., and Harrison, S. (1967). *J. Am. Chem. Soc.* **89,** 5723.

Harrison, C. R., and Hodge, P. (1974). *J. Chem. Soc. Chem. Commun.*, 1009.
Hayatsu, H., and Khorana, H. G. (1966). *J. Am. Chem. Soc.* **88**, 3182.
Hayatsu, H., and Khorana, H. G. (1967). *J. Am. Chem. Soc.* **89**, 3880.
Heitz, W., and Michels, R. (1971). *Makromol. Chem.* **148**, 9.
Heitz, W., and Michels, R. (1972). *Angew. Chem. Int. Ed.* **11**, 298.
Helfferich, F. (1962). "Ion Exchange." McGraw-Hill, New York.
Helfferich, F., and Luten, D. B. (1964). *J. Appl. Pol. Sci.* **8**, 2899.
Issleib, K., and Tzschach, A. (1959). *Chem. Ber.* **92**, 1118.
Kabachink, M. M., Polyakova, I. A., Potapov, V. K., Shaborova, Z. A., and Prokof'ev, M. A. (1970). *Dokl. Akad. Nauk. SSSR Ser. Khim.* **195**, 1344.
Kagan, H. B. (1976a). *Pure Appl. Chem.* **46**, 177.
Kagan, H. B. (1976b). *Chem. Tech.* 510.
Kakis, F. J., Fetizon, M., Douchkine, N., Golfier, M., Mourges, P., and Prange T. (1974). *J. Org. Chem.* **39**, 523.
Kalir, R., Warshawsky, A., Fridkin, M., and Patchornik, A., (1975). *J. Org. Chem.* **39**, 55.
Keinan, E., and Mazur, Y. (1977a). *J. Am. Chem. Soc.* **99**, 3861.
Keinan, E., and Mazur, Y. (1977b). *J. Org. Chem.* **42**, 844.
Keinan, E., and Mazur, Y. (1978). *J. Org. Chem.* **42**, 1020.
Kenner, G. W., McDermott, J. R., and Sheppard, R. C. (1971). *J. Chem. Soc. Chem. Commun.*, 636.
Koplow, W., Esch, T. E. S., and Smid, J. (1973). *Macromolecules* **6**, 133.
Koster, H. (1972a). *Tetrahedron Lett.*, 1527.
Koster, H. (1972b). *Tetrahedron Lett.*, 1535.
Koster, H., and Heyns, K. (1972). *Tetrahedron Lett.*, 1531.
Kurimura, Y., Tsuchida, E., Kaneko, M. (1971). *J. Pol. Sci. A-1*, **9**, 3511.
Kusama, T., and Hayatsu, H. (1970). *Chem. Pharm. Bull.* **18**, 319.
Lalancette, J. M., Rollin, G., and Dumas, P. (1972). *Can. J. Chem.* **50**, 3058.
Lalancette, J. M., Fournier-Breault, M. J., and Thiffault, R. (1974). *Can. J. Chem.* **52**, 589.
Laufer, D. A., Chapman, T. M., Morlborough, D. I., Vaidya, V. M., and Blout, E. R. (1968). *J. Am. Chem. Soc.* **90**, 2696.
Laursen, R. A. (1971). *Eur. J. Biochem.* **20**, 89.
Leebrick, J. R., and Ramsden, H. E. (1958). *J. Org. Chem.* **23**, 235.
Letsinger, R. L., and Hamilton, S. B. (1959). *J. Am. Chem. Soc.* **81**, 3009.
Letsinger, R. L., and Kalus, I. (1964). *J. Am. Chem. Soc.* **86**, 3884.
Letsinger, R. L., and Mahadevan, V. (1966). *J. Am. Chem. Soc.* **88**, 5319.
Letsinger, R. L., Kornet, M. J., Mahadevan, V., and Jerina, D. M. (1964). *J. Am. Chem. Soc.* **86**, 5163.
Leznoff, C. C., and Wong, J. Y. (1973). *Can. J. Chem.* **51**, 3756.
Livshits, A., and Vasil'ev, A. E. (1973). *Zh. Obshch. Khim.* **43**, 219 [*C.A.* 1973, **78**, 111711].
Liu, KT., and Tong, YC. (1978). *J. Org. Chem.* **43**, 2717.
Luche, J. L., Bertin, J., and Kagan, H. B. (1974). *Tetrahedron Lett.*, 759.
McKinley, S. V., and Rakshys, Jr., J. W. (1972). *J. Chem. Soc. Chem. Commun.*, 134.
Manecke, G., and Haake, E. (1968). *Naturwissenschaften.*, **55**, 343.
Marshall, G. R., and Liener, I. E. (1970). *J. Org. Chem.* **35**, 867.
Melby, L. R., and Strobach, D. R. (1967). *J. Am. Chem. Soc.* **87**, 450.
Merrifield, R. B. (1963). *J. Am. Chem. Soc.* **85**, 2149.
Michalska, Z. M., and Webster, D. E. (1975). *Chem. Tech.* 117.
Mitchell, A. R., Kent, S. B. H., Erickson, B. W., and Merrifield, R. B. (1976). *Tetrahedron Lett.*, 3795.
Mutter, N., and Bayer, E. (1974). *Chem. Ber.* **107**, 1344.
Mutter, M., Hagenmaier, H., and Bayer, E. (1971). *Angew. Chem. Int. Ed.* **10**, 811.

References

Narang, C. K., Brunfeldt, K., and Norris, K. E. (1977). *Tetrahedron Lett.,* 1819.
Neckers, D. C., Kooistra, D. A., and Green, G. W. (1972). *J. Am. Chem. Soc.* **94,** 9284.
Overberger, C. G., and Salomone, J. C. (1969a). *Macromolecules* **2,** 533.
Overberger, C. G., and Salomone, J. C. (1969b). *Acc. Chem. Res.* **2,** 217.
Page-Lecuyer, A., Luche, J. L., Kagan, H. B., and Mazieres, C. (1973). *Bull. Soc. Chim. Fr.* 1690.
Panse, G. T., and Laufer, D. A. (1970). *Tetrahedron Lett.,* 4181.
Parr, W., Grohmann, K. (1972). *Tetrahedron Lett.,* 2633.
Parr, W., Grohmann, K., and Hagele, K. (1974). *Annalen,* 655.
Patterson, J. A. (1971). "Biochemical Aspects of Reactions on Solid Supports" (G. R. Stark, ed.), p. 189. Academic Press, New York.
Pfaender, P., Pratzel, H., Blecher, H., Gorka, G., and Hausen, G. (1975). *Pept. Proc. Eur. Pept. Symp., 13th,* 1974, p. 137. Wiley, New York.
Posner, G. H., (1978). *Angew. Chem. Int. Ed.* **17,** 487.
Posner, G. H., Runquist, A. W., and Chapdelaine, M. J. (1977). *J. Org. Chem.* **42,** 1202.
Potapov, V. K., Chekhmakhcheva, O. G., Shabarova, Z. A., and Prokof'ev, M. A. (1971). *Dokl. Akad. Nauk. SSSR Ser. Khim.* **196,** 360.
Potapov, V. K., Turkin, S. I., and Shabarova, Z. A. (1972). *Zh. Obshch. Khim.,* **42,** 2349.
Rebek, J., and Gavina, F. (1974). *J. Am. Chem. Soc.* **96,** 7112.
Regen, S. L., and Koteel, C. (1977). *J. Am. Chem. Soc.* **99,** 3837.
Regen, S. L., and Lee, D. P. (1974). *J. Am. Chem. Soc.* **96,** 294.
Rogozhin, S. V., Davidovich, Yu. A., Andrev, S. N., and Yurtanov, A. I. (1973). *Doklady Akad. Nauk. SSSR Ser. Khim.* **212,** 108. [*C.A.* 1974, **80,** 60192a].
Rosenthal, I., and Archer, J. (1974). *Israel J. Chem.* **12,** 897.
Rousch, W. R., Feitler, D., and Rebek, J. (1974). *Tetrahedron Lett.,* 1391.
Rubinstein, M., and Patchornik, A. (1972). *Tetrahedron Lett.,* 2881.
Rubenstein, M., and Patchornik, A. (1975). *Tetrahedron* **31,** 517.
Sahni, M. K. (1977). Ph.D. Thesis, University of Jodhpur, Jodhpur, India.
Sahni, M. K., Jain, J. C., Narang, C. K., and Mathur, N. K. (1977a). *Indian J. Chem.* **15,** 481.
Sahni, M. K., Sharma, I. K., Narang, C. K., and Mathur, N. K. (1977b). *Synth. Commun.* **7,** 57.
San Filippo, J., Jr., and Chern, C. I. (1977). *J. Org. Chem.* **42,** 2182.
Schott, H., Brandstetter, F., and Bayer, E. (1973). *Makromol. Chem.* **173,** 247.
Seliger, H., and Aumann, G. (1973). *Tetrahedron Lett.,* 2911.
Shambhu, M. B., and Digenis, G. A. (1973). *Tetrahedron Lett.,* 1627.
Shambhu, M. B., and Digenis, G. A. (1974). *J. Chem. Soc. Chem. Commun.* 610.
Sklyarov, L. Yu., Gorbunov, V. I., and Schukina, L. A. (1966).*J. Gen. Chem. U.S.S.R.* **36,** 2217.
Snyder, R. V., Angelici, R. J., and Meck, R. B. (1972). *J. Am. Chem. Soc.* **94,** 2660.
Southard, G. L., Brooke, G. S., and Pettee, J. M. (1969). *Tetrahedron Lett.,* 2505.
Sparrow, J. T. (1975). *Tetrahedron Lett.,* 4637.
Stewart, J. M., and Young, J. D. (1969). "Solid-Phase Peptide Synthesis." Freeman, San Francisco, California.
Takagi, T. (1967). *Poly. Lett.* **5,** 1031.
Takaishi, N., Imai, H., Bertelo, C. A., and Stille, J. K. (1976).*J. Am. Chem. Soc.* **98,** 5440.
Tanimoto, J., Horikawa, J., and Oda, R. (1967). *Kogyo Kagaku Zasshi* **70,** 1269.
Taylor, E. C., Chiang, C.-S., McKillop, A., and White, J. F. (1976).*J. Am. Chem. Soc.,* **98,** 6750.
Tilak, M. A. (1968). *Tetrahedron Lett.,* 6323.
Tsou, K. C., and Yip, K. F. (1973). *J. Macromol. Sci. Chem.* **7**(5), 1097.

Vlasov, G. P., Bilibin, A. Yu., Kuznetsova, N. Yu., Kitkovskaya, I., and Lashkov, V. N. (1973). *Chem. Ztg.* **97,** 236 [*C.A.* 1973, **79,** 42833].
Wachter, E., Hofner, H., and Machleidt, W. (1975). "Solid-Phase Methods in Protein Sequence Analysis" (R. Laursen, ed.), p. 31. Pierce Chem. Co. Rockford, Illinois.
Warshawsky, A., Kalir, R., and Patchornik, A. (1978). *J. Org. Chem.* **43,** 3151.
Weinshenker, N. M., and Shen, C. M. (1972). *Tetrahedron Lett.,* 3281, 3285.
Weinshenker, N. M., Crosby, G. A., and Wong, J. Y. (1975). *J. Org. Chem.* **40,** 1966.
Weygand, F. (1968). *Pept. Proc. Eur. Pept. Symp., 9th, 1968,* p. 183. North-Holland, Amsterdam.
Wieland, Th., and Birr, Ch. (1966). *Angew. Chem. Int. Ed.* **5,** 310.
Wieland, Th., Birr, Ch., and Fleckenstein, P. (1972). *Annalen,* **756,** 14.
Williams, R. E., Brown, J., and Lauren, D. R. (1972). *Polym. Prepr. Am. Chem. Soc., Div. Polym. Chem.* **13,** 823.
Wolman, Y., Kivity, S., and Frankel, M. (1967). *J. Chem. Soc. Chem. Commun.,* 629.
Yaroslavsky, C., and Katchalski, E. (1972). *Tetrahedron Lett.,* 5173.
Yaroslavsky, C., Patchornik, A., and Katchalski, E. (1970). *Tetrahedron Lett.,* 3629.

3 Determination of Functionalization in Polymer Supports

I.	Introduction	37
II.	Chemical Methods for Functional Group Analysis of Polymers	38
III.	Physical and Physicochemical Methods of Determining Functional Groups in Polymers	41
	A. Infrared Spectroscopy	41
	B. Ultraviolet Spectroscopy and Optical Rotation Measurements	42
	C. Nuclear Magnetic Resonance (nmr) Spectroscopy	44
	D. Electron Nuclear Double Resonance (ENDOR) and Raman Spectroscopy	44
	E. Applications of Electron Spin Resonance Spectroscopy	45
	F. Determination of Distribution of Functional Groups in Resin Beads by X-Ray Back Scattering and Autoradiography of Radiolabeled Beads	45
IV.	Physical and Chemical Nature of Immobilization of Reactive Sites on Polymers	46
V.	Use of Radiolabeled Reagents to Follow the Changes in Resin Functionalities	49
VI.	Reporting of Results	50
	References	51

I. INTRODUCTION

The concentration of the functional groups in a polymeric reagent used for synthesis has to be determined so that the maximum amount of reagent available in the polymer can be known with certainty. For a polymer-bound substrate (or its transformed product), the determination must be carried out to discover either the loading of the resin or the "yield" in a reaction.

Unlike the situation with low-molecular-weight compounds, the quantitative determination of polymer-bound functional groups is, for various reasons, restricted. Letsinger *et al.* (1964) showed that the determination of functional groups in popcorn polymers, in the presence of solvents that can swell the polymer, could readily be made, since they were generally available for normal chemical reactions. Chemical methods could, therefore, be applied for determination of functional groups in polymers. Even so, chemically-based analytical methods are generally slow when compared with the many physical methods that might be used to determine functionalities in polymers. They also suffer from the fact that nothing is known of the degree of accessibility of the sites in the polymer. In the following paragraphs, the highlights of the two different approaches will be discussed.

II. CHEMICAL METHODS FOR FUNCTIONAL GROUP ANALYSIS OF POLYMERS

There are several functional groups which, when present in polymers, can be determined by chemical methods. These methods may further be classified as follows.

1. Methods based on determination of specific elements: In certain cases, the polymers contain functional groups of some specific element. For these polymers, it is sufficient to analyze the polymers for that particular element, and the degree of functionalization can be determined from the percentage of that element. In this case, it is assumed that no low-molecular-weight substances are adsorbed or entrapped in the polymer. Table 3-1 lists typical examples of the functional groups in polymers that can be determined by elemental analysis.

2. Chemical methods based on the reactions of functional groups: Chemical methods for the determination of a functional group in a polymer are applicable where the functional group can undergo straightforward reaction and if all of the functional groups, i.e., the reactive sites of the polymer, are readily available to the reagents in solution. Only such straightforward chemical reactions are suitable for functional group determination in which either one of the reactants or the products in the reaction is measureable.

When the functionalized polymer contains acidic or basic groups, such as those in ion-exchange resins, the acid–base titration methods are the most suitable for functional group determination. Such methods are well-established (Helfferich, 1962). In one report (Schou *et al.*, 1975), the free amino group in an unblocked resin–peptide was determined by nonaqueous titration with acid. Similarly, histidine incorporation could be deter-

TABLE 3-1

Examples of Polymers in Which the Degree of Functionalization Can Be Determined from the Analysis of Specific Elements

Polymer	Functional group	Specific element determined	References
Merrifield resin	—CH$_2$Cl	Chlorine	Merrifield, 1963
Polystyrene incorporating boronic acid groups	—B(OH)$_2$	Boron	Seymour and Frechet, 1976
Polymeric organotin dihydride	—Sn(H)$_2$—n-C$_4$H$_9$	Tin	Weinshenker et al., 1975
Polymeric trisubstituted phosphine	—P(C$_6$H$_5$)$_2$	Phosphorus	Camps et al., 1971
Polymeric thiol or sulfide	—SH(Me)	Sulfur	Farrall and Frechet, 1976
Poly(4-vinylpyridine-borane)	H$_2$C=HC—⟨N⟩—BH$_3$	Boron	Hallensleben, 1974
Silylated polystyrene	—SiMe$_2$Cl	Silicon	Farrall and Frechet, 1976
Copolystyrene–DVB)–AlCl$_3$	—AlCl$_3$	Aluminum	Neckers et al., 1972
Transition metal complexes bound to phosphenated and other polymers	—Rh(PPh$_3$)$_n$ —PtCl$_2$	Respective metal	Michalska and Webster, 1975
	H$_2$C=⟨⟩—Ti—⟨⟩		

mined by potentiometric titration with perchloric acid (Brunfeldt et al., 1969).

Hydroxyl and amino groups in polymers are easily determined by acylation methods (Sahni et al., 1977). Ester or amide groups can be determined by saponification. These methods, however, require longer reaction times than for their low-molecular-weight counterparts and are not very suitable for monitoring a solid-phase reaction.

Many more chemical methods of determining polymer-bound functional groups have been developed for specific cases. Since many of the reagents are bound to the polymer reversibly, the chemical methods for their determination can always be used after cleaving them from the polymers. But such methods are, again, time-consuming and unsuitable for rapid monitoring of a reaction. In Table 3-2 are listed some of the functional groups that can be determined by chemical methods.

TABLE 3-2

Functional Groups in Polymers That Can Be Determined by Chemical Methods

Functional group	Recommended method	References
Acidic groups —SO_3H, —CO_2H	Acid–base titration for determination of ion exchange capacity	Helfferich, 1962
Basic groups —$N^+R_3O^-H$, —NH_2, imidazole groups	Acid–base titrations including those in nonaqueous medium	Helfferich, 1962; Schou et al., 1975; Brunfeldt et al., 1969
Hydroxyl group	Acylation	Sahni et al., 1977; Farrall and Frechet, 1976
Ester or amide groups	Saponification method	Alfrey, 1964
Peracid group	Iodometric titration	Harrison and Hodge, 1974
N-haloimides or amides	Iodometric titration	Sahni, 1977
Organotin hydride	Reaction of iodooctane, followed by glc detn. of octane formed	Weinshenker et al., 1975
Sulfoxide in peptidyl resin	Iodometric titration	Norris et al., 1971
Lipoic acid	Reaction of reduced polymer with di-5,5'-thionitrobenzoic acid (DTNB) followed by determination of released 5-thionitrobenzoic acid in supernatant solution	Gorecki and Patchornik, 1973; Ellman, 1959

III. PHYSICAL AND PHYSICOCHEMICAL METHODS OF DETERMINING FUNCTIONAL GROUPS IN POLYMERS

No single physical method can find general application for the analysis of all types of resin-bound reagents, and the detailed description of each of the methods is beyond the scope of this book. These methods are quick and nondestructive and hence preferred over chemical methods. With suitable modifications, many physical methods can be adopted for the analysis of functional groups in solid, insoluble polymers. Crowley and Rapoport (1976) have discussed the possibliities of using physical methods of analysis in solid-phase synthesis. Some of the specific applications of such methods are now briefly dealt with.

A. Infrared Spectroscopy

An extensive literature on the infrared (ir) spectra of polymers exists (Zbinden, 1964; Henniker, 1967). Fortunately the ir spectra of functional groups anchored on polymers do not differ appreciably from those in small molecules, and the technique of taking the spectra of polymeric solids (film, mull, or KBr pellet) is well-developed. The sensitivity attained is approximately the same as that with the small molecules. Infrared spectroscopy has been particularly useful in following polymeric transformations. The characteristic absorption due to a particular functional group often disappears completely on chemical transformation, with a simultaneous appearance of the characteristic absorption of the new group(s). Thus, the completion of a reaction can be easily followed by scanning the ir spectra of reactant and product. Letsinger *et al.* (1964), Blackburn *et al.* (1969), and Farrall and Frechet (1976) have made extensive use of ir spectra to follow chemical transformations in polymers.

Infrared spectroscopy has mainly been used as a qualitative tool to show the presence of certain functional groups in a polymer or to determine the extent to which a chemical transformation has taken place. Quantitative correlations have been made in some cases, by transfer of the spectra from a logarithmic to a linear absorption ordinate chart. Beer's law plots could then be made. Letsinger *et al.* (1964) plotted the ratio of absorption due to the hydrogen bonded hydroxyl to the total hydroxyl vs loading of hydroxyl groups (in percentage) in a polymer. A nearly linear correlation was found, indicating that the extent of hydrogen bonding was directly proportional to the loading. Thus the ir spectra of polymers containing hydroxyl or carboxyl groups can provide additional information about the extent of hydrogen bonding in the resins, which purely chemical methods fail to do.

Crowley and Rapoport (1976) used difference spectra taken with purified unreacted co(polystyrene–2% DVB) pellets (10 mg of resin and 190 mg KBr) in the reference beam and pellets of the functionalized sample resin of the same composition in the other beam. This dramatically simplified the spectra of many solid-phase compounds. As a typical example, the difference spectra of chloromethylated resins having different degrees of substitution were obtained. Satisfactory Beer's law plots could be obtained for chloromethylation based upon the H—C—Cl bending vibration at 1250 cm^{-1}. In a number of other cases, the difference spectra were essentially superimposable on the spectra of model compounds.

Reactive site isolation within the resin matrix and intraresin reactions within cross-linked polystyrenes have been studied by ir spectroscopy. Rapoport and co-workers have demonstrated that complete site isolation does not occur. (Crowley *et al.*, 1973; Crowley and Rapoport, 1976). By reacting carboxylic acid-containing resins with dicyclohexylcarbodiimide and measuring the amount of anhydride formed, they found that even at low levels of loading (0.5%) intraresin reactions occurred. Varying the amount of cross-linking in the resin, and, thus the resin rigidity, did not affect the amount of intraresin reaction (Scott *et al.*, 1977). A selection of ir frequencies of various functional groups, compiled from papers on polymer-mediated syntheses, is shown in Table 3-3.

B. Ultraviolet Spectroscopy and Optical Rotation Measurements

These methods do not find wide, general applications in monitoring of solid-phase synthesis. Some of the polymer-mediated syntheses have been carried out using a soluble polymer support. Nucleotides show characteristic uv absorption, and the attachment of a nucleotide residue to a soluble polymer is easily followed by separating it by ultrafiltration or precipitation. The polymer-bound nucleotide is then redissolved and the concentration of the bound nucleotide determined by uv absorption (Hayatsu and Khorana, 1967).

Even when the polymer is insoluble, the determination can be made. For example, the extent of reaction in a nucleotide synthesis can be monitored by determining the decrease in nucleotide concentration in the supernatant solution by uv absorption. Alternatively, the polymer-bound nucleotide concentration in the supernatant solution can be determined by uv absorption. Alternatively, the polymer-bound nucleotide concentration in the supernatant solution can be determined by uv absorption.

TABLE 3-3

Infrared Frequencies of Some Functionalities on the Polymer

Polymer/functional group	Absorption (cm^{-1})	Assignment	References
Benzene ring	Many characteristic frequencies, used for calibration of ir spectra		Dyer, 1965
—CH_2Cl	1250	C$\underset{Cl}{\overset{H}{\diagdown}}$	Crowley and Rapoport, 1976
—$\overset{O}{\overset{\|}{C}}CH_3$	1685	C=O	Letsinger et al., 1964 Blackburn et al., 1969
—CO_2^-	1550	CO_2	Blackburn et al., 1969
—CO_2H	1725 1685	C=O——H C=O	Letsinger et al., 1964
—CH_2OH	3600 3420	HO———H HO	Letsinger et al., 1964
—CH_2OCOCl	1779	C=O	Letsinger et al., 1964
—CH_2CH_2OH	3600 3100	H—O H—O——H	Blackburn et al., 1969
—CH_2CH_2OAc	1735 1750	C=O	Blackburn et al., 1969
—$CH_2OCONHNH_2$	1715 1665	C=O	Letsinger et al., 1964
—CHO	1695	C=O	Frechet and Pelle, 1975
—SH	2560	S—H	Farrall and Frechet, 1976
—$CONHC_6H_5$	1650 3380, 3300	C=O N—H	Farrall and Frechet, 1976
—$Si(CH_3)_2Cl$	3600, 3380 1250, 770 820	O—H Si—C Si—OH	Farrall and Frechet, 1976

Alternatively, the polymer-bound nucleotide from a small sample can be cleaved, followed by its determination by uv spectroscopy.

In the case of oligosaccharide synthesis on soluble polymers, the attachment of the saccharide residue to the polymer has been determined by measuring the optical rotation. For example, when a preformed monomer incorporating a sugar residue was copolymerized with styrene, the incorporation of sugar in the polymer was determined by measuring optical rotation (Guthrie et al., 1971, 1973). In general, the unreacted oligosaccharide in a reaction mixture can also be determined by measuring the optical rotation. Finally, as in nucleotide syntheses, the saccharide moiety can also be cleaved from the polymer and determined by measurement of optical rotation.

C. Nuclear Magnetic Resonance (nmr) Spectroscopy

An extensive literature on the application of high-resolution nmr to macromolecules has been published (Bovey, 1972). ^{1}H-, ^{13}C-, and ^{19}F-nmr spectroscopy have all been used in monitoring solid-phase reactions. In one case (Relles and Schluenz, 1974), the reaction of a diphenylphosphinite (**2**) with a chloromethylated polymer (**1**) (Arbuzov reaction) was followed by observing the loss of phosphinite methyl group from the solution (Scheme 3-1). In the same studies, the conversion of phenylacetic acid into the corresponding acid chloride by phosgenated polymer **3** was also monitored by observing the change in nmr spectra of benzylic protons.

^{19}F-nmr spectroscopy of trifluoracetylated solid-phase peptide products has been found to be a very sensitive method for the detection of error peptides (Bayer *et al.*, 1972). The ^{19}F-chemical shift of acylated peptides is very sensitive to the nature of the modified residue.

The application of ^{13}C nmr to polymers is a more recent development. The method is based on identifying the ^{13}C atom in natural abundance in different types of resins. Although the ^{1}H-nmr spectra of the polymer tends to be very complex, it has been possible to identify aliphatic and aromatic carbons in polymers by ^{13}C nmr (Crowley and Rapoport, 1976). The natural abundance of ^{13}C can be easily detected using Allerhand probes (Allerhand *et al.*, 1973). The use of specifically enriched ^{13}C atoms in groups on the resin could permit observation of concentrations of the magnitude now being used.

D. Electron Nuclear Double Resonance (ENDOR) and Raman Spectroscopy

According to Crowley and Rapoport (1976), Raman spectroscopy and the ENDOR technique should prove useful for identifying functionalities in polymers. As developed by Feher (1956), the nmr transitions are observed by esr (electron spin resonance) and possess the high sensitivity of esr while retaining the advantage in resolution associated with nmr. The sensitivity of conventional nmr, including that of ^{13}C in natural abundance, can thus be increased by several orders by using a paramagnetic probe in the neighborhood of the moieties of interest. If such a probe could be diffused into the polymer or covalently bound to it,

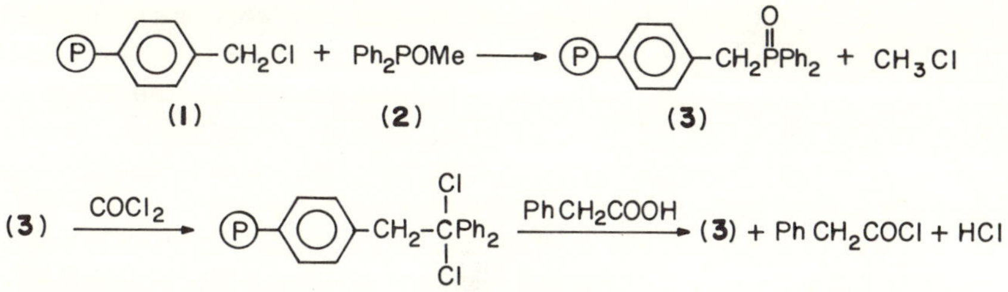

III. Physical and Physicochemical Methods

$$\text{(P)}-\text{C}_6\text{H}_4-\text{CH}_2\text{Cl} + \text{Ph}_2\text{POMe} \longrightarrow \text{(P)}-\text{C}_6\text{H}_4-\text{CH}_2\overset{\overset{\text{O}}{\|}}{\text{P}}\text{Ph}_2 + \text{CH}_3\text{Cl}$$
$$\quad\quad\quad (1) \quad\quad\quad\quad (2) \quad\quad\quad\quad\quad\quad\quad (3)$$

$$(3) \xrightarrow{\text{COCl}_2} \text{(P)}-\text{C}_6\text{H}_4-\text{CH}_2-\underset{\underset{\text{Cl}}{|}}{\overset{\overset{\text{Cl}}{|}}{\text{C}}}\text{Ph}_2 \xrightarrow{\text{PhCH}_2\text{COOH}} (3) + \text{PhCH}_2\text{COCl} + \text{HCl}$$

Scheme 3-1

then ENDOR could prove a very useful method of polymer analysis (Crowley and Rapoport, 1976).

E. Applications of Electron Spin Resonance Spectroscopy

Applications of esr spectroscopy for monitoring the degree of functionalization of a polymer are limited, primarily because esr-active groups are mostly used as probes rather than as reactive functionalities. Electron spin resonance spectroscopy has, however, been used to estimate the proximity of titanium groups in a titanocene polymer (Grubbs, *et al.*, 1973; Bonds *et al.*, 1975). It has also been used to demonstrate the presence of copper prophyrins in polymer-bound metalloporphyrins (Rollmann, 1975).

Kumler and Boyer (1976) used esr spectroscopy to study the glass transition of several spin-labeled polymers and copolymers, and spin-labeled polymeric matrices have been used to study solvent effects on chain flexibility in polystyrenes usually used in peptide solid-phase syntheses (see Section IV).

F. Determination of Distribution of Functional Groups in Resin Beads by X-Ray Back Scattering and Autoradiography of Radiolabeled Beads

Chemists working in the field of solid-phase synthesis have speculated about the uniformity of distribution of the reactive sites in resin particles. The distribution of the functionalities in a resin bead can be found either by autoradiography of the resin beads containing a suitable radiolabel [e.g., tritium-labeled peptide (Merrifield and Littau, 1968)] or by X-ray back scattering from an ion microprobe. Grubbs and Sweet (1975) used X-ray back scattering from an ion microprobe beam to scan sections of the resin bead containing a rhodium-phosphinated polystyrene. X-ray intensities were taken as a measure of the relative concentrations of a given element in different sections of the resin particles. It was established that whereas phosphenation was uniform throughout the bead, the metal distribution could be controlled by the choice of conditions for attachment. Under metal-deficient conditions, the metal distribution in the first 180 μm of the bead was uniform, with no metal in the center of the bead. Metal was uniformly distributed throughout the beads when stoichiometric or excess of the metal complex was used for the synthesis. Thus, X-ray back scattering from an ion microprobe established that the functional groups were uniformly distributed throughout the bead. It follows that the solid-phase reaction is not confined to the bead surface only. Furthermore, it imposes restrictions on the size of the soluble reactants that can diffuse into the bead to these reactive sites, and also upon the size of the product molecule. In such cases, the access to a linear molecule growing in a cavity may be restricted at a certain stage, resulting in truncated sequences.

Some restrictions arising from the distribution of the reactive sites and the pores in the resin beads may also increase the selectivity of the substrate. This has been observed in the case of hydrogenation using polymer-supported catalysts (Grubbs and Kroll, 1971).

IV. PHYSICAL AND CHEMICAL NATURE OF IMMOBILIZATION OF REACTIVE SITES ON POLYMERS

Despite the voluminous literature appearing on solid-phase reactions, relatively little attention has been paid to understanding the physical and chemical nature of immobilizing a reagent on a polymer matrix. Chemical reactions have indicated that reagents bound to 2% cross-linked styrene–DVB polymers are not fully immobilized. This has been attributed not only to the conformational mobility of the bound reagent but also to the fact that, when properly swelled, the polymer strands are translocated to and from the bead surface (Collman and Reed, 1973; Collman et al., 1974).

Regen (1974) described a method of assessing the mobility of the resin sites by covalent attachment of a spin-label probe [2,2,6,6-tetramethyl-4-piperidinyl-1-oxy group (Tempo) (Fig. 3-1. R = (P)—)] to the chloromethylated co(polystyrene–DVB). Rotational correlation times (τ) were calculated from observed room-temperature electron paramagnetic resonance (epr) spectra. The degree of swelling, q, (swelled volume/dry volume) was determined from the measured density of the dry resin and the weight of the imbibed solvent. The data indicate that those

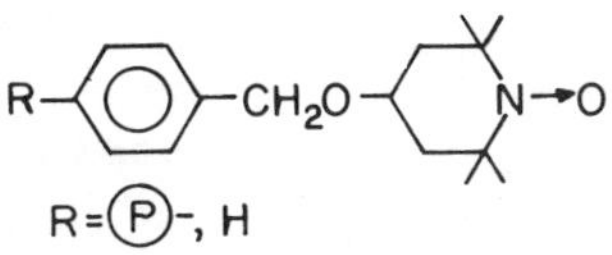

IV. Nature of Immobilization of Reactive Sites on Polymers

R—⟨O⟩—CH₂O—⟨N→O⟩

R = (P)—, H

Fig. 3-1

solvents which swell the resin matrix most will allow for the greatest mobility of the polymer-bound substrate (the nitroxide probe in this case).

It was also found that by increasing the cross-link density of the nitroxide-labeled polymer to 4% and 12%, the degree of swelling was decreased, resulting in an increase of rotational correlation time, i.e., decrease in the mobility of probe.

Regen (1974) interpreted the data to mean that the physical nature of the resin molecules varied substantially with the swelling of the polymer lattice, and that the chemical accessibility and reactivity of groups attached to it may be similarly influenced. In a set of similar experiments, the effect of solvents on spin-labeled polystyrene–poly(methyl methacrylate) polymers has been studied (Verksli and Miller, 1977). It was found that two populations of label were present in the polymers. One, a fast moving set, was being relaxed by terminal-bond motion while the other, a slower moving set, was being relaxed by local mode main chain motion. This latter motion was thought to correlate with the effect of solvent on the monomeric friction coefficient.

In another study (Stegmann et al., 1972), the probe consisted of a paramagnetically marked, mixed resin ester (Fig. 3-2) of 2,6-di-*tert*-butyl-4-hydroxyquinol. An interesting aspect of this latter study was that in addition to the mobility arising from translocation of the polymer strands by swelling, the mobility due to conformational flexibility of the attached group could also be studied. The conformational flexibility of the 2,6-di-*tert*-butyl-4-hydroxyquinol marker was varied by changing the length of the link between the marker and the resin. It was found that the line broadening of the paramagnetic marker has greater dependence on resin loading than on the length of the link.

Electron paramagnetic resonance studies of a nitroxide label have also been used (Regen, 1975) to compare the relative motional freedom of

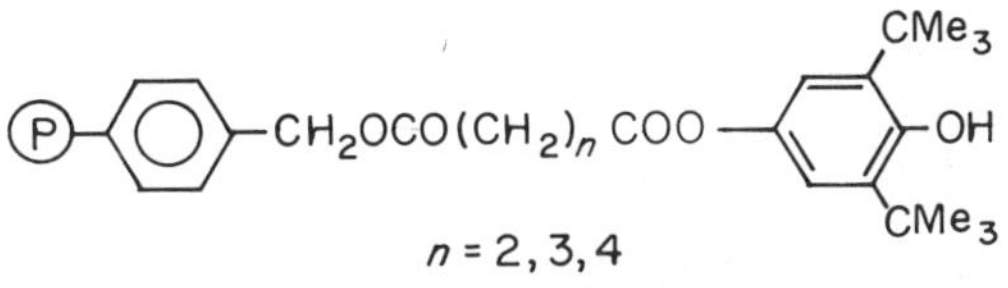

(P)—⟨O⟩—CH₂OCO(CH₂)$_n$COO—⟨O⟩(CMe₃)(OH)(CMe₃)

$n = 2, 3, 4$

Fig. 3-2

resin-bound molecular species and the corresponding free molecule imbibed in the solvent channels of the polymer. The rotational correlation times (τ) of a 2,2,6,6-tetramethyl-4-piperidiol oxyl substituent on resin benzyl groups (Fig. 3-1, R = (P)—) were compared with those of the benzyl ether (Fig. 3-1, R = H) of the nitroxide when it was imbibed into unfunctionalized resin at nearly the same concentration as the loading of functionalized resin. It was observed that the imbibed nitroxide had nearly 10 times more motional freedom than the immobilized label at the same degree of cross-linking. Furthermore, the motional freedom of the immobilized label and the imbibed label decreased in nearly the same order with increasing cross-linking or decreasing degree of swelling.

Regen (1977) has extended these studies to include the interaction of water with a polystyrene matrix containing pendant poly(ethylene glycol monomethyl ether) groups. This type of polymer has been used in triphase catalysis (see Chapter 13). It was demonstrated that "relatively fluid polar and nonpolar zones" existed within the polymer and that, upon changes in solvent composition, either a contraction of the polar–nonpolar interface occurred, or the polarity and viscosity in the polar and nonpolar zones were changed.

It is evident from these results that the degree of swelling of polystyrene matrix in benzene, as defined by the cross-link density, influences not only the mobility of polymer-bound molecules, but also the motion of molecules located within the solvent channels.

In addition to the degree of cross-linking and swelling of the resin, mobility and distribution of the functional groups on a polymer are also affected by ion-clustering. It has been reported that, in the case of polymers functionalized with charged groups (polyelectrolytes), extensive ion-cluster formation takes place when the polymer groups are anionic and the counter cations are small, e.g., lithium. This results in local, high concentrations of the functional groups in certain polymer regions. Due to such ion-clustering, the functional group isolation on the polymer does not remain effective, resulting in site aggregation and bis-reaction (Weinshenker *et al.*, 1975; Crosby and Kato, 1977). In contrast to this, cationic functional groups on the polymer (with anionic counter ions) cause repulsion of the polymer strands and favor site separation and mono reaction (Regen and Lee, 1974).

One of the major problems arising from chain mobility within the polymer matrix is lack of site isolation. Several uses of polymers as aids in organic chemistry have been based on the premise that sites within the polymer are isolated from one other. Experiments designed to test site isolation within polymer matrices have been numerous, but none has conclusively demonstrated *complete* and long-lasting isolation of sites

V. Use of Radiolabeled Reagents

within the polymer. The site-isolation experiments have relied on chemical reactions, ir spectroscopy, polymer-chelating ability, and reactive intermediate trapping (see Crowley and Rapoport, 1976; Neckers, 1978; Wulff and Schulze, 1978). Increased yields of cyclic peptides (Fridkin et al., 1965a,b; Patchornik and Kraus, 1970) and improved yields of unidirectional Dieckmann products from triethyl pimelate esters have been obtained (Crowley and Rapoport, 1970). Both results suggest that sites within the resin are isolated. One set of ir spectroscopic results suggest that complete site isolation does not occur (Letsinger et al., 1964) while another, more recent, study suggests that it does (Scott et al., 1977). The chelation of cobalt, copper, and iron by polymers containing pyridine and imidazole residues has been studied (Melby, 1975), and they suggest that site isolation was only partially achieved. Finally, reactive intermediate trapping experiments (Jayalekshmy and Mazur, 1976) suggest that the benzyne intermediate generated from the benzotriazole (Scheme 3-2) was present in an isolated state for about one minute after generation. Thus, site isolation in this case was partially achieved.

It is apparent from the foregoing that complete and long-lasting site isolation within polymer supports has not been observed. Before complete use of the hyperentropic efficacy (the high dilution principle) (Crowley and Rapoport, 1976) of polymers can be made in solid-phase synthesis, more information about site isolation will be necessary, especially on those polymer supports designed to make use of this factor.

V. USE OF RADIOLABELED REAGENTS TO FOLLOW THE CHANGES IN RESIN FUNCTIONALITIES

Wide possibilities exist for following the reactions of functionalized resins if radiolabeled reactants are employed. Groups incorporated into the resin or those remaining in the solution could be determined by scintillation counting of the solution or the resin. Many such examples have been reported. A few of them will now be discussed to illustrate the applications and scope of the method.

^{14}C-labeled Boc-amino acids have been used for following the coupling reaction of Boc-amino acids in solid-phase synthesis (Beyermann et al., 1973). In this case, the radioactivity incorporated on the resin was used as

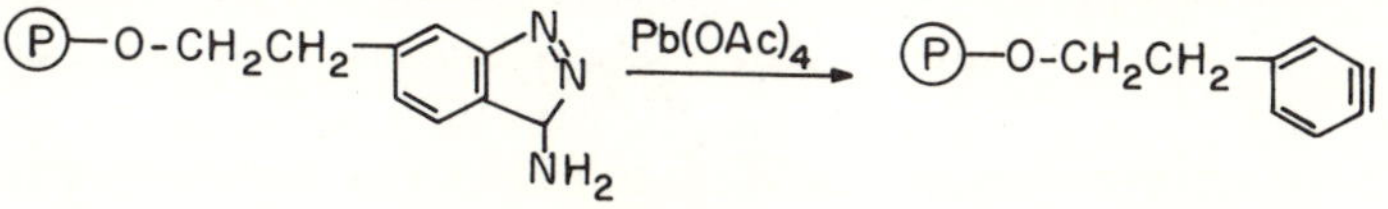

Scheme 3-2

a measure of the coupling reaction. Similarly, the N-deblocking of resin–Boc–peptides can also be followed by measuring the loss of radioactivity from the peptide–resin.

The method was also used to follow Dieckmann cyclization of a resin-bound pimelate ester. In loading of the resin and the subsequent reactions, scintillation counting was used to determine the fate of the radiolabel (Crowley and Rapoport, 1970). In oligonucleotide synthesis on polymer supports, thymidine [(^{3}H)methyl] was loaded on the polymer. This aided in determination of molar ratio of the products after cleavage, by a combination of scintillation counting and uv spectrophotometry (Narang et al., 1977).

The main limitation of this method is the availability of the required labeled compounds and their cost in a routine synthesis. In addition, quenching of the radioactivity by the polymer also has to be taken into account.

VI. REPORTING OF RESULTS

The molecular weight of most polymers is not uniform. Thus, the usual methods of expressing the concentration of a functional group in a molecule, e.g., equivalents per mole, percentage, or gram equivalent weight, cannot be used for polymeric molecules. It is generally convenient to express the concentration of a functional group in a polymer in terms of mEq/gm. This is generally referred to as the "capacity" of the resin. This expression, which was originally used in the case of ion-exchange resins, has also been applied to other functionalized resins.

When a polymer with a known capacity undergoes chemical transformation, the capacity of the newly introduced functional groups in the transformed polymer can also be determined. From this the "yield" of the reaction is readily calculated. When a functional group in a polymer is used to bind a substrate, the fraction of the functional groups reacting with the substrate is referred to as the "degree of loading of a functionalized polymer." In the case of functionalized styrene polymers, the percentage substitution (or degree of substitution) of the functional groups can be easily calculated from the known "resin capacity" in mEq/gm. As an example, a styrene–DVB copolymer, when completely monosubstituted with chloromethyl groups, should consist predominantly of units each having a formula weight of 152.5, and with each ring containing one chloromethyl group (one equivalent or 1000 mEq). Hence, the total capacity of such a polymer should be 1000/152.5 or about 6.5 mEq/gm. In typical chloromethylated co(polystyrene–DVB) polymers (Merrifield resins), however, the degree of chloromethylation is purposely kept low (0.75–1.25 mEq/gm corresponding to the chloromethylation of one out of every 7.8 phenyl groups) in order to avoid the possibility of residual unreacted groups after the initial loading.

REFERENCES

Alfrey, T. (1964). "Chemical Reactions of Polymers," (E. M. Fettes, ed.). Wiley (Interscience), New York.
Allerhand, A., Childers, R. F., and Oldfield, E. (1973). *J. Magn. Reson.* **11,** 272.
Bayer, E., Hunziker, P., Mutter, M., Sievers, R. E., and Uhmann, R., (1972).*J. Am. Chem. Soc.* **94,** 265.
Beyerman, H. C., vander Kemp, P. R. M., de Leer, E. W. B., van den Brink, W. M., Paramentier, J. H., and Westerling, J. (1973). *Pept. Proc. Eur. Pept. Symp., 11th, 1971,* p. 138. North-Holland Publ., Amsterdam.
Blackburn, G. M., Brown, M. J., and Harris, M. R. (1969). *J. Chem. Soc.* **C,** 676.
Bonds, Jr., W. D., Brubaker, Jr., C. H., Chandrasekaran, E. S., Gibbons, C., Grubbs, R. H., and Kroll, L. C. (1975) *J. Am. Chem. Soc.* **97,** 2128.
Bovey, F. A. (1972). "High Resolution N. M. R. of Macromolecules" Academic Press, New York.
Brunfeldt, K., Roppestorff, P., and Thomson, J. (1969). *Acta, Chem. Scand.* **23,** 2906.
Camps, F., Castells, J., Font, J., and Vela, F. (1971). *Tetrahedron Lett.,* 1715.
Collman, J. P., and Reed, C. A. (1973). *J. Am. Chem. Soc.* **95,** 2048.
Collman, J. P., Gagne, R. R., Kouba, J., and Ljusberg-Wahren, H. (1974).*J. Am. Chem. Soc.* **96,** 6800.
Crosby, G. A., and Kato, M. (1977). *J. Am. Chem. Soc.***99,** 278.
Crowley, J. I., and Rapoport, H. (1970). *J. Am. Chem. Soc.* **92,** 6363.
Crowley, J. I., and Rapoport, H. (1976). *Acc. Chem. Res.* **9,** 135.
Crowley, J. I., Harvey III, B., and Rapoport, H. (1973). *J. Macromol. Sci. Chem.* **7,** 1118.
Dyer, J. R. (1965). "Applications of absorption spectroscopy of organic compounds." Prentice-Hall, Englewood Cliffs, New Jersey.
Ellman, G. L. (1959). *Arch. Biochim. Biophys.* **82,** 70.
Farrall, M. J., and Frechet, J. M. J. (1976). *J. Org. Chem.* **41,** 3877.
Feher, G. (1956). *Phys. Rev.* **103,** 500.
Frechet, J. M. J., and Pelle, G. (1975). *J. Chem. Soc. Chem. Commun.,* 225.
Fridkin, M., Patchornik, A., and Katchalski, E. (1965a). *Israel J. Chem.* **3,** 69.
Fridkin, M., Patchornik, A., and Katchalski, E. (1956b). *J. Am. Chem. Soc.* **87,** 4646.
Gorecki, M., and Patchornik, A. (1973). *Biochim. Biophys. Acta* **303,** 36.
Grubbs, R. H., and Kroll, L. C. (1971). *J. Am. Chem. Soc.* **93,** 3062.
Grubbs, R. H., and Sweet, E. M. (1975). *Macromolecules,* **8,** 241.
Grubbs, R. H., Gibbons, C., Kroll, L. C., Bonds, W. D., and Brubaker, C. H. (1973).*J. Am. Chem. Soc.* **95,** 2373.
Guthrie, R. D., Jenkin, A. D., and Stehlicek, J. (1971). *J. Chem. Soc.* **C** 2690.
Guthrie, R. D., Jenkin, A. D., and Roberts, G., (1973). *J. Chem. Soc., Perkin 1,* 2414.
Hallensleben, M. L., (1974). *J. Polym. Sci. Polym. Symp.* **47,** 1.
Harrison, C. R., and Hodge, P. (1974). *J. Chem. Soc. Chem. Commun.,* 1009.
Hayatsu, H., and Khorana, H. G. (1967). *J. Am. Chem. Soc.* **89,** 3880.

Helfferich, F., (1962). "Ion Exchange." McGraw-Hill, New York.
Henniker, J. C. (1967). "Infra-red Spectrometry of Industrial Polymers." Academic Press, New York.
Jayalekshmy, P., and Mazur, S. (1976). *J. Am. Chem. Soc.* **98,** 6710.
Kumler, P. H., and Boyer, R. F. (1976). *Macromolecules* **9,** 903.
Letsinger, R. L., Kornet, M. J., Mahadevan, V., and Jerina, D. M. (1964). *J. Am. Chem. Soc.* **86,** 5163.
Melby, L. R. (1975). *J. Am. Chem. Soc.* **97,** 4044.
Merrifield, R. B. (1963). *J. Am. Chem. Soc.* **85,** 2149.
Merrifield, R. B., and Littau, V. (1968). *Pept. Proc. Eur. Pept. Symp., 9th, 1968,* p. 179. North-Holland Publ., Amsterdam.
Michalska, Z. M., and Webster, D. E. (1975). *Chem. Tech.* 117.
Narang, C. K., Brunfeldt, K., and Norris, K. E. (1977). *Tetrahedron Lett.,* 1819.
Neckers, D. C. (1978). *Chem. Tech.* 108.
Neckers, D. C., Kooistra, D. A., and Green, G. W. (1972). *J. Am. Chem. Soc.* **94,** 9284.
Norris, K. E., Halstrom, J., and Brunfeldt, K. (1971). *Acta. Chem. Scand.* **25,** 945.
Patchornik, A., and Kraus, M. A. (1970). *J. Am. Chem. Soc.* **92,** 7587.
Regen, S. L. (1974). *J. Am. Chem. Soc.* **96,** 5275.
Regen, S. L. (1975). *J. Am. Chem. Soc.* **97,** 3108.
Regen, S. L., (1977). *J. Am. Chem. Soc.* **99,** 3838.
Regen, S. L., and Lee, D. P. (1974). *J. Am. Chem. Soc.,* **96,** 294.
Relles, H. M., and Schluenz, R. W. (1974). *J. Am. Chem. Soc.* **96,** 6469.
Rollmann, L. D. (1975). *J. Am. Chem. Soc.* **97,** 2132.
Sahni, M. K. (1977). Ph.D. Thesis, University of Jodhpur, India.
Sahni, M. K., Sharma, I. K., Narang, C. K., and Mathur, N. K. (1977). *Synth. Commun.* **7,** 57.
Schou, O., Brunfeldt, K., Rubin, I., and Hansen, L. (1975). *Z. Physiol.* **356,** 1451.
Scott, L. T., Rebek, J., Ovsyanko, L., and Sims, C. L. (1977). *J. Am. Chem. Soc.* **99,** 625.
Seymour, E., and Frechet, J. M. J. (1976). *Tetrahedron Lett.,* 1149.
Stegmann, H. B., Breuninger, H., Scheffler, K. (1972). *Tetrahedron Lett.,* 3793.
Veksli, Z., and Miller, W. G. (1977). *Macromolecules* **10,** 686.
Weinshenker, N. M., Crosby, G. A., and Wong, J. W. (1975). *J. Org. Chem.* **40,** 1966.
Wulff, G., and Schulze, I. (1978). *Angew. Chem. Int. Ed.* **17,** 537.
Zbinden, R., ed. (1964). "Infra-red Spectroscopy of High Polymers." Academic Press, New York.

4

Polypeptide Synthesis on Polymer Supports

I.	Introduction and History	53
II.	Basic Principles of Merrifield's Solid-Phase Peptide Synthesis	55
III.	Supports for Solid-Phase Peptide Synthesis	56
IV.	Linkage of the First Amino Acid to the Polymer	62
V.	Protecting Groups Used in Solid-Phase Peptide Synthesis	63
VI.	Coupling of Successive Amino Acids to Resin-Bound Amino Acids	64
VII.	Cleavage of the Resin–Peptide Bond	65
VIII.	Monitoring of Solid-Phase Peptide Synthesis	67
IX.	Automation in Solid-Phase Peptide Synthesis	69
X.	Racemization Problems in Solid-Phase Peptide Products	70
XI.	Purification of Solid-Phase Peptide Products	70
XII.	Problems in Solid-Phase Synthesis	71
XIII.	Solid-Phase Coupling of N-Carboxylanhydrides (NCA)	73
XIV.	Solid-Phase Synthesis Using Side-Chain Functionalities for Attachment to Polymers and Bidirectional Extension of Peptide Chains	73
XV.	Fragment Condensation Strategy in Solid-Phase Peptide Synthesis	74
	References	75

I. INTRODUCTION AND HISTORY

In 1963, R. B. Merrifield reported an unconventional method of peptide synthesis for which the name "solid-phase peptide synthesis" was coined. This method differed from general organic synthetic methods in that one of the reactants was reversibly and covalently bound to an insoluble, solid polymer support which was then reacted with the reagent to give a resin-bound product. After filtering the latter from the reaction mixture and after repetition of as many steps as necessary to achieve the synthesis, the product was obtained by a suitable cleavage reaction (Merrifield, 1963; Marshall and Merrifield, 1965). It may be recalled that, in contrast, ordinary synthetic methods involve reactions of low-molecular-weight substances in solution, followed by multistage purification and recovery of the desired product.

Although solid-phase peptide synthesis is a relatively new technique, it has attracted the attention of many workers in the fields of chemistry, biochemistry, pharmaceutical and medical sciences, and certain other allied fields. During the short span of about 15 years, hundreds of research papers have appeared on this and allied topics.

Almost simultaneously with Merrifield, Letsinger's group (1963) reported the synthesis of a dipeptide on a "popcorn polymer support." A basic difference in the strategies employed by the two groups existed. In classical methods of peptide synthesis, it had been well established that chain elongation by successive coupling of amino acids at the NH_2-terminus of the growing peptide chain was more convenient than working from the COOH-terminus. Merrifield's group, following the "normal" strategy, attached the polymer to the first N-protected amino acid residue at the carboxylate ester. After N-deprotection, the peptide chain was built up by successive addition of amino acid residues to the NH_2-terminus. Letsinger's group, however, bound the NH_2-terminus of a C-protected amino acid to the polymer via a urethane bond and the chain was elongated from the COOH-terminus.

In the present monograph, we have dealt with solid-phase peptide synthesis as one of the several types of organic syntheses that can be achieved using polymeric reagents. The extensive literature on this topic will not be discussed here since it has been reviewed by Erickson and Merrifield (1976). Only important references, which cover the progress of the work in this field, are given. The three important sources available which survey the annual progress in this field are:

1. *Amino Acids, Peptides and Proteins, Specialist Periodical Reports* published by The Chemical Society, London, Vols. 1–10 (covering work during 1968–1976).

2. *Proceedings of European Peptide Symposia.* Recent *Proceedings* have been published by North-Holland Publishing Co., Amsterdam (1967–1973); Keter Press, Jerusalem (1975); Éditions de l'Université Bruxelles, Brussels (1976).

3. *Proceedings of American Peptide Symposia* published by Ann Arbor Science Publishers, Ann Arbor, Michigan.

In addition to these sources, a large number of reviews and books contain extensive coverage of work on solid-phase peptide synthesis (Merrifield, 1965a,b, 1966, 1967, 1968, 1969, 1971, 1972; Stewart and Young, 1969; Marshall and Merrifield, 1970; Stark, 1971; Stewart, 1972;

Young, 1972; Katsoyannis, 1973; Meienhofer, 1973; Hey and John, 1973; Fridkin and Patchornik, 1974; Stewart, 1976; Erickson and Merrifield, 1976).

II. BASIC PRINCIPLES OF MERRIFIELD'S SOLID-PHASE PEPTIDE SYNTHESIS

The principle of Merrifield's solid-phase synthesis of peptides is illustrated in Scheme 4-1 (Merrifield, 1963; Stewart and Young, 1969). The most commonly employed resin support is chloromethylated, 1% crosslinked co(polystyrene–divinylbenzene). The first N-protected amino acid is bound by its carboxyl group to the polymer via a benzyl ester. The second step involves the deprotection of the amino group (NH_2 group) under conditions that do not cleave the resin–amino acid ester bond. In the third step, a second N-protected amino acid is coupled to the amino group of the polymer-bound amino acid using dicyclohexylcarbodiimide (DCC) or via an active ester coupling. The N-deblocking (ii) and coupling (iii) steps are repeated until the desired sequence is formed (see Scheme 4-1). Finally,

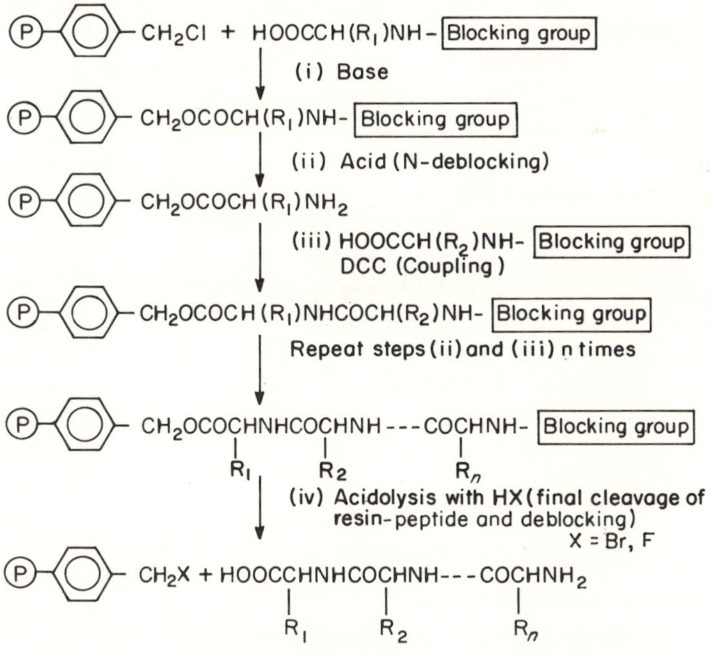

(P)–⟨◯⟩–CH_2Cl + $HOOCCH(R_1)NH$– Blocking group

        ↓ (i) Base

(P)–⟨◯⟩–$CH_2OCOCH(R_1)NH$– Blocking group

        ↓ (ii) Acid (N-deblocking)

(P)–⟨◯⟩–$CH_2OCOCH(R_1)NH_2$

        ↓ (iii) $HOOCCH(R_2)NH$– Blocking group
            DCC (Coupling)

(P)–⟨◯⟩–$CH_2OCOCH(R_1)NHCOCH(R_2)NH$– Blocking group

        ↓ Repeat steps (ii) and (iii) n times

(P)–⟨◯⟩–$CH_2OCOCHNHCOCHNH$---$COCHNH$– Blocking group
               | | |
               R_1 R_2 R_n

        ↓ (iv) Acidolysis with HX (final cleavage of resin-peptide and deblocking)
              X = Br, F

(P)–⟨◯⟩–CH_2X + $HOOCCHNHCOCHNH$---$COCHNH_2$
                       | | |
                       R_1 R_2 R_n

Scheme 4-1

the resin–peptide bond is cleaved by a suitable acid-catalyzed cleavage reaction (HBr–AcOH or TFA, HF), which results in simultaneous N-deblocking and deblocking of most of the side-chain functionalities (iv). Since only the final peptide is obtained by cleavage from the polymer, the purity of the final product depends upon the coupling efficiency of each step. This efficiency should be quantitative, or at least better than 99%, for the synthesis of long chains.

Initial experiments indicated reasonable success with this new technique for small and medium-sized (5–15 amino acid residues) peptides. However, many limitations of the new method were observed subsequently. Most of these limitations have since been overcome but some still remain (Stewart, 1976).

III. SUPPORTS FOR SOLID-PHASE PEPTIDE SYNTHESIS

The choice of chloromethylated polystyrene as the resin support to bind the amino acid substrate by benzyl ester is amply justified on the basis of the frequent use of benzyl esters of amino acids in solution and methods of peptide synthesis (Greenstein and Winitz, 1961; Schroder and Lubke, 1965; Bodanszky and Ondetti, 1966). Such a bond is stable during the various reactions of peptide synthesis but is readily cleaved by anhydrous hydrogen bromide at the end of synthesis. Details for the improved synthesis of chloromethylated co(polystyrene–DVB) have been reported (Feinberg and Merrifield, 1974; see also Chapter 2). Both microporous (swellable) and macroreticular or macroporous polymers have been examined as support material (Tilak and Hollinden, 1971; Sano et al., 1971), but in most synthetic reactions the swellable resins have an edge over the nonswellable resins.

In addition to the benzyl chloride group, certain other supports incorporating other functionalities have been used. These are summarized in Table 4-1. Although polystyrene resins can be synthesized readily from monomers and can be functionalized (Stewart and Young, 1969), Merrifield resins are now commercially available. The commercial products meet uniform specifications with respect to characteristics such as the particle size and shape, degree of cross-linking (which determines the degree of swelling and accessibility to reagents), and degree of functionalization.

The resin-bound substrates are assumed to be present on the surface of the resin beads as well as in the cavities or pores of the beads (Merrifield and Littau, 1968). The rate of coupling reactions in the solid-phase method is usually slower compared with those in solution. The reaction

TABLE 4-1

Polymer-Supported Peptide Synthesis

Polymer type/active site of bonding	Nature of the Ⓟ—first amino acid bond	End cleavage and remarks	References
Merrifield resins: chloromethylated, nitro-, and bromochloromethylated			
Bromochloromethylated co(polystyrene–DVB)	Ⓟ—bromobenzyl ester bond	HBr/AcOH	Merrifield, 1963
Chloromethylated co(polystyrene–DVB)			
microreticular, swelling type	Ⓟ—benzyl ester bond	HBr/AcOH	Marshall and Merrifield, 1965
macroreticular, nonswelling type	Ⓟ—benzyl ester bond	HBr/AcOH	Tilak and Hollinden, 1971
Linear, soluble	Ⓟ—benzyl ester bond	Same as above/product separation by ultrafiltration or dialysis	Shemyakin et al., 1965; Ovchinnikov et al., 1968; Green and Garson, 1969
Modified Merrifield resins, with more stable resin–peptide linkages			
Nitrochloromethylated polystyrene–DVB	Ⓟ—nitrobenzyl ester bond	HBr/HoAc	Merrifield, 1963
Aminoacyl-OCH$_2$–PAM polystyrene–DVB	Ⓟ—phenylacetamidomethylester bond	HF	Mitchell et al., 1976; Blake and Li, 1976; Mitchell et al., 1978, c.f. Blake and Li, 1976

(*Continued*)

TABLE 4-1 (Continued)

Polymer type/active site of bonding	Nature of the ⓟ—first amino acid bond	End cleavage and remarks	References
Modified Merrifield resins, useful for linking the first amino acid to the polymer, under mild conditions			
Bromomethylated co(polystyrene–DVB)	ⓟ—benzyl ester bond, coupling at room temperature	HBr/AcOH	Tilak, 1968
Merrifield resin, incorporating sulfonium groups	ⓟ—benzyl ester bond	HBr/AcOH	Dorman and Love, 1969
Hydroxymethylated co(polystyrene–DVB)	ⓟ—benzyl ester bond, coupling by carbonyldiimidazole, or by redox coupling or by DCC/4-dimethylaminopyridine	HBr/AcOH or NH_2NH_2/DMF	Bodansky and Sheehan, 1966; Lasse and Neubert, 1968; Ueki and Ikeda, 1974; Wang, 1975; Chang et al., 1976
Pellicularized resins, i.e., glass beads coated with co(polystyrene–DVB), and functionalized (benzyl halides) glass or silica beads. The swelling of the support is independent of the solvent, and eliminates failure sequences, but the resin capacity is low			
Glass beads coated with Merrifield resin	ⓟ—benzyl ester bond	HBr/AcOH	Bayer et al., 1970c
Silanized glass incorporating benzyl halide groups	ⓟ—benzyl ester bond		Bayer et al., 1970c; Parr and Grohmann, 1971, 1972; Parr et al., 1974
Polymers incorporating "safety catch" device for cleavage of ⓟ—peptide bond			
Co(polystyrene–DVB), incorporating (p-hydroxyphenyl)-thioether groups	ⓟ—phenyl ester bond, coupling by Carbonyldiimidazole	Sulfide oxidized to sulfone, with H_2O_2–AcOH or m-chloroperoxybenzoic acid–CH_2Cl_2, and cleavage by base	Marshall and Liener, 1970

Resin	Bond/coupling	Cleavage	Reference
Co(polystyrene–DVB), incorporating sulfonamido groups	Ⓟ—N-acylsufonamide bond, coupling by DCC	N-Methylation with CH_2N_2, and cleavage by N_2H_4, NH_3, or NaOH	Kenner et al., 1971
Merrifield resin, incorporating hydrazino groups	Ⓟ—N-acylhydrazine, coupling by DCC	N-Bromosuccinimide oxidative cleavage	Wieland et al., 1970

Resins incorporating spacer arms for peptide fragment coupling or stepwise peptide synthesis. The nonbenzylic resin–ester is cleaved by reactions other than protolysis, i.e., ammonolysis, hydrazinolysis, or mild alkaline saponification. Acid-labile side-chain protecting groups are not cleaved, so that the protected peptides can subsequently be used for fragment condensation by solution methods

Resin	Bond/coupling	Cleavage	Reference
Bromochloroacetylated co(polystyrene–DVB)	Ⓟ—phenacyl ester bond	Sodium thiophenolate	Weygand, 1968; Mizoguchi et al., 1969, 1970
Merrifield resin incorporating 2-hydroxyethylsulfone groups	Ⓟ—alkyl ester, coupling by DCC	Dilute alkali involving elimination	Tesser and Ellenbroek, 1967; Tesser et al., 1976
Polystyrene, incorporating hydroxyalkyl side chains	Ⓟ—alkyl ester bond, coupling by carbonyldiimidazole	Ammonolysis or alkaline saponification	Tilak and Hollinden, 1968; Parr et al., 1972; Bayer et al., 1971a
Polystyrene, incorporating spacer-arm (N-acetylamino) alcohol	Ⓟ—alkyl ester bond, coupling by carbonyldiimidazole	Alkaline saponification	Tilak and Hollinden, 1968
Polystyrene, incorporating tBoc-hydrazine spacer-arm	Ⓟ—amino acid (N-Bpoc protected) hydrazide bond	50% TFA/CH_2Cl_2	Wang and Merrifield, 1969, 1972
Polystyrene, incorporating p-alkoxybenzylcarbonyl hydrazine groups	Ⓟ—amino acid (N-Bpoc protected) hydrazide bond	50% TFA/CH_2CP_2	Wang, 1973
Polystyrene, incorporating p-alkoxybenzyl alcohol groups	Ⓟ—p-alkoxybenzyl ester bond	Same as above/suitable for the synthesis of protected peptide fragments having free carboxyl group	Wang, 1973
Polystyrene, incorporating tert-butylalcohol groups	Ⓟ—ester bond, by coupling with phthalylglycine, via mixed anhydride	50% TFA	Wang and Merrifield, 1972

(Continued)

TABLE 4-1 (*Continued*)

Polymer type/active site of bonding	Nature of the ⓟ—first amino acid bond	End cleavage and remarks	References
Resins incorporating benzyhydrol and benzhydrylamine groups. ⓟ—benzhydryl ester or benzhydryl–amide bond cleaved by mild acidolysis to give free peptide and peptide–amide, respectively			
Polystyrene, incorporating benzhydrol groups	ⓟ—benzhydryl ester bond, coupled via benzhydryl chloride in the presence of a base	5–50% TFA/CHCl$_3$	Southard et al., 1969
Polystyrene, incorporating diphenyldiazomethane groups	ⓟ—benzhydryl ester bond,	5–50% TFA/CHCl$_3$	Chapman and Walker, 1975
Polystyrene, incorporating benzhydrylamine groups	ⓟ—benzhydryl ester bond	Anhydrous HF	Pietta et al., 1973, 1974; Orlowski et al., 1976
ⓟ—CH$_2$O—⟨ ⟩—CH$_2$NH$_2$	ⓟ—benzyloxybenzylamine	TFA	Pietta and Brenna, 1975
Resins incorporating benzylchloroformate (benzyloxycarbonyl chloride) group for peptide synthesis from COOH-terminus of the polymer-bound amino acid (Letsinger's method and Merrifield's azide method). The polymer can also be bound through ε-NH$_2$ group of lysine			
Polystyrene, incorporating benzylchloroformate groups	ⓟ—urethane bond, amino acid ester coupled in presence of base	HBr/AcOH	Letsinger and Kornet, 1963
Polystyrene, incorporating benzylchloroformate groups	ⓟ—urethane bond, amino acid tBoc-hydrazide coupled in presence of base, peptide synthesis from COOH-terminus via azide	HBr/TFA	Felix and Merrifield, 1970

Polymer supports linking the first amino acid by side-chain functionalities, for bidirectional extension of the peptide chain

Polymer support	Bond	Cleavage	References
Polystyrene, incorporating benzyl chloroformate groups	(P)—urethane bond with ε-amino-Lys, or δ-amino Orn	HBr/dioxane or TFA/bidirectional synthesis of Orn- or Lys-peptides	Skylarov and Shaskova, 1969; Meienhofer and Trzeciak, 1971
Polystyrene, incorporating N-(5-fluoro-2,4-dinitrophenyl)glycyl groups	(P)—histidine N_{im}-(5-cysteine)-2,4-dinitrophenyl bond, coupling by base	Thiolysis	Glass et al., 1972, 1974
Merrifield resin	(P)—γ-glutamic acid ester or β-aspartic acid ester bond	HBr/AcOH	Mishin and Shvachkin, 1971; Krumdieck and Baugh, 1969
Carboxymethyl dextran	(P)—ε-lys-amide, by DCC coupling	—	Livshits and Vasil'ev, 1973
Miscellaneous polymer supports			
Polystyrene, incorporating p-hydroxyphenyl groups	(P)—phenyl ester bond	Mild selective cleavage by MeOH/NH$_3$	Inukai et al., 1968
Polystyrene, incorporating o-nitrobenzyl groups	(P)—o-nitrobenzylester bond	Photolysis in anhydrous methanol or ethanol	Rich and Gurwara, 1973, 1975
Poly(ethylene glycol) H(OCH$_2$CH$_2$)$_n$-OH MW = 2,000–20,000 (soluble)	(P)—amino acid ester bond	Saponification/peptide separation by ultrafiltration	Mutter and Bayer, 1974; Bayer and Mutter, 1974; Bayer et al., 1974
Copolymer of N-vinylpyrrolidinone and vinyl alcohol	(P)—amino acid ester bond	Saponification/peptide separation by ultrafiltration	Bayer and Geckeler, 1974
Hydroxypropylated dextran (Sephadex LH20)	(P)—amide of amino acid coupling by carbonyldiimidazole	Saponification	Vlasov et al., 1973
Copolydimethylacylamide, incorporating β-alanine	(P)—amide of amino acid derivative (Boc-Gly-OCH$_2$-COOT$_{cp}$)	Amino acid benzyl ester bond cleaved by HF	Atherton et al., 1975
Co(polyisobutylenene–maleic anhydride), functionalized to certain N-chloromethyl groups	(P)—amino acid ester bond,	Acidolysis	Wildi and Johnson, 1968

rates are controlled by such factors as the rate of diffusion of the low-molecular-weight reactants to the sites of polymer-bound substrate and the reagents. The nature of the solvent also plays an important role in solid-phase synthesis.

A list of polymer supports, including others than those employed in the conventional Merrifield method, is given in Table 4-1. Relevant information such as the binding site on the resin and the conditions of the reaction, as well as the advantages or limitations compared to Merrifield's original procedure, is also given. None of the alternative polymer supports appears to present a vast improvement over the Merrifield resin. It may be seen from Table 4-1 that the newer resins have been developed with different aims, such as improving resin–peptide bond stability, solvent–resin–product compatibility, support loading, coupling efficiency, cleavage of finished resin–peptide bond, and synthesis of protected peptide fragments including peptide esters, amides or hydrazides. Functionalized resins incorporating "safety catch" devices have also been developed to eliminate side reactions, provide a means of driving the reactions to completion under more vigorous conditions and, yet, provide an easy means of removing the product from the support. Soluble resins for liquid-phase synthesis and pellicularized resins based on silica have also been developed to ensure high efficiency in coupling. One of the latest trends is to develop polyamide resins which are more compatible with the reagents and solvents.

IV. LINKAGE OF THE FIRST AMINO ACID TO THE POLYMER

In Merrifield's original procedure, the linkage of the first amino acid to the support polymer was achieved by nucleophilic substitution of the N-blocked aminoacylate group in presence of a base (Et_3N). Although no racemization of the COOH-terminal amino acid was observed, 48 hours of refluxing with stirring was considered to be rather a drastic treatment. It has also been reported that quaternization of the chloromethyl group in the polymer (by triethylamine) may occur, lowering the loading capacity of the polymer. The use of tetramethylammonium (Loffet, 1971), benzyltrimethylammonium (Andreatta and Rink, 1973), or cesium (Gisin, 1973) salts of Boc-amino acids is reported to eliminate these side reactions. Crown ethers (18-Crown-6) have also been used to improve esterification of the chloromethylated resin by Boc-amino acid potassium salts (Roeske and Gesellchen, 1976).

Loading efficiency of the first amino acid to the polymer has also been improved by such modifications of the polymer as the incorporation of bromomethyl, hydroxymethyl, or sulfonium groups (Table 4-1). Loading of the hydroxymethyl polymer is accomplished by oxidation–reduction coupling, by carbonyldiimidazole coupling (Matsueda et al., 1972; Ueki and Ikeda, 1974), and by ester transesterification (Khan and Sivanandaiah 1976; Bodanszky and Fagan, 1977). Loading is also improved by employing certain resins incorporating spacer-arms and supports based on glass beads, where the reactive groups are confined to the surface of the beads only (Table 4-1).

After linking the first amino acid to the polymer support, the unreacted chloromethyl groups on the resin should be masked as acetoxy groups; otherwise, these groups may become reactive sites in the subsequent steps of synthesis, giving rise to peptides of shorter sequences. If the chloromethyl groups are not blocked, they lead to several problems during the synthesis of long chains. For example, they can be quaternized by the amine used at the neutralization step. This leads to the presence of ion-exchange sites on the resin and, thus, the presence of unwanted acetic and trifluoroacetic acid groups during coupling steps. In addition, chain termination by the chloromethyl groups is also possible during the neutralization step of the synthesis.

Difficulties in attachment and lower yields have been encountered with certain specific amino acids such as N^α-Boc-nitroarginine and N-Boc-proline. Difficulties in linking the first amino acids to the polymer support have also been observed in case of N^α-Boc-methionine and N^α-Boc-N^{im}-benzylhistidine. For methionine, the difficulty is due to the formation of sulfonium derivatives. This was overcome by using methionine, as its sulfoxide, and then reducing the sulfoxide to its thioether by sodium iodide and acetyl chloride in DMF (Norris et al., 1971). The two last-named amino acids can be linked without any difficulty to the hydroxymethyl polymers and the problem circumvented.

V. PROTECTING GROUPS USED IN SOLID-PHASE PEPTIDE SYNTHESIS

Merrifield, in his first experiments, employed the benzyloxycarbonyl group for protection of the α-amino group. This protecting group was selected because (i) it was stable during the linking of the first amino acid, as well, during the coupling of the subsequent amino acids and (ii) it could be easily and completely removed by hydrogen bromide in glacial acetic acid. However, in order to protect the resin–amino acid benzyl ester linkage from cleaving during the N-deblocking reaction, it was necessary to use a nitrobenzyl or bromobenzyl polymer. In most subsequent solid-phase methods, N^α-Boc-amino acids have been employed, because this group can be cleaved by differential acidolysis (HCl–acetic acid or HCl–dioxane or 50% TFA–CH$_2$Cl$_2$) leaving the resin–benzyl ester bond intact. In addition to the Boc group, o-nitrophenylsulfenyl (Losse and Neubert, 1968) and 2-(biphenyl)isopropoxycarbonyl groups (Bpoc) (Wang and Merrifield, 1969) have been used for N-protection.

As in the case of solution methods, appropriate protecting groups for the third functions (side chains) must be employed. Such protecting groups should be capable of resisting cleavage under the conditions of acidolysis of N-Boc group. Thus, N^α-Boc-ϵ-Z-Lys is a suitable lysine derivative for solid-phase synthesis. The N^α-Boc group can be deblocked in preference to the ϵ-amino-Z group. During N^α-deblocking, even slight deblocking of the third function may result in exposure of side-chain nucleophiles, which could lead to the formation of branched peptides. This was indeed detected by Yaron and Schlossmann (1968) who overcame the difficulty by employing the p-NO$_2$-Z group for ϵ-amino-Lys, which is more resistant to TFA–CH$_2$Cl$_2$ than the Z group. The choice of protecting groups for the third function in solid-phase peptide synthesis is not basically different from solution methods. Developments in one field are soon adopted by the other. Lists of protected amino acids that have been used for peptide synthesis are available (Pettit, 1976; Fletcher and Jones, 1972, 1975; cf. Erickson and Merrifield, 1976).

VI. COUPLING OF SUCCESSIVE AMINO ACIDS TO RESIN-BOUND AMINO ACIDS

As already mentioned, the resin–amino acid ester bond remains intact during N-deblocking. Successive N-blocked amino acids are then coupled to the N-deblocked amino group of the polymer-bound amino acid. DCC has been the reagent of choice for coupling, and methylene chloride or DMF are commonly employed solvents. In contrast to the solution methods of peptide synthesis, no detectable racemization has been observed in coupling reactions using DCC. In an improved coupling procedure (Rebek and Feitler, 1974) using DCC, six equivalents of the carboxyl component and three equivalents of DCC are reacted and the precipitated dicyclohexylurea (DCU) filtered off. The filtrate containing the symmetrical anhydride is then added to the amino–resin. Thus, the precipitation of insoluble DCU on the polymer matrix is avoided.

According to the recent view (Rebek and Feitler, 1974) on the mechanism of DCC coupling in solid-phase peptide synthesis, symmetrical anhydrides are the sole acylating species, whereas O-acylurea has been shown to be the major acylating agent in solution methods. This is presumably due to the fact that the rate of diffusion of the reagents to the polymer-bound amino component is slower than the rate of formation of symmetrical anhydride.

Hagenmaier and Frank (1972) have reported the possibility of "diacylamine formation" using DCC and a large excess of carboxyl component that ultimately may result in either chain shortening or chain lengthening. However, very few cases of this have been reported (Mitchell and Roeske, 1970).

Active esters, which have been suggested as an alternative to DCC coupling, certainly employ milder conditions. Use of active esters also eliminates the possibility of dehydration of β- or γ-amides in the cases of asparagine and glutamine, respectively. Although active esters have been used in the case of special amino acids, DCC coupling is simpler in most cases.

In addition to the active esters, symmetrical anhydrides (Wieland et al., 1971) and EEDQ (N-ethoxycarbonyl- 2- ethoxy-1,2-dihydroquinoline) (Yajima and Kawatani, 1971; Sipos and Gaston, 1973) have been used for coupling of successive amino acids in solid-phase synthesis. The combination of two polymers—one, a linear polymer carrier and the other an insoluble acylating reagent—has also been tried (Heusel et al., 1977).

VII. CLEAVAGE OF THE RESIN–PEPTIDE BOND

In the original Merrifield synthesis, HBr–AcOH was used for the cleavage of the resin–peptide bond. During the cleavage, the other acid-labile protecting groups (benzyl-based protecting groups for third functions, e.g., ω-carboxyl, ϵ-amino, N^{im}, hydroxyl) on the resin-bound peptide are also cleaved. Since then, many alternative methods for acidolytic cleavage of resin–peptide bond have been suggested. These include HBr–TFA (Merrifield, 1964) and anhydrous HF (Lenard and Robinson, 1967).

Some mild acidolytic cleavage methods have also been proposed. These include boron tris-(trifluoracetate) in TFA or $CH_2 Cl_2$ (Pless and Bauer, 1973; Bauer and Pless, 1975). Another reagent, trifluoromethanesulfonic acid, may also prove as useful as HF for this purpose (Yajima and Kiso, 1974; Yajima et al., 1974a).

Hydrogenation of the loaded resin using palladium acetate/DMF/H_2 has also been used to remove completely the finished peptide from the resin (Schlatter et al., 1977).

Peptide-removal by base-catalyzed transesterification (Ohno and

Anfinsen, 1967; Halpern *et al.*, 1968; Beyerman *et al.*, 1968) and hydrazinolysis or ammonolysis (Bodanszky and Sheehan, 1964; Chang *et al.*, 1976) of the loaded resin have also been demonstrated. The acid-labile third-function protecting groups can survive these treatments and the acylpeptide–hydrazine fragment can be used in classical azide coupling reactions of fragments. Diazomethane-catalyzed transesterification has also been reported for cleavage of the peptide from the resin (Rebek and Feitler, 1974). However, because of the basic conditions in the above reactions, there is always danger of racemization. Cyanide-catalyzed transesterification of peptide resins has been accomplished with KCN, either alone (Moore and McMaster, 1978) or in the presence of the crown ether, 18-Crown-6 (Tam *et al.*, 1977).

Self-catalyzed transesterification with 2-dimethylaminoethanol in DMF has been found effective in cleaving protected peptides from the Merrifield resin, in the absence of side-chain benzyl ester protecting groups (Barton *et al.*, 1973). This mild, virtually racemization-free procedure may prove extremely useful for fragment condensation because it yields N-Z- and N-Boc-protected peptides. Since the reagent, 2-dimethyl-aminoethanol, specifically cleaves of the benzylic ester group, the benzyl esters of glutamic acid and aspartic acid, if present, will also be hydrolyzed (Barton *et al.*, 1973; Savoie and Barton, 1974).

So far we have discussed the cleavage of the resin–peptide bond in Merrifield-type resins. In order to effect cleavage using milder acidolytic conditions, alternative resins (e.g., benzyhydrylester-forming resins) have been developed (Southard *et al.*, 1969, 1971). Similarly, resins which yield on cleavage, acylpeptide hydrazides or acylpeptide amides, have also been developed (Wang and Merrifield, 1969). In addition, a large number of nonbenzylic ester resins can be cleaved from peptides by transesterification, ammonolysis, alkaline hydrolysis, hydrazinolysis, and certain other nucleophilic esterolytic catalytic reactions (Table 4-1). In an unconventional method, using either an o-nitrobenzyl group-containing resin or an α-methylphenacyl group-containing resin, the peptide was cleaved by photolysis (12–17 hr irradiation at 300 nm in methanol and in the absence of air) (Rich and Gurwara, 1973, 1975; Tjoeng *et al.*, 1977, 1978; Wang, 1976). No racemization occurred and most of the side-chain protecting groups were left intact.

A large number of resins incorporating safety catch devices have been developed (Table 4-1). Many side-chain protecting groups remain intact under the conditions of safety catch cleavage. To illustrate (Scheme 4-2), the peptide chain is built up on a resin incorporating sulfonamido groups. The first amino acid is bound to the polymer via its carboxyl group (acylamide). Alkaline hydrolysis of the bond does not occur due to the

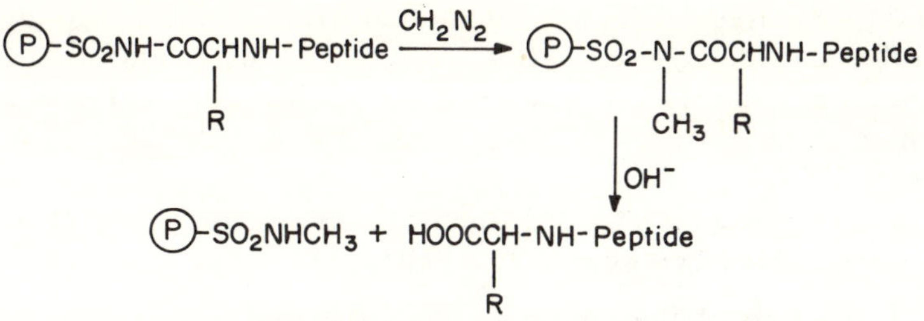

VIII. Monitoring of Solid-Phase Peptide Synthesis

$$\text{(P)}-SO_2NH-COCHNH-\text{Peptide} \xrightarrow{CH_2N_2} \text{(P)}-SO_2-N-COCHNH-\text{Peptide}$$
$$|||$$
$$RCH_3R$$
$$\Big\downarrow OH^-$$
$$\text{(P)}-SO_2NHCH_3 + HOOCCH-NH-\text{Peptide}$$
$$|$$
$$R$$

Scheme 4-2

formation of the sulfonamido anion. On *N*-methylation of the sulfonamide with diazomethane, the resin–peptide bond becomes vulnerable to alkaline hydrolysis.

VIII. MONITORING OF SOLID-PHASE PEPTIDE SYNTHESIS

An important feature of Merrifield's method of sequential synthesis on the polymer support is that the synthesis goes unchecked. Unless the coupling reaction proceeds to completion in every step, the final product obtained after cleavage is bound to be contaminated with peptides differing from the desire sequence by one or more amino acid residues. Thus, because of the multiple uncertainties associated with solid-phase synthesis, it is highly desirable to have rapid analytical control of the two major synthetic operations, i.e., coupling and deprotection, in order to achieve unambiguous synthesis of the desired peptide.

Direct determination of the functional groups on the polymer, with sufficient accuracy, is generally difficult. The simplest method of monitoring consists of cleaving-off the peptide from a portion of the resin after each reaction sequence, and carrying out amino acid analysis on that portion. This method is, however, time-consuming, and the successive synthetic operations are delayed in absence of rapid analytical results.

Mass spectrometry has been used on the cleaved peptide (Bayer *et al.*, 1970a) to detect the formation of error sequences. Weygand and Obermeier (1968) coupled mass spectrometry with Edman degradation to determine the acylation of free amino groups. The Edman degradation has been further improved by using radiolabeled phenyl isothiocyanate and isotope dilution analysis or scintillation counting (Hornle and Geising, 1971). The use of fluorescamine has been suggested (Tometsko and Vogelstein, 1975) as a reagent for measurement of the free amino groups on the cleaved peptide. The peptide is cleaved with sodium methoxide, the cleavage product reacted with fluorescamine, and the fluorescence concentration measured at 480 nm. In another method, the amino groups of the polymer are dansylated. After thorough washing with methylamine solution, the resin is subjected to hydrazinolysis and the dansyl-containing material released determined fluorimetrically (Garden and Tometsko, 1972).

A number of other methods, which do not require prior cleavage of resin–peptide, have been developed. These are summarized as follows.

1. The reaction of fluorescamine with the free amino groups on the polymer has been used for visualizing their presence without resorting to cleavage of resin–peptides (Felix and Jimenez, 1973).

2. The determination of residual free amino acids on the resin (after N-deblocking) can be made by employing 2-hydroxy-1-naphthaldehyde (Esko *et al.*, 1968; Esko and Karisson, 1970). The resin is treated with an excess of the reagent, which forms a Schiff base with the free amino groups. The excess reagent is washed off, and the aldehyde, after displacement with a soluble amine (benzylamine), is determined spectrophotometrically. This method is slow and the ultimate results do not accurately reflect the coupling efficiency of a reaction. 5-Phenylazosalicylaldehyde and 5-nitrophenylazosalicylaldehyde (Losse and Ulbrich, 1973) have been suggested as alternate reagents for monitoring polymer-bound free amino groups. The resin turns yellow and red, respectively, with these two reagents when free amino groups are present. Thus the Schiff base formation can be followed visually.

3. When the coupling reaction is carried out using certain active esters, the progress of the reaction can be followed by spectroscopic measurements of either the released 5-chloro-8-hydroxyquinoline (Rudinger and Gut, 1967) or *p*-nitrophenol (Bodanszky and Sheehan, 1966). Similarly, the disappearance of *N*-(*o*-nitrophenylsulfenyl)amino acids during the coupling reaction can also be followed (Weitzel *et al.*, 1968; Gut and Rudinger, 1968).

4. The determination of free amino groups in peptide-containing resins by nonaqueous titration using perchloric acid in acetic acid or acetic acid/methylene chloride is possible (Brunfeldt *et al.*, 1969, 1972a,b). The method is rapid and has been automated and used as a feed-back signal for N-deblocking, as well as for coupling reactions, in automated peptide synthesis.

5. A rapid method for the determination of free amino groups on an insoluble polymer has been described (Gisin, 1972; Gisin and Merrifield, 1972). The polymer is treated with 0.1 N picric acid in methylene chloride and, after removal of the excess reagent, the picric acid is eluted out by means of a solution of diisopropylethylamine or 0.1 M pyridine hy-

drochloride and determined spectrophotometrically. The method has been used for automatic monitoring of solid-phase peptide synthesis (Hodges and Merrifield, 1975).

6. The qualitative ninhydrin method of Kaiser et al. (1970) has been worked into a quantitative method by Chou et al. (1971). A known amount of the resin is treated with the ninhydrin reagent and the absorption of the supernatant is measured after dilution, at 570 nm, with relation to a ninhydrin blank. Bayer et al. (1976), however, noticed that the method does not give absolute values since a correction factor, dependent on the relevant amino acid and its position in the peptide chain, is needed.

7. A rapid analytical method for determining the percentage reaction (coupling and deblocking) based on the use of ^{14}C-labeled Boc-amino acids has been suggested (Beyerman et al., 1973). Incorporation of labeled Boc groups into the resin can be easily followed by measuring the decrease in radioactivity in aliquots of the supernatant layer. Similarly, the deblocking of the resin-bound peptide can be followed by measuring the loss in radioactivity of the resin. Though considered to be one of the most suitable methods for monitoring, the method is expensive in the long-term view, and errors due to adsorption of Boc-amino acids on the resin are reported.

8. In another method, the free amino groups in the peptide–resin are converted to the corresponding hydrochloride or hydrobromide by reaction with the respective pyridinium halide (Dorman and Love, 1969; Losse and Ulbrich, 1971; Losse, 1971). The halide ion is then displaced from the resin by triethylamine and measured titrimetrically. The speed and sensitivity of the method has been improved (Hancock et al., 1973) by using pyridine H^{36}Cl, which enables the released chloride to be determined by measurement of radioactivity and which allows the method to be automated. Just as the free amino groups in the resin can be converted into their hydrochlorides, salt formation with other radiolabeled acids, e.g., 1-[^{14}C]chloroacetic acid or [^{35}S] sulfuric acid (Beyerman et al., 1973), has been used to determine the amino group content of the resin.

IX. AUTOMATION IN SOLID-PHASE PEPTIDE SYNTHESIS

The most appealing feature of solid-phase peptide synthesis is its amenability to automation. A review on the subject has been published (Brunfeldt, 1973; Erickson and Merrifield, 1976). The credit for introducing automation into peptide synthesis goes again to the inventors of the technique, R. B. Merrifield and co-workers (Merrifield, 1966; Merrifield et *al.*, 1966). Since then, a large number of reports have appeared on the design and instrumentation of automated solid-phase peptide synthesis. It has been possible to carry out complete reaction sequences, including filtration and washing, with the help of a machine under computer control (Brunfeldt *et al.*, 1969, 1972a,b; Hruby *et al.*, 1972). Most components for the automated system are standard; however, specially designed reaction vessels have been used in all the machines set up to date. In the case of pellicular resins, the synthesis is performed in a chromatographic column wherein all synthetic operations are carried out (Scott *et al.*, 1972). For swellable resins, Merrifield *et al.* (1966) used a glass cylinder with a fritted glass disc, with mixing accomplished by revolving the vessel 180°. Boni *et al.* (1971) used a system where the resin is kept between two teflon filters and where mixing is done by moving the liquid phase up and down through the filters by gas pressure. Mixing by shaking was made use of by Brunfeldt *et al.* (1972b) in a specially designed reactor, combined with a recirculation system. Mixing by stirring is used in one of the commercially available units (Beckman, model 990's).

X. RACEMIZATION PROBLEMS IN SOLID-PHASE PEPTIDE PRODUCTS

One great merit of the conventional solid-phase peptide synthesis has been its virtual freedom from racemization. With DCC or active esters coupling, virtually racemization-free products are obtained. As a test case, Bayer *et al.* (1971b) carried out the solid-phase peptide synthesis of a model dodecapeptide ($_L$Leu-$_L$Ala)$_6$. After the peptide was completely hydrolyzed, the resulting amino acids were analyzed for racemization by gas chromatography. No detectable recemization occurred.

Acidolytic cleavage is also free from racemization. However, the danger of racemization exists when the resin–peptide bond is cleaved by ammonolysis or base-catalyzed transesterification. These cleavage reactions are sometimes used with the modified resin supports or when the protected fragment is desired for fragment condensation.

XI. PURIFICATION OF SOLID-PHASE PEPTIDE PRODUCTS

In the case of smaller peptides, highly pure products have been obtained in good yields by simple evaporation of filtrate from the resin–peptide cleavage product. If necessary, the product peptide can be purified by ion-exchange chromatography to separate the error-peptides ef-

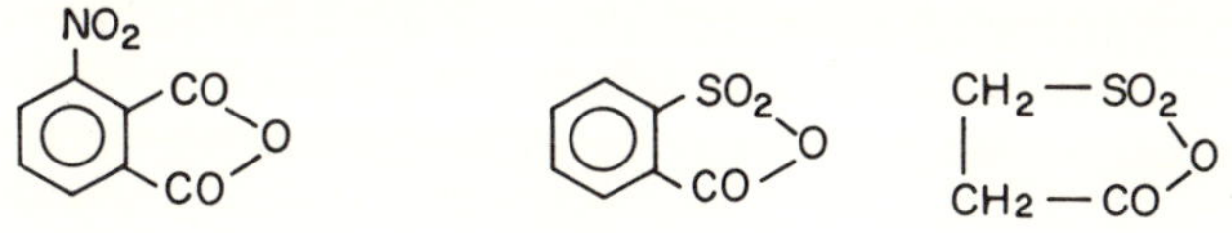

Fig. 4-1

fectively. Wieland *et al.* (1969) suggested that, after each coupling stage, the peptide–resin be treated with one of the cyclic anhydrides (Fig. 4-1) in order to acylate the error-peptides on the resin. After cleavage, the strongly anionic acidic derivatives of the error-peptides can easily be separated on ion-exchangers.

In the case of large peptides, it is necessary to carry out purification by methods such as counter-current extraction or chromatography (including high-pressure-liquid, ion-exchange, affinity, and gel-filtration) in order to get the pure major fraction. The purification methods are checked by biological assay. For large biopolymers (proteins), however, the capacity of any of these methods to detect small variations in sequence remains doubtful. Thus, the amount of the desired material present in the final product may be very low and, even though biological activity may be present in the preparation, the preparation is one of many related molecules whose structures differ at one or more sites within the molecule.

XII. PROBLEMS IN SOLID-PHASE SYNTHESIS

Problems and limitations of solid-phase peptide synthesis have been critically discussed by Wunsch (1971), Blossey and Neckers (1975), and Ericksen and Merrifield (1976) in their compilations of selected papers on solid-phase synthesis. According to Wunsch, there were certain defects inherent in the conventional Merrifield method, but some of the limitations have now been reduced to insignificant levels. Major limitations and difficulties of solid-phase methods fall into the following categories.

1. Noncompatibility of the resin: The compatibility of polystyrene resins in peptide synthesis has been questioned. According to Sheppard (1973), as the growing polar peptide chain grafts onto the nonpolar styrene backbone, one can envisage collapse of either peptide chains or of the styrene backbone, depending on the nature of the solvating medium. Alternative polar resins such as polyacrylamides, however, have yet to establish themselves as better than styrene resins.

2. Blocking groups: Acidolytic cleavage of N^α-protecting groups (α-Boc and -NPS) is bound to cleave the blocking groups used for side-

chain functionalities, to some extent exposing side-chain nucleophiles. Subsequent coupling steps may result in formation of branched-chain peptides at these sites.

3. Linkage of first amino acid to the resin: The problems related to incomplete esterification in general and difficulties with specific amino acids in particular have been discussed (Section IV). The problem has been reduced to an insignificant level by using improved methods of linking and blocking of the unreacted groups.

4. Peptide–resin linkage stability: In long syntheses, problems have arisen because of the instability of the peptide to resin linkage during the deblocking steps. Recently, attempts have been made to overcome this problem.

5. Formation of error-peptides: Weygand and Obermeier (1968) and Bayer et al. (1970a) have divided the error-peptides (peptides other than those desired) into two classes: (a) truncated sequences and (b) failure sequences. When a peptide chain on the resin fails to grow beyond a certain length, it is termed a "truncated sequence." In a "failure sequence," one or more amino acids may be missing from the desired peptide sequences.

Whereas in the solution method, the intermediate peptides are purified after each coupling reaction, thus eliminating the unreacted material and by-products, in the solid-phase method, the error-peptides keep accumulating on the resin making their separation after cleavage very difficult.

The formation of truncated sequences has been attributed to steric factors and alkylation of chain ends by residual chloromethyl groups. The most common situation, however, is one in which the growth of the peptide chain in some of the sites may stop prematurely due to the limiting size of the cavity in the polymer. To overcome this problem, low cross-linked resins with low functionalization have been employed. Truncated sequences do not pose a problem either, when pellicular resins are used. It is difficult to assign a proper explanation for the formation of failure sequences other than to propose a statistical failure of some amino acids being incorporated into certain peptide chains. The formation of failure sequences can only be minimized by choosing proper, i.e., different, sets of experimental conditions for coupling different amino acids.

6. Formation of error-peptides due to interchain aminolysis: It has been established (Crowley and Rapoport, 1976) that low ($<2\%$) cross-linked resins are not very rigid in the solvated state. The interchain aminolysis of N-blocked peptides has been observed (Beyerman et al., 1972) and results in the formation of error-peptides by chain dimerization. In certain spe-

cific cases, such as a DVal-LPro-resin, carboxyl-catalyzed intramolecular aminolysis (70%) was observed, giving rise to diketopiperazine (Gisin and Merrifield, 1972). Although this phenomenon does not take place to an appreciable extent with most peptides, such observations have been reported (Brunfeldt *et al.*, 1972a; Rothe and Mazanek, 1974).

Many other difficulties arise in specific cases of solid-phase peptide synthesis, e.g., destruction of the tryptophan residue during acidolytic cleavage of resin–peptide bond, pyroglutamyl derivative formation, chain termination by acetylation with acetic acid leaching from teflon components, and formation of γ-glutamyl peptides. Solutions to these problems provide the basis for much of the continuing work in the field.

XIII. SOLID-PHASE COUPLING OF *N*-CARBOXYLANHYDRIDES (NCA)

The application to peptide synthesis of the stepwise addition of NCAs at controlled pH in solution has also been carried out on a polymer support (Scheme 4-3). The addition of the NCA was controlled by the formation of the resulting carbamate at basic pH (Blecher and Pfaender, 1973). Another way of controlling the addition of the NCA is to use a suitably N-blocked NCA derived from *N-o*-nitrophenylsulfenyl (Kricheldorf, 1973) or *N*-(4,4'-dimethoxybenzhydryl) (Halstrom and Kovacs, 1973) amino acids. After each NCA addition, the N-blocking group is cleaved and addition of the successive N-blocked NCA is repeated. Maher *et al.*, (1972) reported repetitive peptide synthesis using unsubstituted NCA, but the resin-bound amino acid was *N*-silylated (Scheme 4-3).

XIV. SOLID-PHASE SYNTHESIS USING SIDE-CHAIN FUNCTIONALITIES FOR ATTACHMENT TO POLYMERS AND BIDIRECTIONAL EXTENSION OF PEPTIDE CHAINS

It has been suggested that the scope of solid-phase synthesis could be extended by attachment of the amino acid or a peptide to the polymer by side-chain functionalities. This has been done with ornithine (Skylarov and Shaskova, 1969) and lysine (Meienhofer and Trzeciak, 1971) using a benzyloxycarbonyl chloride polymer and the δ- or ϵ-amino group; with the glutamic acid γ-carboxylic group, using benzyl chloride polymer (Mishin and Shvachkin, 1971); with histidine-N^{im} using a 2,4-

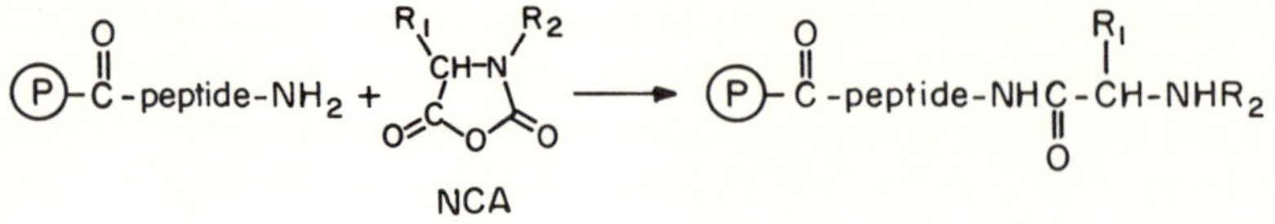

$$\text{(P)}-\overset{\overset{O}{\|}}{C}-\text{peptide-NH}_2 + \underset{NCA}{\begin{array}{c}R_1R_2\\ \text{CH-N}\\ O=CC=O\\ \diagdown O \diagup\end{array}} \longrightarrow \text{(P)}-\overset{\overset{O}{\|}}{C}-\text{peptide-NH}\overset{R_1}{\underset{\overset{\|}{O}}{C}}-\overset{R_1}{\underset{}{CH}}-\text{NHR}_2$$

R_1 = amino acid side chain; R_2 = H, - O-nitrosulfenyl,
 - N- (4,4'- dimethoxybenzhydryl),
 - trimethylsilyl

Scheme 4-3

dinitrofluorophenyl polymer (Glass *et al.*, 1972); and with cysteine using an S-2,4-dinitrofluorophenyl polymer (Glass *et al.*, 1974) (Table 4-1).

The obvious limitation of this type of polymeric side-chain protection technique is its limited application to peptides containing these specific amino acids only. Such a scheme, however, appears to provide considerably more latitude in planning a synthesis; e.g., the cyclization of a peptide can be carried out while it is still bound to the support. More work in this field will have to be done before the method is fully established.

XV. FRAGMENT CONDENSATION STRATEGY IN SOLID-PHASE PEPTIDE SYNTHESIS

An important tactical variation of the conventional stepwise "one-amino acid-at-a-time solid-phase peptide synthesis" involves the coupling of larger peptide fragments to an already polymer-bound peptide fragment. As early as 1966, Weygand and Ragnarson reported the use of a phenacyl bromide resin for coupling peptide fragments on solid support. The peptide fragments used did not have side-chain functionalities, and the resin–peptide cleavage was brought about by sodium thiophenolate, which certainly involved the danger of racemization.

As in Merrifield's method, the amino component peptide is polymer-immobilized while the carboxylic component peptide is kept in solution. Although small and simpler peptide fragments can be coupled by the active ester method in the solid-phase method (Weygand and Ragnarson, 1966), azide coupling (cf. solution method) is generally preferred for coupling larger protected peptide fragments. It has been possible to make the two fragments separately by the solid-phase method, then cleave one of them (the carboxyl-component) as a peptide–hydrazine from the resin. The third function protecting groups should remain intact during the cleavage of peptide–hydrazide. This fragment can then be condensed with the amino-component peptide immobilized on the polymer, via azide coupling (Yajima *et al.*, 1970; Wang and Merrifield, 1972).

It is obvious that the carboxyl-component peptide fragment should be cleaved from the resin as a hydrazide. A *tert*-alkoxycarbonylhydrazine–co(polystyrene–DVB) resin can be used for making such a hydrazine–peptide. If the first amino acid is glycine, the polymeric derivative of *tert*-butyl alcohol may be used as the resin support, since alkaline cleavage of the glycine–resin bond does not cause racemization. Polymeric derivatives of *p*-alkoxybenzyl alcohol and the corresponding carboxy hydrazine derivatives can similarly be used for making the carboxy peptide fragment (Wang, 1973).

Bovine basic inhibitor (Kunitz's inhibitor), which contains 58 amino acids, was obtained (Yajima and Kiso, 1974; Yajima *et al.*, 1974b) by this method by condensation of five fragments. Coupling was brought about using DCC over a period of nine days, and the yields in each step were reported to be 75–100%. The purified inhibitor showed 50% of the biological activity of the native inhibitor. Among the other peptides synthesized by fragment condensation, are human corticotrophin-like intermediate lobe peptide, (CLIP) coresponding to ACTH sequence 18–39 (Kawatani *et al.*, 1974); insulin and proinsulin (Weber and Andre, 1975); and the ACTH sequence 1–24, employing an oxidation–reduction method on solid support. A number of sequential polypeptides have been synthesized by coupling of fragments on solid supports (for example, Bhatnagar and Rapaka, 1975; Rapaka and Bhatnagar, 1975).

Fragment condensation strategy provides a means of minimizing the number of acidolytic deprotection steps which can result in partial deprotection of side-chain functional groups, in addition to causing destruction of tryptophan residues. Furthermore, the possibility of formation of a large variety of truncated and failure sequences is also reduced; and, thus, the purification of the product is very much simplified. The economic aspect is a great limitation to the fragment condensation method since a liberal excess of the acylpeptide (carboxy-component) is necessary to maximize the coupling yield and since the recovery of the unused portion at the end of the reaction is very difficult.

REFERENCES

Andreatta, R. H., and Rink, H. (1973). *Helv. Chim. Acta.* **56**, 1205.
Atherton, E., Clive, D. L. J., and Sheppard, R. C. (1975). *J. Am. Chem. Soc.* **97**, 6584.
Barton, M. A., Lemieux, R. U., and Savoie, J. Y. (1973). *J. Am. Chem. Soc.* **95**, 4501.

Bauer, W., and Pless, J. (1975). *In* "Peptides: Chemistry, Structure and Biology" (R. Walter and J. Meienhofer, eds.), p. 341. Ann Arbor Press, Ann Arbor, Michigan.
Bayer, E., and Geckeler, K. (1974). *Annalen,* 1671.
Bayer, E., and Mutter, M. (1974). *Chem. Ber.* **107,** 1344.
Bayer, E., Eckstein, H., Hagele, K., Konig, W. A., Bruning, W., Hagenmaier, H., and Parr, W. (1970a). *J. Am. Chem. Soc.* **92,** 1735.
Bayer, E., Gil-Av., E., Konig, W. A., Nakaparksin, S., Ore, J., and Parr, W. (1970b). *J. Am. Chem. Soc.* **92,** 1738.
Bayer, E., Jung, G., Halsaz, I., and Sebastian, I. (1970c). *Tetrahedron Lett.,* 4503.
Bayer, E., Breitmaier, E., Jung, G., and Parr, W. (1971a). *Z. Physiol. Chem.* **352,** 759.
Bayer, E., Hagenmaier, H., Jung, G., Parr, W., Eckstein, H., Hunziker, P., and Sievers, R. E. (1971b). *Pept. Proc. Eur. Pept. Symp., 9th, 1969,* p. 65. North-Holland Publ., Amsterdam.
Bayer, E., Mutter, M., Uhmann, E., Polster, J., and Mauser, H. (1974). *J. Am. Chem. Soc.* **96,** 7333.
Beyerman, H. C., Hindricks, H., and de Leer, E. W. B. (1968). *J. Chem. Soc. Chem. Commun.,* 1668.
Beyerman, H. C., de Leer, E. W. B., and van Voseen, W. (1972). *J. Chem. Soc. Chem. Commun.,* 929.
Beyerman, H. C., van den Brink, W. M., Parmentier, J. H., and Westerling, J. (1973). *Pept. Proc. Eur. Pept. Symp. 11th, 1971,* p. 138. North-Holland Publ., Amsterdam.
Bhatnagar, R. S., and Rapaka, R. S. (1975). *Biopolymers,* **14,** 597.
Blake, J. and Li, C. H. (1976). *J. Chem. Soc. Chem. Commun.,* 504.
Blecher, H., and Pfaender, P. (1973). *Annalen,* 1263.
Blossey, E. C., and Neckers, D. C. (eds.), (1975). "Solid-Phase Synthesis." Dowden, Hutchinson and Ross Inc., Stroudsberg, Pennsylvania.
Bodanszky, M., and Fagan, T. D. (1977). *Int. J. Pept. Protein Res.* **10,** 375.
Bodanszky, M., and Ondetti, M. A. (1966). "Peptide Synthesis." Wiley (Interscience), New York.
Bodanszky, M., and Sheehan, J. T. (1964). *Chem. Ind.* 1423.
Bodanszky, M., and Sheehan, J. T. (1966). *Chem. Ind.* 1597.
Boni, R., Bonora, G. M., Ciceri, L., Gambini, A., Scatturin, A., and Scoffone, E. (1971). *Chim. Ind.* **53,** 10.
Brunfeldt, K. (1973). *Pept. Proc. Eur. Pept. Symp., 12th, 1972,* p. 141. North-Holland Publ., Amsterdam.
Brunfeldt, K., Roepstorff, P., and Thomsen, J. (1969). *Acta Chem. Scand.* **23,** 2906.
Brunfeldt, K., Bucher, D., Christensen, T., Roepstorff, P., Rubin, I., Schou, O., and Villemoes, P. (1972a). *In* "Chemistry and Biology of Peptides " (J. Meienhofer, ed.), p. 183. Ann Arbor Press, Ann Arbor.
Brunfeldt, K., Christensen, T., and Villemoes, P. (1972b). *FEBS Lett.* **22,** 238.
Chang, J. K., Shimizu, M. and Wang S. S. (1976). *J. Org. Chem.* **41,** 3255.
Chapman, P. H., and Walker, D. (1975). *J. Chem. Soc. Chem. Commun.,* 690.
Chou, F. C. H., Chawla, R. K., Kibler, R. F., and Shapira, R. (1971). *J. Am. Chem. Soc.* **93,** 267.
Crowley, J. I., and Rapoport, H. (1976). *Acc. Chem. Res.* **9,** 135.
Dorman, L. C., and Love, J. (1969). *J. Org. Chem.* **34,** 158.
Erickson, B. W., and Merrifield, R. B. (1976). *In* "The Proteins" 3rd ed. (H. Neurath, R. L. Hill, C. L. Boeder, eds.). Academic Press, New York.
Esko, K., and Karlsson, S., (1970). *Acta. Chem. Scand.* **24,** 1415.
Esko, K., Karlsson, S. and Porath, J. (1968). *Acta. Chem. Scand.* **22,** 3342.

References

Feinberg, R. S., and Merrifield, R. B. (1974). *Tetrahedron* **30**, 3209.
Felix, A. M., and Jimenez, M. H. (1973). *Anal. Biochem.* **52**, 377.
Felix, A. M., and Merrifield, R. B. (1970). *J. Am. Chem. Soc.* **92**, 1385.
Fletcher, G. A., and Jones, J. H. (1972). *Int. J. Pept. Protein Res.* **4**, 347.
Fletcher, G. A., and Jones, J. H. (1975). *Int. J. Pept. Protein Res.* **7**, 91.
Fridkin, M., and Patchornik, A. (1974). *Ann. Rev. Biochem.* **43**, 419.
Garden, J., and Tometsko, A. M. (1972). *Anal. Biochem.* **46**, 216.
Gisin, B. F. (1972). *Anal. Chim. Acta.* **58**, 248.
Gisin, B. F. (1973). *Helv. Chim. Acta.* **56**, 1476.
Gisin, B. F., and Merrifield, R. B. (1972). *J. Am. Chem. Soc.* **94**, 3102.
Glass, J. D., Schwartz, I. L., and Walter, R. (1972). *J. Am. Chem. Soc.* **94**, 6209.
Glass, J. D., Talansky, A., Grzonka, Z., Schwartz, I. L., and Walter, R. (1974). *J. Am. Chem. Soc.* **96**, 6476.
Green, B., and Garson, R. (1969). *J. Chem. Soc.* **C** 401.
Greenstein, J. P., and Winitz, M. (1961). "Chemistry of the Amino Acids." Vols. 1–3. Wiley, New York.
Gut, V., and Rudinger, J. (1968). *Pept. Proc. Eur. Pept. Symp., 9th, 1968*, p. 185. North-Holland Publ., Amsterdam.
Hagenmaier, H., and Frank, H. (1972). *Z. Physiol. Chem.* **353**, 1973.
Halpern, B., Chew, L., Close, V., and Patton, W. (1968). *Tetrahedron Lett.*, 5163.
Halstrom, J., and Kovacs, K. (1973). *Pept. Proc. Eur. Pept. Symp., 12th, 1972*, p. 173. North-Holland Publ., Amsterdam.
Hancock, W. S., Prescot, D. J., Vagelos, P. R., and Marshall, G. R. (1973). *J. Org. Chem.* **38**, 774.
Heusel, G., Bovermann, G., Gohring, W., and Jung, G. (1977). *Angew. Chem. Int. Ed.* **16**, 642.
Hey, D. H., and John, D. I., eds. (1973). "Amino acids, peptides and related compounds" *MTP Int. Rev. Sci., Organic Chemistry, Ser. One*, **6**.
Hodges, R. S., and Merrifield, R. B. (1975). *Analyt. Biochem.* **65**, 241.
Hornle, S., and Geising, W. (1971). *Physiol. Chem. Phys.* **352**, 5.
Hruby, V. J., Barstow, L. E., and Linhart, T. (1972). *Anal. Chem.* **44**, 343.
Inukai, N., Nakano, K., and Murakami, M. (1968). *Bull. Chem. Soc. Jpn.* **41**, 182.
Kaiser, E., Colescott, R. L., Bossinger, C. D., and Cook, P. I. (1970). *Anal. Biochem.* **34**, 595.
Katsoyannis, P. G. (ed.). (1973). "The Chemistry of Polypeptides." Plenum, New York.
Kawatani, H., Tamura, F., and Yamija, J. (1974). *Chem. Pharm. Bull.* (Japan) **22**, 1879.
Kenner, G. W., McDermoth, J. R., and Sheppard, R. C. (1971). *J. Chem. Soc. Chem. Commun.*, 636.
Khan, S. A. and Sivanandaiah, K. M. (1976). *J. Chem. Soc. Chem. Commun.*, 614.
Kricheldorf, H. R. (1973). *Agnew. Chem. Int. Ed.* **12**, 73.
Krumdieck, C. L., and Baugh, C. M. (1969). *Biochemistry* **8**, 1568.
Lenard, J., and Robinson, A. B. (1967). *J. Am. Chem. Soc.* **89**, 181.
Letsinger, R. L., and Kornet, J. (1963). *J. Am. Chem. Soc.* **85**, 3045.
Livshits, A., and Vasil'ev, A. E. (1973). *Zh. Obshch. Khim.* **43**, 219.
Loffet, A. (1971). *Int. J. Protein Res.* **3**, 297.
Losse, A. (1971). *Tetrahedron Lett.*, 4989.
Losse, G., and Neubert, K. (1968). *Z. Chem.* **8**, 387.
Losse, G., and Ulbrich, R. (1971). *Z. Chem.* **11**, 346.
Losse, G., and Ulbrich, R. (1972). *Tetrahedron* **28**, 5823.
Losse, G., and Ulbrich, R. (1973). East German patent 83 529 [*C. A.* 1973, **78**, 84822].

Maher, J. J., Furey, M. E., and Greenberg, L. J. (1972). *Tetrahedron Lett.*, 1581.
Marshall, D. L., and Liener, I. E. (1970). *J. Org. Chem.* **35**, 867.
Marshall, G. R., and Merrifield, R. B. (1965). *Biochemistry* **4**, 2394.
Marshall, G. R., and Merrifield, R. B. (1970). *In* "Handbook for Biochemistry," 2nd ed. (H. A. Sober, ed.), p. C-145. The Chemical Rubber Co., Cleveland, Ohio.
Matsueda, R., Kitazawa, E., Maruyama, H., Takahagi, H., and Mukaiyama, T. (1972). *Chem. Lett.* 379.
Meienhofer, J. (1973). *In* "Hormonal Proteins and Peptides" (C. H. Li, ed.), Vol. 2, p. 45. Academic Press, New York.
Meienhofer, J., and Trzeciak, A. (1971). *Proc. Nat. Acad. Sci. U. S. A.* **68**, 1006.
Merrifield, R. B. (1963). *J. Am. Chem. Soc.* **85**, 2149.
Merrifield, R. B. (1964). *J. Am. Chem. Soc.* **86**, 304.
Merrifield, R. B. (1965a). *Endeavour* **24**, 3.
Merrifield, R. B. (1965b). *Science* **150**, 178.
Merrifield, R. B. (1966). *In* "Hypotensive Peptides" (E. G. Erdos, N. Back, and F. Sicuteri, eds.), p. 1. Springer-Verlag, Berlin and New York.
Merrifield, R. B. (1967). *Recent Prog. Horm. Res.* **23**, 451.
Merrifield, R. B. (1968). *Sci. Am.* **218**, 56.
Merrifield, R. B. (1969). *Advances in Enzymology*, **32**, 221.
Merrifield, R. B. (1971). *Harvey Lect.* **67**, 143.
Merrifield, R. B. (1972). *Beckman Rep.* (**1**) 3 [*C. A.* 1972, **77** 102, 155m].
Merrifield, R. B., and Littau, V. (1968). *Pept. Proc. Eur. Pept. Symp., 9th, 1968*, p. 179. North-Holland Publ., Amsterdam.
Merrifield, R. B., Stewart, J. M., and Jernberg, N. (1966). *Anal. Chem.* **38**, 1905.
Mishin, G. P., and Shvachkin, Yu. P. (1971). *Zh. Obshch. Khim*, **41**, 234.
Mitchell, A. R., and Roeske, R. W. (1970). *J. Org. Chem.* **35**, 1171.
Mitchell, A. R., Erickson, B. W., Ryabtsev, M. N., Hodges, R. S., and Merrifield, R. B. (1976). *J. Am. Chem. Soc.* **98**, 7357.
Mitchell, A. R., Kent, S. B. H., Engelhard, M., and Merrifield, R. B. (1978). *J. Org. Chem.* **43**, 2845.
Mizoguchi, T., Shigezane, K., and Takamura, N., (1969). *Chem. Pharm. Bull. Jpn.* **17**, 411.
Mizoguchi, T., Shigezane, K., and Takamura, N. (1970). *Chem. Pharm. Bull. Jpn.* **18**, 1465.
Moore, G., and McMaster, D. (1978). *Int. J. Pept. Protein Res.* **11**, 140).
Mutter, M., and Bayer, E. (1974). *Angew. Chem. Int. Ed.* **13**, 88.
Norris, K. E., Halstrom, J., and Brunfeldt, K. (1971). *Acta Chem. Scand.* **25**, 945.
Ohno, M., and Anfinsen, C. B. (1967). *J. Am. Chem. Soc.* **84**, 5994.
Orlowski, R. C., Walter, R., and Winkler, D. (1976). *J. Org. Chem.* **41**, 3701.
Ovchinnikov, Yu. A., Kiryushkin, A. A., and Kozhevnikova, I. V. (1968). *Zh. Obshch. Khim.* **38**, 2631, 2636.
Parr, W., and Grohmann, K. (1971). *Tetrahedron Lett.*, 2633.
Parr, W., and Grohmann, K. (1972). *Angew. Chem. Int. Ed.* **11**, 314.
Parr, W., Yang, C., and Holzer, G. (1972). *Tetrahedron Lett.* 101.
Parr, W., Grohmann, K., and Hagele, K. (1974). *Annalen*, 655.
Pettit, G. R. (1976). "Synthetic Peptides," Vol. 4. Academic Press, New York.
Pietta, P., and Brenna, O. (1975). *J. Org. Chem.* **40**, 2995.
Pietta, P. G., Cavallo, P. F., and Marshall, G. R. (1973). *Pept. Proc. Eur. Pept. Symp., 11th, 1971*, p. 172. North-Holland Publ., Amsterdam.
Pietta, P. G., Cavallo, P. F., Takahashi, K., and Marshall G. R. (1974). *J. Org. Chem.* **39**, 44.
Pless, J., and Bauer, W. (1973). *Angew. Chem. Int. Ed.* **12**, 147.
Rapaka, R. S., and Bhatnagar, R. S. (1975). *Int. J. Pept. Protein Res.* **7**, 119, 475.

References

Rebek, J., and Feitler, D. (1974). *J. Am. Chem. Soc.* **96,** 1606.
Rich, D. H., and Gurwara, S. K. (1973). *J. Chem. Soc. Chem. Commun.,* 610.
Rich, D. H., and Gurwara, S. K. (1975). *J. Am. Chem. Soc.* **97,** 1575.
Roeske, R. W., and Gesellchen, P. D. (1976). *Tetrahedron Lett.,* 3369.
Rothe, M., and Mazanek, J. (1974). *Annalen,* 439.
Rudinger, J., and Gut, V. (1967). *Pept. Proc. Eur. Pept. Symp., 8th, 1966,* p. 89. North-Holland Publ., Amsterdam.
Sano, S., Tokunaga, R., and Kun, K. A. (1971). *Biochim. Biophys. Acta.* **244,** 201.
Savoie, J. V., and Barton, M. A. (1974). *Can. J. Chem.* **52,** 2832.
Schlatter, J. M., Mazur, R. H., and Goodmonson, O. (1977). *Tetrahedron Lett.,* 2851.
Schröder, E., and Lübke, K. (1965). "The Peptides," Vols. 1 and 2. Academic Press, New York.
Scott, R. P. W., Zolty, S., and Chan, K. K. (1972). *J. Chromatogr. Sci.* **10,** 384.
Shemyakin, M. M., Ovchinnikov, Yu., A., Kiryushkin, A. A., and Kazhevnikova, I. V. (1965). *Tetrahedron Lett.,* 2323.
Sheppard, R. C. (1973). *Pept. Proc. Eur. Pept. Symp., 11th, 1971,* p. 111. North-Holland Publ., Amsterdam.
Sipos, F., and Gaston, D. W. (1973). *Pept. Proc. Eur. Pept. Symp., 11th, 1971,* p. 165. North-Holland Publ., Amsterdam.
Skylarov, L. Yu., and Shaskova, I. V. (1969). *Zh. Obshch. Khim.* **39,** 2778.
Southard, G. L., Brooke, G. S., and Pettee, J. M. (1969). *Tetrahedron Lett.,* 3505.
Southard, G. L., Brooke, G. S., and Pettee, J. M. (1971). *Tetrahedron* **27,** 2701.
Stark, G. R. (1971). "Biochemical Aspects of Reactions on Solid-Supports." Academic Press, New York.
Stewart, J. M. (1972). *Annu. Rep. Med. Chem.* **7,** 289.
Stewart, J. M. (1976). *J. Macromol. Sci. Chem.* **A10,** 259.
Stewart, J. M., and Young, J. D. (1969). "Solid-Phase Peptide Synthesis." Freeman, San Francisco, California.
Tam, J. P., Cunningham-Rundles, W. F., Erickson, B. W., and Merrifield, R. B. (1977). *Tetrahedron Lett.,* 4001.
Tesser, G. I., and Ellenbroek, B. W. J. (1967). *Pept. Proc. Eur. Pept. Symp., 8th, 1966,* p. 124. North-Holland Publ., Amsterdam.
Tesser, G. I., Buis, J. T. W. A. R. M., Wolters, E. Th. M., and Bothé-Helmes, E. G. A. M. (1976). *Tetrahedron* **32,** 1069.
Tilak, M. A. (1968). *Tetrahedron Lett.,* 6323.
Tilak, M. A., and Hollinden, C. S. (1968). *Tetrahedron Lett.,* 1297.
Tilak, M. A., and Hollinden, C. S. (1971). *Org. Prep. Proced.* **3,** 183.
Tjoeng, F. S., Staines, W., St.-Pierre, S., and Hodges, R. S. (1977). *Biochim. Biophys. Acta.* **490,** 489.
Tjoeng, F. S., Tong, E. K., and Hodges, R. S. (1978). *J. Org. Chem.* **43,** 4190.
Tometsko, A. M., and Vogelstein, E. (1975). *Anal. Biochem.* **64,** 438.
Ueki, M., and Ideda, S. (1974). *Chem. Lett.* 2b.
Vlasov, G. P., Bilibin, A. Yu., Kuznetsova, N. Yu., Ditkovskaya, I., and Lashkov, V. N. (1973). *Chem. Ztg.* **97,** 236 [*C. A.* 1973, **79,** 42 833].
Wang, S. S. (1973). *J. Am. Chem. Soc.* **95,** 1328.
Wang, S. S., (1975). *J. Org. Chem.* **40,** 1235.
Wang, S. S., (1976). *J. Org. Chem.* **41,** 3258.
Wang, S. S., and Merrifield, R. B. (1969). *J. Am. Chem. Soc.* **91,** 6488.
Wang, S. S., and Merrifield, R. B. (1972). *Int. J. Pept. Protein Res.* **4,** 309.
Weber, U., and Andre, M. (1975). *Z. Physiol. Chem.* **356,** 701.

Weitzel, G., Weber, U., Hornle, S., and Schneider, F. (1968). *Pept. Proc. Eur. Pept. Symp., 9th, 1968,* p. 171. North-Holland Publ., Amsterdam.
Weygand, F. (1968). *Pept. Proc. Eur. Pept. Symp., 9th, 1968,* p. 183. North-Holland Publ., Amsterdam.
Weygand, F., and Obermeier, R. (1968). *Z. Naturforsch Teil B.* **23,** 1390.
Weygand, F., and Ragnarsson, U. (1966). *Z. Naturforsch Teil B.* **21,** 1141.
Wieland, Th., Birr, C., and Wissenbach, H. (1969). *Angew. Chem.* **81,** 782.
Wieland, Th., Lewalter, J., and Birr, C. (1970). *Annalen,* **740,** 31.
Wieland, Th., Birr, C., and Flor, F. (1971). *Angew. Chem. Int. Ed.* **10,** 336.
Wildi, B. S., and Johnson, J. H. (1968). *Am. Chem. Soc. Natl. Meet. 155th,* A-8.
Wunsch, E. (1971). *Angew. Chem. Int. Ed.* **10,** 786.
Yajima, H., and Kawatani, H. (1971). *Chem. Pharm. Bull. Jpn.* **19,** 1905.
Yajima, H., and Kiso, Y. (1974). *Chem. Pharm. Bull. Jpn.* **22,** 1087.
Yajima, H., Kawatani, H., and Watanabe, H. (1970). *Chem. Pharm. Bull. Jpn.* **18,** 1333.
Yajima, H., Fujii, N., Ogawa, H., and Kawatani, H. (1974a). *J. Chem. Soc. Chem. Commun.,* 107.
Yajima, H., Kiso, Y., Okada, Y., and Watanabe, H. (1974b). *J. Chem. Soc. Chem. Commun.,* 106.
Yaron, A., and Schlossmann, S. F. (1968). *Biochemistry* **7,** 2673.
Young, G. T. (1972). *Essays Chem.* **4,** 115.

5

Oligonucleotide Synthesis on Polymer Supports

I.	Introduction	81
II.	General Principles of Solid-Phase Oligonucleotide Synthesis	84
III.	Polymer Supports	85
	A. Styrene-Based Polymer Supports	85
	B. Other Hydrophilic Supports	94
IV.	Functionalization of Polymer Supports	95
V.	Strategies Used for Oligonucleotide Synthesis on Polymer Supports	95
VI.	Protection of Reactive Groups	96
VII.	Cleavage of the Protecting Groups	97
VIII.	Attachment of the Polymeric Carrier to the Nucleotide or Nucleoside	97
IX.	Elongation of the Nucleotide Chain on the Polymer Support	98
X.	Cleavage of the Polymer–Nucleoside/Nucleotide Bond	98
XI.	Monitoring in Polymer-Supported Oligonucleotide Synthesis	100
XII.	Purification of Synthetic Oligonucleotides	100
XIII.	Synthesis of Oligoribonucleotides on Polymer Supports	100
XIV.	Miscellaneous Application of Polymers in Polynucleotide Synthesis	101
XV.	Advantages and Limitations	101
	References	102

I. INTRODUCTION

After the synthesis of peptides, oligonucleotide synthesis is the field in which the solid-phase method has been exploited with the most reasonable success.

Although the first solid-phase synthesis of a trinucleotide (Letsinger and Mahadevan, 1965) was attempted only 2 years after Merrifield (1963) announced the first synthesis of a peptide on a polymer support, the last 12 years have seen relatively little activity in this field.

Finding a suitable polymer support which has good solvating properties and which is mechanically and chemically compatible with the reactants and solvents used in the synthesis has been a major problem. The variety of groups (carbohydrates, bases, and phosphate moieties) in nucleotides that need to be protected during oligonucleotide synthesis and solubility problems introduced by the very large and polar nucleotide molecules pose other difficulties. It has been found that the yield of the reaction products falls in every step of synthesis, and often yields of oligonucleotides with more than three units are so low that they are not major products (Kossel and Seliger, 1975). Although it has not been possible to build up chains larger than an octanucleotide on a polymer support, the simplicity, convenience, and rapidity of synthetic procedures are too tempting for the oligonucleotide synthetic chemist to ignore. The ever-growing demand for oligonucleotides with a known sequence, especially in the field of genetics and genetic engineering, can be met only if they can be synthesized quickly. Indeed, the present world-wide state of research in this field is an indication that it may soon be possible to build chains up to the decanucleotide level in quite satisfactory yields, making automation possible (Norris, 1977). The method may serve as a good supplement to syntheses in solution in that it could provide blocks for subsequent fragment synthesis.

In solution, two types of approaches have been followed for the synthesis of oligonucleotides: the so-called "phosphodiester" and the "phosphotriester" approaches. The basic principles of the phosphodiester approach, pioneered by Khorana and co-workers, are well-established. Details of the method are easily accessible through several monographs (Chargaff and Davidson, 1955, 1960; Khorana, 1961; Michelson, 1963; Kochetkov and Budovskii, 1971) and review articles (Cramer, 1961, 1966, 1969; Khorana, 1967, 1968; Agarwal *et al.*, 1972; Zhdanov and Zhenodarova, 1975; Kossel and Seliger, 1975; Amarnath and Broom, 1977; Van Boom, 1977). Briefly, the first step in this approach is comprised of suitable protection of the groups likely to interfere in synthesis. This includes the amino groups of the heterocyclic bases and the hydroxy and phosphate groups of deoxyribonucleotides and ribonucleotides. The second step consists of coupling a nucleotide with a free 3'-OH with another having a 5'-phospha $\circ$, thus forming the 3'–5' phosphodiester bond. This internucleotide bond formation is brought about with the help of either dicyclohexylcarbodiimide (DCC) (Gilham and Khorana, 1958; Jacob and Khorana, 1964) or hindered arylsulfonyl chlorides [e.g., mesitylenesulfonyl chloride (MDC1) (Khorana *et al.*, 1962); triisopropyl-

I. Introduction

benzenesulfonyl chloride (TPS) (Lohrmann and Khorana, 1966)]. Sulfonyl chlorides afford higher yields without significant side reactions and are generally preferred for most condensations. Rubinstein and Patchornik (1972, 1975) developed poly(3,5-diethylstyrene)sulfonyl chloride as a solid-phase coupling reagent. The subsequent steps in the diester synthesis consist of isolation, deprotection, purification, and characterization of the oligonucleotide. A typical synthesis, following this approach is illustrated in Scheme 5-1.

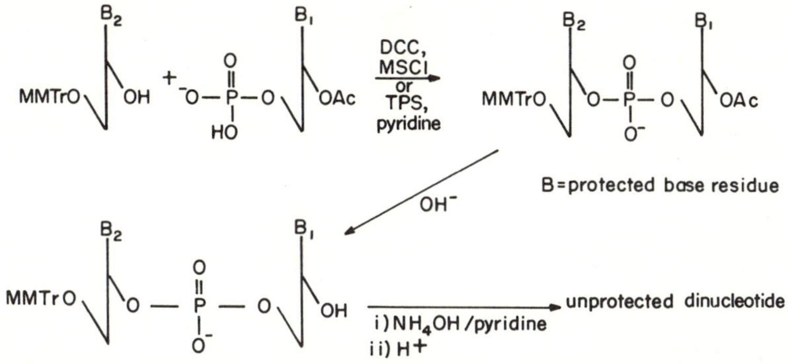

Scheme 5-1

The triester approach, for synthesizing oligodeoxyribonucleotides (Van Boom, 1977), involves the protection of the internucleotide phosphate by such groups as β-cyanoethyl, trichloroethyl, phenyl, or substituted phenyl. The key intermediate (4) outlined in Scheme 5-2 is formed by direct esterification of the monophosphate functions (1). Alternatively, 4 may also be formed from 2 by reaction with β-cyanoethyl dihydrogen phosphate or phenyl dihydrogen phosphate in the presence of TPS. The intermediate 4 can also be formed by first reacting 2 with trichloroethyl or phenylphosphorodichloridate to give 3, which on subsequent hydrolysis yields 4. Finally, the internucleotide linkage is established by condensation of 4 with a suitably protected 5'-hydroxyl bearing component 5 in the presence of TPS. Compound 3 may also be directly condensed with 5 to give 6 without the use of condensing agent. Although certain side reactions reported in the diester approach are prevented in the triester approach, the rate of internucleotide bond formation is lowered by a factor of 10. Hence, the reaction period is usually prolonged. In a modification of this approach (Scheme 5-2), Katagiri *et al.* (1975) have employed 7 as the starting material. This compound, by employing β-cyanoethyl and *p*-chlorophenyl groups, has a fully protected 3'-phosphate group. After deblocking the β-cyanoethyl group by treatment with mild alkali, the

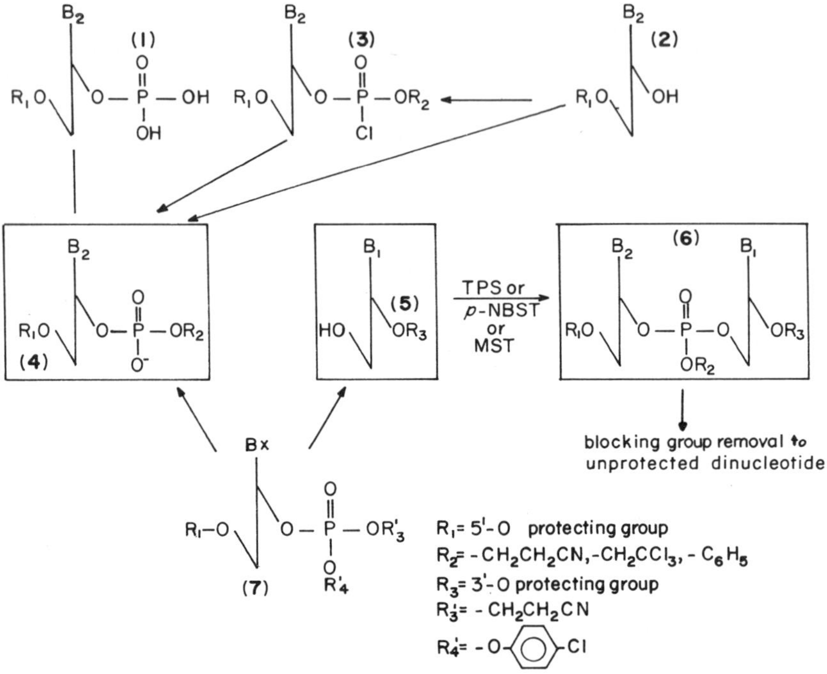

Scheme 5-2

product was condensed with 5 ($R_3 = PO_3R_3'R_4'$) in the presence of p-nitrobenzenesulfonyltriazolide (p-NBST) or mesitylenesulfonyltriazolide (MST) as the condensing agent. Using this approach and silica gel, instead of the conventional DEAE-cellulose chromatography, substantially higher yields have been reported.

II. GENERAL PRINCIPLES OF SOLID-PHASE OLIGONUCLEOTIDE SYNTHESIS

The solid-phase synthesis of oligonucleotides consists basically of the sequential repetition of reaction steps similar to those previously described, with the properly protected monomeric nucleotides (for reviews, see Kossel and Seliger, 1975; Amarnath and Broom, 1977). The essential steps are like those used in solid–phase peptide synthesis.

1. Suitable functionalization of the polymeric carrier.
2. Protection of the groups in the initial monomeric nucleotide or nucleoside that are likely to interfere in synthesis.

3. Attachment of the first nucleotide of the projected sequence to the polymeric support.
4. Blocking of the unreacted functional groups on the support and deblocking of the grafted monomer.
5. Coupling with a suitable protected second nucleotide in the presence of certain condensing agents to form the phosphodiester internucleotide bond (diester approach).
6. Blocking of the unreacted hydroxyl groups in order to prevent lengthening of the undesirable chains (attempted by some workers only).
7. Deblocking of the protective group on the reactive hydroxyl group.
8. Repetition of steps 5, 6, and 7 until the desired sequence is achieved.
9. Cleavage of the product from the support.
10. Deprotection of groups (step 10 in some cases may precede step 9).
11. Final purification and characterization.

III. POLYMER SUPPORTS

The choice of polymer support is the most crucial factor in solid-phase oligonucleotide synthesis. Various types of supports used by different workers are summarized in Tables 5-1 and 5-2. Although most reported uses of polymers in nucleotide syntheses are styrene-based supports (both insoluble and soluble), other polymers such as polyamides, poly(ethylene glycol), vinylacetate–N-vinylpyrrolidone copolymer, poly(vinyl alcohol), Sephadex LH20, and silica gel have also been used.

A. Styrene-Based Polymer Supports

Although polystyrene-type resins possess chemical stability and ease of functionalization, there is agreement among oligonucleotide synthetic chemists that styrene-based polymers are not compatible with the synthesis of oligonucleotide chains larger than a tri- or tetranucleotide.

Three main types of styrene-based polymers have been used: (a) insoluble copolymers of styrene cross-linked with divinylbenzene; (b) linear, "soluble" polystyrenes of different molecular weights; and (c) isotactic polystyrenes.

Among the insoluble types, polystyrenes with different amounts of cross-linking (0.1–40%) have been tried. Polystyrenes with a small degree of cross-linking (0.1–2%) swell considerably in organic solvents and provide higher condensation rates than the more highly cross-linked polymers. As the oligonucleotide chain grows, however, the hydrophobic nature of polystyrene becomes increasingly incompatible with the increasing

TABLE 5-1

Synthesis of Oligodeoxynucleotides on Polymer Supports Using the "Diester" Approach[a]

Polymer support and cross-linking (X)	Functional group	Nature of Ⓟ—Nu bond and conditions of formation	Conditions for successive coupling	Conditions for cleavage of Ⓟ—Nu bond	Synthesis achieved (yields %)	References
Polystyrene (soluble), $X = 0$	Ⓟ—MMTrCl	Nucleoside-5'-trityl ether, resin, excess dT/dry Py	pdTOAc, MSCl; pdA^{bz}OAc, MSCl; pdC^{an}OAc, MSCl; pdG^{ib}OAc, MSCl	1% TFA/CHCl$_3$ at 0°C for 15 min	dT pdT (96%); dT(pdT)$_2$, (83%); dT pdC, (91%); dT pdA, (76%)	Hayatsu and Khorana, 1966, 1967
Polystyrene (soluble), $X = 0$		Nucleoside-5'-trityl ether, resin, dTOAc/dry Py	(i) Deblock at 3'-O (ii) pdTOAc	50% TFA/dioxane, 1:100 for 1 hr at R.T.	dT(pdT)$_2$ (11%)	Cramer et al., 1966
Polystyrene (soluble), $X = 0$	Ⓟ—⟨benzene⟩—NH$_2$	Nucleotide-5'-phosphoramidate and resin	pdTOAc	80% HOAc	—	Cramer et al., 1966; Seliger, 1973
Popcorn polystyrene, $X = 0.2\%$	Ⓟ—COCl	Nucleoside-3'-ester	Refer to Scheme 5-3	0.5 N NaOH in water/dioxane 1:1, for 12 hr at R.T.	dTpdTpdT (51%)	Letsinger et al., 1967a
Popcorn polystyrene, $X = 0.2\%$	Ⓟ—DMTrCl	Nucleoside-5'-trityl ether, resin, dTOAc/dry Py, 48 hr at 70°C	(i) Deblock at 3'-O (ii) pdTOAc	0.1% TFA/CH$_2$Cl$_2$ 1 hr at R.T.	dTpdT (64%; dT(pdT)$_5$ (3.5%)	Koster and Cramer, 1972

Support	X =	Procedure	Reagents	Deprotection	Product (yield)	Reference
Polystyrene, 1%	ⓟ—MMTrCl	Nucleoside-5′-trityl ether, resin, excess dT in abs. pyridine or DMF/Py or C_6H_6/Py for 48 hr at R.T.	pdTOAc, DCC, or TPS	$AcOH-H_2O-C_6H_6$, 16:4:5 v/v for 5 hr at R.T. or 1% $AcOH-C_6H_6$ for 18–24 hr at R.T.	dTpdT, (59%); dT(pdT)$_2$, (52%); dT(pdT)$_3$, (39%); dT(pdT)$_4$, (49%); (yields reported based on the next lower member in the series)	Melby and Strobach, 1967, 1969
Polystyrene, 2%	ⓟ—TrCl	Nucleoside-5′-trityl ether, resin, excess dT, dry Py, 1 hr, 60°C Keep for a day at R.T.	dpG^{ac}OAc, DCC coupling repeated at each step	2% TFA in dry C_6H_6 for 15 min at 14°C	dTpdG, (27%); dT(pdG)$_3$, 2%	Zarytova et al., 1970
Polystyrene, 2%	ⓟ—CH$_2$O—⟨⟩—NH$_2$	Nucleotide-5′-phosphoramidate, resin	pdTOAc, DCC	80% AcOH for 24 hr	(pdT)$_2$, (40%); (pdT)$_3$, (13%); (of the di-)	Blackburn et al., 1967, 1969; Ohtsuka et al., 1972
Polystyrene, 2%	ⓟ—⟨⟩—CH$_2$COCl	Nucleoside-5′-ester, resin dT $\xrightarrow{\text{anhydrous pyridine}}$ for 22 hr at R.T.	pdT, MSCl or DCC	Conc. NH$_4$OH/dioxane, 1:1 v/v 120 hr at R.T.	dT(pdT)$_n$, n = 1–5 polycondensation dTpdT (85%)	Kusama and Hayatsu, 1970

[a] an = anisoyl; bp = β-benzoylpropionyl; bz = benzoyl; DMTr = dimethoxytrityl; ib = isobutyryl; IUdR = 5′-iododeoxyuridine; MMTr = monomethoxytrityl; MSCl = mesitylenesulfonyl chloride; Nu = nucleotide(oside); Obp = β-benzoylpropionyl; Py = pyridine; R.T. = room temperature; TFA = trifluoroacetic acid; TPS = 2,4,6-triisopropylbenzenesulfonyl chloride; Tr = trityl.

(Continued)

TABLE 5-1 (Continued)

Polymer support and cross-linking (X)	Functional group	Nature of Ⓟ—Nu bond and conditions of formation	Conditions for successive coupling	Conditions for cleavage of Ⓟ—Nu bond	Synthesis achieved (yields %)	References
Polystyrene, X = 3%	Ⓟ—MMTrCl	Nucleoside-5'-trityl ether, dT $\xrightarrow{\text{anhydrous pyridine}}$, 48 hr	pdTOAc, MSCl	80% AcOH	dTpdT, (73%); dT(pdT)$_2$, (40%)	Glaser et al., 1973
Polystyrene, X = 5%	Ⓟ—⟨⟩—CH$_2$Cl	Nucleotide-5'-phosphorthioate, resin, 5'-phosphorthioate-dT	pdTOAc, TPS; pdC^{an}OAc, TPS; pdA^{bz}OAc, TPS	Iodine (10 mg/ml in pyridine/water 3:1; 20 hr at R.T.	(pdT)$_2$, (39%); (pdT)$_3$, (25%); (pdT)$_4$, (19%); (pdT)$_5$, (21%); yields based on the next lower member some other di- and trinucleotides also reported	Sommer and Cramer, 1972; cf. Seliger and Cramer, 1969
Polystyrene, X = 6%	(structure with OH, C, CH$_2$, N, HOH$_2$C, CH$_2$C)	Nucleotide-5'-phosphodiester, resin, pdTOAc DCC	pdTOAc	2 N NaOCH$_3$ in CH$_3$OH/pyridine 1:1	(pdT)$_2$,(35%); (pdT)$_6$, (2%)	Freist and Cramer, 1970
Polystyrene, X = 20% (microgel)	Ⓟ—MMTrCl	Nucleoside-5'-trityl ether, resin, dTOAc	—	80% AcOH	—	Koster and Geussenhainer, 1972
Polystyrene, X = high	Ⓟ—MMTrCl	Nucleoside-5'-trityl ether, dTOAc, Py, 48 hr	(i) Deblock at 3'-O (ii) pdTOAc, TPS; pdTpdTOAc, TPS; pdC^{an}OAc, TPS; pdA^{bz}OAc, TPS	80% AcOH for 1 hr at 70% or for 6 hr at R.T. or Py/AcOH/water 3:10:3 v/v/v for 48 hr at R.T.	dT(pdT)$_5$	Cramer and Koster, 1968; Koster and Cramer, 1973

Support	Linker	Procedure	Conditions	Products	Reference
Polystyrene, highly porous, nonswellable, X = high	ⓟ—MMTrCl	Nucleoside-5′-trityl ether, resin, dTOAc or dTpdTOAc, Py, 48 hr, 70°C	80% AcOH	dT(pdT)$_2$, dT(pdT)$_3$, dT(pdT)$_5$, dTpdTpdC, dTpdTpdA, dTpdTpdG	Koster and Cramer, 1974a
Polystyrene, X = high	ⓟ—MMTrCl ⓟ—TrCl	Nucleoside-5′-trityl ether, resin, excess dT, Py	(i) Deblock at 3′-O (ii) pdTOAc, TPS; pdA^{bz}OAc, TPS; pdG^{an}OAc, TPS; pdC^{an}OAc, TPS; pdC^{an}OAc, TPS	2% TFA in dry C_6H_6 or $CHCl_3$ at 15°C for 30 min	dTpdA, (80%); dT(PdA)$_2$, (64%); dT(pdA)$_3$ (43%)	Potapov et al., 1971; Kabachink et al., 1973; Levina et al., 1972
Isotactic polystyrene	ⓟ—MMTrCl	Nucleoside-5′-trityl ether, resin, excess dT, Py	pdA^{bz}OAc, MSCl	2% TFA in dry C_6H_6 or $CHCl_3$ at 15°C for 30 min	dTpdA, (80%); dT(pdA)$_2$, (56%); dT(pdA)$_2$, (28%)	Potapov et al., 1971
Isotactic polystyrene	ⓟ—MMTrCl	Nucleoside-5′-trityl ether: (i) Resin + dT dry pyridine 45 hr, 80°C (ii) Resin + IUdR, dry Py, 25 hr, 80°C (iii) Resin + dA, dry Py, 36 hr, R.T.	pdTOAc, MSCl pIUdR-3″-OAc, MSCl	10% TFA/CHCl at 0°C for 10 min	dTpT, (50%); dT(pdT)$_2$, (56.8%); dTpIUdR, (63%); dApdT, (58%); IUdRpdT, (36%); IUdR(pdT)$_2$, (25%)	Tsou and Yip, 1973
Poly-L-lysine cross-linked N$^\epsilon$-(p-amino benzoylated)	ⓟ—NH—C(=O)—C$_6$H$_4$—NH$_2$	Nucleoside-5′-phosphoramidate, resin, pdTOAc, DCC/Py	(i) Deblocked at 3′-O (ii) pdTOAc, TPS (iii) Naphthyl-isocyanate in order to block the unreacted hydroxyl groups	Isoamylnitrite in Py/ACOH Py/ACOH 1:1 v/v	(pdT)$_2$, (43%); (pdT)$_3$, (14%)	Chapman and Kleid, 1973

(Continued)

TABLE 5-1 (Continued)

Polymer support cross-linking (X)	Functional group	Nature of Ⓟ—NU bond and conditions of formation.	Conditions for successive coupling	Conditions for cleavage of Ⓟ—Nu bond	Synthesis achieved (yields %)	References
Polydimethyl-acrylamide, cross-linked, insoluble	—S(CH₂)OH with (CH₂)₂—CO—NH—Ⓟ benzene ring	Nucleotide-5′-phosphodiester resin, pdTOAc/TPS	(i) Deblocked at 3′-O (ii) pdTOAc, TPS; pdC^{an}OAc, TPS; pdA^{bz}OAc, TPS; pdG^{ib}OAc, TPS (iii) C^{6}H^5—N=C=O in order to block the unreacted hydroxyl	(i) 0.2 N N-chloro-succinimide in buffer 0.1 M K₃PO₄ (pH 7.5) in dioxane/water 1:1, 2 times, 15 min each (ii) 0.2 M NaOH in dioxane/water/MeOH (5:4:1)	(pdT)₅ (47%); (pdT)₅pdC (24%); d(pC-A-G-T-G-A-T)	Gait and Sheppard, 1976, 1977
Polyacrylmorpholide (Enzacryl K), cross-linked insoluble	Ⓟ—C(=O)—NH(CH₂)₂NH₂	Nucleoside-5′-trityl ether, resin, 5-O-(p-carboxymethyloxy-TrO thymidine $\xrightarrow{\text{DCC}}$ Pyridine)	pdTOAc, TPS	1% CF₃COOH in CH₃CN/H₂O, 1:1	dT(pdT)₄ 84–90% in each step of condensation	Narang et al., 1977
Macroporous polyacrylic acid Bio-Rex 70	Ⓟ—C(=O)—OCH₂CHCH₃—O—COCl	Nucleoside-5′-ester	pdTMMTr, TPS Py, 30 hr (50°C)	0.1 M Toluene sulfonic acid CH₃CN	dTpdT, (18%); dTpdA, (8%), and dTpdApdApdC, (13%); dTpdTpdApdC, (6%)	Seliger et al., 1978

Support	Functional group	Reaction	Conditions	Cleavage	Products	Reference
Copolymer of vinyl acetate and N-vinyl pyrrolidone, non-cross-linked	Ⓟ—OH	Nucleoside-5′-carbonate, resin, ClCOdT-Obp $\xrightarrow{\text{DMF, Et}_3\text{N}}_{24\text{ hr, R.T.}}$	(i) Deblock at 3′-Obp with N_2H_4 (ii) pdT, TPS; pdTOMMTr, TPS; prU(OAc)$_2$, TPS	Conc. NH_4OH	dT(pdT)$_n$ polycondensation products (dT(pdT)$_n$rU n = 1–2 dTpdT, (56%)	Seliger and Aumann, 1973, 1975
α-ω-Diamino poly(ethylene glycol)	Ⓟ—NH$_2$	Nucleotide-5′-phosphoramidate, resin, pdA^{bz}OAc $\xrightarrow{\text{DCC, Py}}_{2\text{ days}}$	(i) Deblock at 3′-O (ii) pdA^{bz}OAc, TPS; pdTOAc, TPS	Isoamyl nitrite in AcOH/Py 1:1	d(pdA)$_3$, (14%); d(pdT)$_5$, (8%)	Brandstetter et al., 1973
Poly(ethylene glycol)	Ⓟ—(NH—C$_6$H$_4$—S—(CH$_2$)$_2$OH)$_2$	Nucleotide-5′-phosphodiester bond, resin, pdTOAc, TPS/Py	(i) Deblock at 3′-O (ii) pdTOAc, TPS	(i) N-Chlorosuccinimide in 0.1 M phosphate buffer for 10 min at R.T. (ii) 1 N NaOH for 5 min at 0°C	—	Brandstetter et al., 1974
Poly(ethylene glycol)	Ⓟ—MMTrCl	Nucleoside-5′-tritylether, resin, dTOAc	—	—	—	Koster, 1972b
Poly(ethylene glycol)	Ⓟ—OH	Nucleotide-5′-phosphodiester bond, resin, $\xrightarrow{\text{TPS, dry Py}}_{22°C, 20\text{ hr}}$	—	—	—	
Sephadex LH20	Ⓟ—OH Obp	Nucleotide-5′-phosphodiester bond, resin, $\xrightarrow{\text{TPS, dry Py}}_{22°C, 24\text{ hr}}$	pdTOAc	0.1 N NaOH for 20 min at 22°C	—	Koster and Heyns, 1972
Silica gel	Ⓟ—TrCl	Nucleoside-5′-tritylether		80% AcOH	dTpdT (54%)	Koster, 1972a

TABLE 5-2

Synthesis of Oligodeoxynucleotides on Polymer Supports Using the "Triester" Approach[a]

Polymer support	Functional group	Type of P—Nu bond and conditions of formation	Conditions for successive coupling	Cleavage of P—Nu bond	Synthesis achieved (yields are in percent)	References
Polystyrene, popcorn (copolymer of p-vinylbenzoic acid, styrene, DrB) $X = 0.02$–0.2%	ⓟ—COCl	Deoxycytidine N-carbamate, resin, excess 5'-OTr-dC, Pyr	(i) β-Cyanoethyl phosphate, DCC (ii) dT/MSCl	0.2 M NaOH in dioxane/water, 1:1	dC(pdT)$_n$, $n=1$ (61%); $n=2$ (16%); $n=3$ (14%)	Letsinger and Mahadevan, 1965, 1966
as above	ⓟ—COCl	Nucleoside-5'-ester bond, resin excess N-DMTrdG, Py	(i) β-Cyanoethyl phosphate, MSCl (ii) MSCl, N-DMTrdG or MSCl, dT	0.5 M NaOH in dioxane/water, 1:1	dG(pdG)$_n$, $n = 1$–3 dG(pdG)$_2$pdT	Shimidzu and Letsinger, 1968, 1971
as above	ⓟ—COCl	Nucleoside-5'-ester bond, resin excess N-DMTrdG, Py	(i) β-Cyanoethyl phosphate, MSCl (ii) 3'-O-β-benzoylpropionyl-dT, TPS	0.5 M NaOH in dioxane/water 1:1	dT(pdT)$_2$ (78%)	Letsinger et al., 1967a,b

Support	Reagent	Attachment	Procedure	Cleavage	Products	Reference
Polystyrene, $X = 0.2\%$	ⓟ—COCl	Nucleotide-3'-ester, resin, 5'-MMTrdT/Py, 40 hr, R.T.	Refer to Scheme 5-4	0.5 M NaOH in dioxane/water, 1:1; R.T., 30 hr	dTpdT, (64%); dT(pdT)$_2$, (38%); dT(pdT)$_4$, (9%)	Pless and Letsinger, 1975
Polystyrene, biobeads $X = 2\%$	ⓟ—⟨benzene ring⟩—C(=O)—C(CH$_2$)$_2$—COCl	Nucleoside-5'-ester bond, resin, dT-3'-MMTr/Py, 48 hr, R.T.	(i) Deblock at 3'-O (ii) Phosphorylation at 3'-O with CH$_3$OPOCl$_2$ (iii) dT-3'-MMTr	0.5 M NaOH in dioxane/water, 1:1, 48 hr, R.T.	dTpdT; (38%); dT(pdT)$_2$, (10%); dT(pdT)$_3$, (4%)	Ogilvie and Kroeker, 1972
Polystyrene, $X = 2\%$	ⓟ—TrCl	Nucleoside-5'-trityl ether	(i) Phosphorylation at 3'-OH with POCl$_3$ (ii) dTOAc or 2',3'-Isopropylidineuridine	1% AcOH in CHCl$_3$ or C$_6$H$_6$	dTpdT, (52%); dTpU, (67%)	Kabachink et al., 1970a,b
Polystyrene popcorn, $X = 0\%$	ⓟ—TrCl	Nucleoside-5'-tritylether, resin, dT	(i) Phosphorylation at 3'-OH NO$_2$—⟨benzene⟩—OPOCl$_2$ (ii) dTOAc	1% TFA in CHCl$_3$ or C$_6$H$_6$	dTpdT, (60%)	Kabachink et al., 1970a,b

[a] Abbreviations used are same as in Table 5-1.

content of strongly polar phosphodiester bonds. In addition, there is a drastic change in the accessibility of the active centers with the change of solvent (steric hindrance). Among the low-level cross-linked polystyrenes, the popcorn type—the one used by Letsinger and Mahadevan (1965) in the first polymer-supported synthesis—appears to be the most attractive. It has very low cross-linking ($\sim 1\%$) and possesses maximum chain flexibility (Breitenbach and Olaj, 1968). At the same time, it is mechanically stable, insoluble, and easily recoverable by filtration.

More highly cross-linked polymers are rigid, nonswellable, and relatively insensitive to differences in the solvation exerted by different media. With these types the immobilized reactant is attached only on the surface of the pores and accessible cavities of the heterogeneous macroreticular gel. A compromise must therefore be found between pore size and mechanical stability. Koster and Geussenhainer (1972) approached this problem by using very small particles with a diameter of 0.1–1 μm—the so-called microgels, in which the ratio of outer to inner surface is greatly reduced. The small size of the particles, however, precludes their filtration. Koster and Cramer (1974a) and Koster *et al.* (1974) have reported the use of a macroporous polystyrene made by copolymerizing styrene with divinylbenzene in the presence of n-dodecane and toluene. Using this support, they could synthesize chains up to the octanucleotide level.

The soluble polystyrenes are linear, without any cross-links, and have molecular weights ranging from 150,000 to 300,000. They are soluble in a number of organic solvents, but are insoluble in alcohol, water, and petroleum ether. These carriers provide condensation yields comparable with those in homogeneous phase syntheses. However, as the oligonucleotide chain grows to a trinucleotide, with the addition of water or alcohol, the polymer tends to become colloidal and therefore not easily filtrable. Thus, each step of condensation and filtration causes a loss of 10–15% of the polymer carrying the nucleotide. The losses arising from incomplete precipitation have been claimed to be minimized by gel filtration using Sephadex-LH20 (Potapov *et al.*, 1972).

Isotactic polystyrenes strike a mean between insoluble and soluble polystyrenes. They have poor solubility in organic solvents and can be recovered completely by washing with polar solvents such as methanol, ethanol, or water.

B. Other Hydrophilic Supports

Considering the limitations of polystyrene, other hydrophilic polymeric supports have been tried recently: poly(ethylene glycol) (Koster, 1972b;

Brandstetter *et al.*, 1973), poly(vinyl alcohol) (Schott, 1973; Schott *et al.*, 1974), vinylacetate–*N*-vinylpyrrolidone copolymer (Seliger and Aumann, 1973), Sephadex LH20 (Koster and Heyns, 1972), and silica gel (Koster, 1972a). Most of them are reported to give yields no better than those of polystyrene, and they suffer from the disadvantage that nucleotides tend to be adsorbed on them. Polyamide supports appear to be more promising supports for nucleotide synthesis (Gait and Sheppard, 1976, 1977). A cross-linked polyacrylmorpholide has also been used (Narang *et al.*, 1977). The cross-linked polyacrylmorpholide resin (Enzacryl gel) is commercially available from Koch Light Labs. Ltd., England. Both the acrylamide and morpholide resins have good handling properties, show excellent swelling in organic solvents, and exhibit very little tendency for adsorbing nucleotides. A poly-L-lysine support has been reported by Chapman and Kleid (1973).

IV. FUNCTIONALIZATION OF POLYMER SUPPORTS

Various functional groups may be incorporated into the polymeric carrier either before or after polymerization. Their prime function is to bind the first nucleoside or nucleotide. It is also possible to introduce a functional group into the nucleoside or nucleotide and then couple it with the suitable derivatized polymer support. In either case, the bond between the support and the nucleotide or nucleoside must be strong enough to resist cleavage during the synthesis. At the end of the synthesis it should be possible to cleave the bond completely under conditions which result in no cleavage of the internucleotide bond and no deamination or depurination of the nucleotides.

The more common functionalities used by various workers have been summarized in Table 5-1, 5-2.

V. STRATEGIES USED FOR OLIGONUCLEOTIDE SYNTHESIS ON POLYMER SUPPORTS

As in solution syntheses, two types of strategies have been used for oligodeoxynucleotide synthesis. These are the "diester" and the "triester" approaches. The earliest attempts by Letsinger and co-workers employed the triester approach. This is illustrated in Schemes 5-3 and 5-4, X = H). Most of the recent work has employed the diester approach. The advantage in this particular approach is that the phosphorylating reagents are more reactive than those leading to phosphotriesters. Most workers

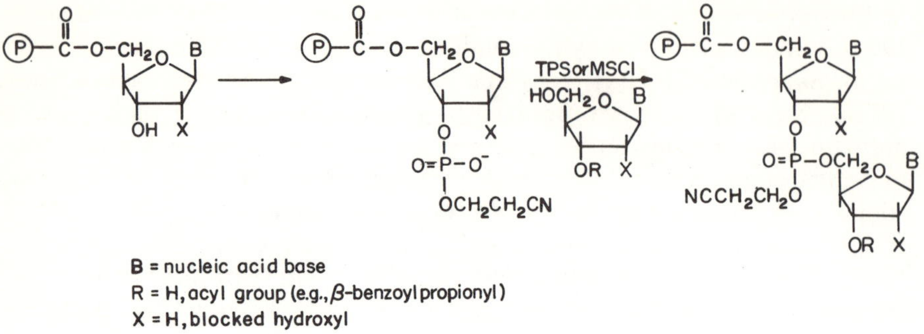

96   5. Oligonucleotide Synthesis on Polymer Supports

[Scheme 5-3 diagram]

B = nucleic acid base
R = H, acyl group (e.g., β-benzoylpropionyl)
X = H, blocked hydroxyl

Scheme 5-3

have preferred blocking the 5'-O position with the functionalized polymer, carrying out elongation of the oligonucleotide chain from the 3'-O end by coupling with another nucleotide having free 5'-phosphate and protected at the 3'-position (Scheme 5-3).

VI. PROTECTION OF REACTIVE GROUPS

The protecting groups for polymer-supported oligonucleotide synthesis have been essentially those adopted from the solution methods. By using a support, one of the groups—either the sugar 5'-OH or 3'-OH, or the corresponding phosphates, or a heterocyclic base amino group—is automatically protected by binding to the polymeric carrier. All other groups which are likely to interfere in the synthesis need to be protected. However, the selection of protecting groups must be appropriate to the strategy employed in the synthesis. The normal rule is that the protective groups used for each of the functional groups should be independently cleavable, like the "lock and key" system. For the heterocyclic base amino group, it is common to use anisoyl, benzoyl, and isobutyryl groups for cytosine, adenosine, and guanosine, respectively. During acylation of

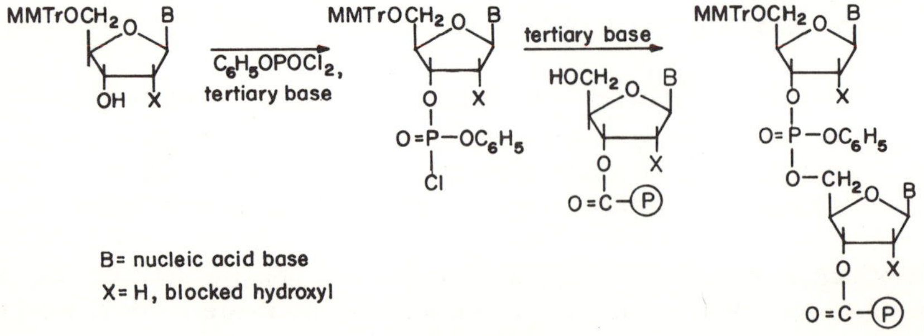

[Scheme 5-4 diagram]

B = nucleic acid base
X = H, blocked hydroxyl

Scheme 5-4

the amino group, the free hydroxyl groups are also acylated, at least partially. They can be readily deacylated by the usual alkaline treatment, which leaves all other protecting groups unaffected. Shimidzu and Letsinger (1968, 1971) employed dimethoxytrityl groups for the protection of the amino group of guanosine. However, this method proved unsatisfactory since some of the protecting groups were cleaved off during phosphorylation reactions.

VII. CLEAVAGE OF THE PROTECTING GROUPS

The N-acyl groups used for the protection of amino groups of the heterocyclic bases can all be cleaved simultaneously by reaction with concentrated ammonium hydroxide in pyridine or methanol. The cleavage conditions of the polymer–nucleoside/nucleotide bond are given in Table 5-1 and 5-2. O-Acetyl groups have been cleaved by alkaline treatment in many different media: 2 N NaOH/pyridine (Py) (1:1 v/v) (Freist and Cramer, 1970); 1 N NaOH for 5 min (Brandstetter et al., 1974); 0.1 N NaOH for 1 hr at 0°C (Koster, 1972b); 3 N NaOH in methanol/Py, 3:1 v/v, for 20 min at 0°C (Koster and Heyns, 1972); 0.1 N $(CH_3)_4$NOH in $(CH_3)_2$CHOH for 3 hr at room temperature, or DMF + Py + 1 M NaOCH$_3$, 5:4:1 v/v/v, for 1–2 hr at room temperature (Cramer and Koster, 1968; Koster and Geussenhainer, 1972; Koster et al., 1974; Koster and Cramer, 1972, 1974b); 1 M NaOCH$_3$ in CH$_3$OH + DMSO + Py, 2:4:4 v/v/v, for 15 min at room temperature (Tsou and Yip, 1973); conc. NH$_4$OH/Py, 3:1 v/v, for 3 days at room temperature (Melby and Strobach, 1967); 0.2 M KOH in CH$_3$OH/dioxane, 1:9 v/v, for 15–30 min at room temperature (Melby and Strobach, 1969); a mixture of 2 M $(CH_3)_4$NOH/Py/EtOH, 2:1:1 v/v/v, for 5 min at 0°C (Narang et al., 1977); 0.2 N NaOCH$_3$ in CH$_3$OH/Py 1:1 v/v (Hayatsu and Khorana, 1967; Blackburn et al., 1967; Gait and Sheppard, 1976, 1977; Chapman and Kleid, 1973); 2 N NaOH/Ch$_3$OH/Py 1:1:1 v/v/v for 3 hr at 20°C (Potapov et al., 1971); 2 N NaOH/CH$_3$OH/Py 10:10:7 v/v/v, for 3 hr at room temperature (Zarytova et al., 1971); 1 N KOH for 1 hr at room temperature (Kabachink et al., 1970a,b).

VIII. ATTACHMENT OF THE POLYMERIC CARRIER TO THE NUCLEOTIDE OR NUCLEOSIDE

The polymer can be linked via the sugar 5'-OH or 3'-OH, via the corresponding phosphates, or at the amino side chain of the heterocyclic base. Most workers have linked the polymer at the 5'-O position through a

5'-nucleoside ether, 5'-nucleoside ester, or 5'-nucleoside carbonate bond. The workers who have used 5'-phosphate nucleotides for binding to the support have used phosphoramidate, phosphorthioate, or phosphodiester bonds. Letsinger *et al.* (1967a) and Pless and Letsinger (1975) have linked the polymer at the 3'-O end with an ester bond. Ohtsuka *et al.* (1972) bound the polymer at the 3'-phosphate with a phosphoramidate linkage. The procedural details for attachment of the support to the first nucleoside or nucleotide are given in Table 5-1, 5-2.

IX. ELONGATION OF THE NUCLEOTIDE CHAIN ON THE POLYMER SUPPORT

After the first nucleotide has been bound to the polymer support, the unreacted functional groups still present on the polymer must be blocked. During this process, the hydroxyl groups on the nucleotide may also be blocked, at least partially; it is therefore important to deblock these groups selectively before coupling with the second nucleotide. The coupling reaction is then made with a suitably protected nucleotide in the presence of a condensing agent to form the phosphodiester internucleotide bond (Schemes 5-3 and 5-4). After deprotection of the reactive hydroxyl group (3'-OH or 5'-OH), the condensation reaction is repeated until the desired sequence of oligonucleotide chain is reached. Since the coupling yield in each synthetic step seldom, if ever, reaches 100%, some of the hydroxyl groups remain unreacted and they are likely to react in the subsequent synthetic stages. The net result is a mixture of products which is difficult to resolve. In order to minimize this difficulty, some workers, following the suggestion by Agarwal and Khorana (1972), have attempted to block the unreacted 3'-OH groups after each coupling reaction. The use of naphthylisocyanate (Chapman and Kleid, 1973) or phenylisocyanate (Gait and Sheppard, 1976, 1977) in this type of blocking has been reported. The latter workers found that phenylisocyanate has an additional advantage in that it reacts with water, thus making it possible to avoid the operation of repeated evaporation with absolute pyridine before the coupling reaction.

X. CLEAVAGE OF THE POLYMER–NUCLEOSIDE/NUCLEOTIDE BOND

Cleavage of the oligonucleotide chain after the desired sequence has been built up depends upon the nature of the polymer–nucleoside/

X. Cleavage of the Polymer–Nucleoside/Nucleotide Bond

nucleotide (Ⓟ—Nu) bond, the nature of the surrounding groups, and the nature of the support. The conditions used by different workers for cleavage of the oligonucleotide chain from the polymer are summarized in Tables 5-1 and 5-2. Since acidic media are not employed during the synthetic cycles, most workers have preferred to use the trityl—Nu ether bond which is acid-labile and can be easily cleaved off. The facility of cleavage, other things remaining the same, falls in the order dimethoxytrityl (DMTr) > monomethoxytrityl (MMTr) > trityl (Tr). The DMTr—Nu bond is, however, so labile that some nucleotide material may be lost during the synthetic cycles. MMTr has been the obvious optimum choice for most workers. Because the nature of the support changes with the growing oligonucleotide chain, it has been found (Narang et al., 1977) that the conditions (1% TFA in CH_2Cl_2 for 1 hr at room temperature) that were quite suitable for cleavage of the Ⓟ — Nu bond up to a trinucleotide stage become unsatisfactory at higher stages. The use of 1% TFA in 50% aqueous acetonitrile has proved more satisfactory. It is important that conditions for acidic cleavage be chosen properly, since prolonged exposure to acid may lead to degradation and depurination of the chain. Other linkages that could be cleaved in neutral or alkaline media have been tried. Phosphoramidate linkages can be cleaved by isoamyl nitrite in pyridine–acetic acid 1:1 (v/v). The alkali-labile ester and carbonate Ⓟ — Nu linkages, however, forbid the use of 3'-O-acetyl protection which is commonly employed for the protection of the growing chain. However, the synthesis is possible if 3'-O-β-benzoylpropionyl group protection is used (Letsinger et al., 1967b). As in peptide synthesis, certain polymer supports incorporating a "safety-catch" device have been utilized in the cleavage of an oligonucleotide chain from the polymer support. A β-mercaptoether was used and cleaved by safety-catch activation via oxidation of the sulfide to the corresponding sulfone followed by base-catalyzed elimination [Eq. (1)] (Brandstetter et al., 1974; Gait and Sheppard, 1977).

$$\text{Ⓟ}-S(CH_2)_2-ONu \xrightarrow{NCS} \text{Ⓟ}-\overset{O}{\underset{O}{\overset{\|}{S}}}-(CH_2)_2-ONu$$

$$\downarrow \text{base}$$

$$\text{Ⓟ}-\overset{O}{\underset{O}{\overset{\|}{S}}}-CH=CH_2 + Nu$$

(NCS = N-chlorosuccinimide) (1)

XI. MONITORING IN POLYMER-SUPPORTED OLIGONUCLEOTIDE SYNTHESIS

Since 100% yields in each coupling reaction are seldom, if ever, attainable, failure sequences and truncated sequences are present on the polymer support. A similar problem occurs in polypeptide synthesis (see Chapter 4). On cleavage all of them pass into the cleavage product. It is important that the yield of the longest desired oligonucleotide formed after each coupling step be determined. Separation of these different products may be affected satisfactorily by paper chromatography, electrophoresis, anion-exchange chromatography, or high-pressure liquid chromatography (HPLC). The concentration of each product can then be determined by uv measurements. If the first nucleotide carries a radiolabeled base, the cleavage product will contain sequences carrying the radiolabeled base. These can be similarly separated and a molar ratio of each found by scintillation counting (Narang et al., 1977).

XII. PURIFICATION OF SYNTHETIC OLIGONUCLEOTIDES

In the polymer-supported synthesis, the product will contain not only the desired longest oligomer, but also species of shorter lengths down to the mononucleotide. Adequate resolution of all these is highly desirable in order to obtain the pure product. Anion-exchange chromatography has been the most widely used technique for purification of oligonucleotides and their base-protected derivatives. DEAE-cellulose has been the most widely employed support. In addition, the use of other ion-exchangers, for example, RPC5 (Egan, 1973) Dowex-1 (Asteriadis et al., 1976), and Partisil-10-SAX (Hartwick and Brown, 1975; Gait and Sheppard, 1977), has been also explored, especially for use in HPLC. It must be emphasized that there is still scope for improvement in the development of rapid and efficient fractionation methods for oligonucleotides, both at the protected and unprotected stages. With these improved methods, solid-phase oligonucleotide synthesis will achieve real breakthroughs.

XIII. SYNTHESIS OF OLIGORIBONUCLEOTIDES ON POLYMER SUPPORTS

In the synthesis of oligoribonucleotides, there is the additional problem of protection of the 2'-OH group in order to prevent internucleotide bond formation at this site. (Schemes 5-3 and 5-4, X = blocked hydroxyl). The participation of the free 2'-OH can also cause hydrolysis of 3', 5'-phosphodiester bonds under basic and acidic conditions. Yip and Tsou (1971) were the first to report the synthesis of the trimer, UpUpU on an isotactic succinylated polystyrene support. 2' - O - β - Benzoylpropionyl-3' - O - methoxyacetyluridine was bound to the polymer through a 5'-ester bond and deblocked at the 3'-O-position. This served as the monomer unit. It was condensed with 2'p O - β - benzoylpropionyl - 3' - O - methoxyacetyluridine-5' - phosphate to form the dinucleotide. Repetition of this last step gave the trinucleotide.

Ohtsuka et al. (1972) have synthesized oligoribonucleotides on an insoluble 4'-aminophenoxymethylpolystyrene support. The first nucleotide, 5'-OMMTr-2'-O-acyl protected, was linked to the polymer via a 3'-phosphoramidate linkage. After deblocking at the 5'-O, it was coupled to the second nucleotide, a 5'-OMMTr-2'-O-acyl-protected 3'-phosphate, to form the dinucleotide.

XIV. MISCELLANEOUS APPLICATION OF POLYMERS IN POLYNUCLEOTIDE SYNTHESIS

Support-aided conversion of nucleosides to nucleotides has been described. Trityl chloride carriers have been used along with various phosphorylating agents to phosphorylate several nucleosides (Kabachnik et al., 1970a,b). An aldehyde-containing polymer has been used to demonstrate the support-aided phosphorylation of deoxyribonucleoside rU (Hanessian et al., 1974).

Polymer supports have also been used in attempts to elucidate the mechanism of carbodiimide- and sulfonyl chloride-aided phosphorylations (Blackburn et al., 1966, 1967, 1969).

Reagents supported on polymeric backbones have been used to catalyze internucleotide bond formation in either diester or triester syntheses (Rubenstein and Patchornik, 1972, 1975). These will be considered again in Chapter 12.

XV. ADVANTAGES AND LIMITATIONS

Like any other solid-phase methods of synthesis, oligonucleotide synthesis on polymer supports offers the advantages of ease and quickness of product isolation. Because the synthesis is a sequential type, the successive coupling operations can be done on intermediates still bound to the polymer. The method appears to be attractive for the synthesis of short oligomers (tri or tetra) for which good to excellent yields have been reported.

The main limitation of the method lies in the rapid decrease in yields seen in each successive coupling step. Accompanying the decreased yield is the formation of failure and truncated sequences. The decreased yields, which seem to stem from nucleotide–support incompatability and from steric hindrance of reaction sites, necessitate elaborate methods for separation of the product from unwanted material. Current methods of separation are considered to be inadequate for this purpose. Part of this problem could be overcome if fragment-type syntheses could be employed so as to maximize the difference between the starting materials and products.

REFERENCES

Agarwal, K. L., and Khorana, H. G. (1972). *J. Am. Chem. Soc.* **94**, 3578.
Agarwal, K. L., Yamazaki, A., Cashion, P. J., and Khorana, H. G. (1972). *Angew. Chem.* **84**, 489.
Amarnath, V., and Broom, A. D. (1977). *Chem. Rev.* **77**, 183.
Asteriadis, G. T., Armbruster, M. A., and Gilham, P. T. (1976). *Anal. Biochem.* **70**, 64.
Blackburn, G. M., Brown, M. J., and Harris, M. R. (1966). *J. Chem. Soc. Chem. Commun.*, 611.
Blackburn, G. M., Brown, M. J., and Harris, M. R. (1967). *J. Chem. Soc.* (C) 2438.
Blackburn, G. M., Brown, M. J., Harris, M. R., and Shire, D. (1969). *J. Chem. Soc.* (C) 676.
Brandstetter, F., Schott, H., and Bayer, E. (1973). *Tetrahedron Lett.*, 2997.
Brandstetter, F., Schott, H., and Bayer, E. (1974). *Tetrahedron Lett.*, 2705.
Breitenbach, J. W., and Olaj, O. F. (1968). *Chimia.* **22**, 157.
Chapman, T. M., and Kleid, D. J. (1973). *J. Chem. Soc. Chem. Commun.*, 193.
Chargaff, E., and Davidson, J. N. (1955, 1960) "The Nucleic Acids," Vol. 1-3. Academic Press, New York.
Cramer, F. (1961). *Angew. Chem.* **73**, 49.
Cramer, F. (1966). *Angew. Chem.* **78**, 186.
Cramer, F. (1969). *Pure and Appl. Chem.* **18**, 197.
Cramer, F., and Koster, H. (1968). *Angew. Chem. Int. Ed.* **7**, 473.
Cramer, F., Helbig, R., Hettler, H., Scheit, K. H., and Seliger, H. (1966). *Angew. Chem. Int. Ed.* **5**, 601.
Egan, Z. B. (1973). *Biochim. Biophys. Acta* **299**, 245.
Freist, W., and Cramer, F. (1970). *Angew. Chem.* **82**, 358.
Gait, M. J., and Sheppard, R. C. (1976). *J. Am. Chem. Soc.* **98**, 8514.
Gait, M. J., and Sheppard, R. C. (1977). *Nucl. Acids Res.* **4**, 1135.
Gilham, P. T., and Khorana, H. G. (1958). *J. Am. Chem. Soc.* **80**, 6212.
Glaser, R., Sequin, U., and Tamm, C. (1973). *Helv. Chim. Acta* **56**, 654.
Hanessian, S., Ogawa, T., Guindon, Y., Kamenof, J. L., and Roy, R. (1974). *Carbohyd. Res.* **38**, C15.
Hartwick, R. A., and Brown, P. R. (1975). *J. Chromatogr.* **112**, 651.
Hayatsu, H., and Khorana, H. G. (1966). *J. Am. Chem. Soc.* **88**, 3182.
Hayatsu, H., and Khorana, H. G. (1967). *J. Am. Chem. Soc.* **89**, 3880.

References

Jacob, T. M., and Khorana, H. G. (1964). *J. Am. Chem. Soc.* **86,** 1630.
Kabachink, M. M., Polyakova, I. A., Potapov, V. K., Shabarova, Z. A., and Prokof'ev, M. A. (1970a). *Dokl. Akad. Nauk. SSSR* **195,** 1344.
Kabachink, M. M., Potapov, V. K., Shabarova, Z. A., and Prokof'ev, M. A. (1970b). *Dokl. Akad. Nauk. SSSR* **195,** 1107.
Kabachink, M. M., Potapov, V. K., Shabarova, Z. A., and Prokof'ev, M. A. (1971). *Dokl. Akad. Nauk. SSSR* **201,** 858.
Kabachink, M. M., Timofeva, N. G., Budanov, M. V., Potapov, V. K., Shabarova, Z. A., and Prokof'ev, M. A. (1973). *Zh. Obshch Khim.* **43,** 379.
Katagiri, N., Itakura, K., and Narang, S. A. (1975). *J. Am. Chem. Soc.* **97,** 7332.
Khorana, H. G. (1961). "Recent Developments in the Chemistry of Phosphate Esters of Biological Interest." Wiley, New York.
Khorana, H. G. (1967). *Proc. Int. Cong. Biochem., 7th, 1967,* p. 17.
Khorana, H. G. (1968). The Harvey Lectures **62,** 79.
Khorana, H. G., Vizsolyi, J. P., and Ralf, R. K. (1962). *J. Am. Chem. Soc.* **84,** 414.
Kochetkov, N. K., and Budovskii, E. J. (1971). "Organic Chemistry of Nucleic Acids." Plenum, New York.
Kossel, H. and Seliger, H. (1975). *Fortschr. Chem. Org. Naturestoffe* **32,** 297.
Koster, H. (1972a). *Tetrahedron Lett.,* 1527.
Koster, H. (1972b). *Tetrahedron Lett.,* 1535.
Koster, H., and Cramer, F. (1972). *Liebigs Ann. Chem.* **766,** 6.
Koster, H., and Cramer, F. (1973). *Makromol. Chem.* **167,** 171.
Koster, H., and Cramer, F. (1974a). *Liebigs Ann. Chem.,* 946.
Koster, H., and Cramer, F. (1974b). *Die Makromol. Chemie.* **168,** 171.
Koster, H., and Geussenhainer, S. (1972). *Angew. Chem.* **84,** 712.
Koster, H., and Heyns, K. (1972). *Tetrahedron Lett.,* 1531.
Koster, H., Pollak, A., and Cramer, F. (1974). *Liebigs Ann. Chem.,* 959.
Kusama, T., and Hayatsu, H. (1970). *Chem. Pharm. Bull. Tokyo* **18,** 319.
Letsinger, R. L., and Mahadevan, V. (1965). *J. Am. Chem. Soc.* **87,** 3526.
Letsinger, R. L., and Mahadevan, V. (1966). *J. Am. Chem. Soc.* **88,** 5319.
Letsinger, R. L., Caruthers, M. H., Miller, P. S., and Ogilvie, K. K. (1967b). *J. Am. Chem. Soc.* **89,** 7146.
Letsinger, R. L., Caruthers, M. H., and Jerina, D. M. (1967a). *Biochemistry* **6,** 1379.
Letsinger, R. L., Kornet, M. S. Mahadevan, V. and Jerina, D. M. (1964). *J. Am. Chem. Soc.* **86,** 5163.
Levina, A. S., Potapov, V. K., Khorre, D. G., Shabarova, Z. A., and Shubina, T. M. (1972). *Izv. Sib. Otd. Akad. Nauk.,* 117.
Lohrmann, R., and Khorana, H. G. (1966). *J. Am. Chem. Soc.* **88,** 829.
Melby, L. R., and Strobach, D. R. (1967). *J. Am. Chem. Soc.* **89,** 450.
Melby, L. R. and Strobach, D. R. (1969). *J. Org. Chem.* **34,** 421, 427.
Merrifield, R. B. (1963). *J. Am. Chem. Soc.* **85,** 2149.
Michelson, A. (1963). "The Chemistry of Nucleosides and Nucleotides." Academic Press, New York.
Narang, C. K., Brunfeldt, K., and Norris, K. E. (1977). *Tetrahedron Lett.,* 1819.
Norris, K. E. (1977). Personal Communication.
Ogilvie, K. K., and Kroeker, K. (1972). *Can. J. Chem.* **50,** 1211.
Ohtsuka, E., Morioka, S., and Ikehara, M. (1972). *J. Am. Chem. Soc.* **94,** 3229.
Pless, R. C., and Letsinger, R. L. (1975). *Nucl. Acid. Res.* **2,** 773.
Potapov, V. K., Chekhmakhcheva, O. G., Shabarova, Z. A., and Prokof'ev, M. A. (1971). *Dokl. Akad. Nauk. SSSR* **196,** 360.
Potapov, V. K., Turkin, S. I., and Shabarova, Z. A. (1972). *Zh. Obshch. Khim.* **42,** 2349.

Rubinstein, M., and Patchornik, A. (1972). *Tetrahedron Lett.,* 2281.
Rubinstein, M., and Patchornik, A. (1975). *Tetrahedron* **31,** 1517.
Schott, H. (1973). *Angew. Chem. Int. Ed.* **12,** 246.
Schott, H., Brandstetter, F., and Bayer, E. (1974). *Makromol. Chem.* **173,** 247.
Seliger, H. (1973). *Makromol.Chem.* **169,** 83.
Seliger, H., and Aumann, G. (1973). *Tetrahedron Lett.,* 2911.
Seliger, H., and Aumann, G. (1975). *Makromol. Chem.* **176,** 609.
Seliger, H., and Cramer, F. (1969). *Angew. Chem. Int. Ed.* **8,** 609.
Seliger, H., Holupirek, M., Gortz, HH. (1978). *Tetrahedron Lett.,* 2115.
Shimidzu, T., and Letsinger, R. L. (1968). *J. Org. Chem.* **33,** 708.
Shimidzu, T., and Letsinger, R. L. (1971). *Bull. Chem. Soc. Jpn.* **44,** 1673.
Sommer, H., and Cramer, F. (1972). *Angew. Chem.* **84,** 710.
Tsou, K. C., and Yip, K. F. (1973). *J. Macromol. Sci. Chem.* **A7** (5) 1097.
Van Boom, J. H. (1977). *Heterocycles* **7,** 1197.
Yip, K. F., and Tsou, K. C. (1971). *J. Am. Chem. Soc.* **93,** 3272.
Zarytova, V. F., Potapov, V. K., Shabarova, Z. A., and Knorre, D. G. (1971). *Dokl. Akad. Nauk. SSSR,* **199,** 1072.
Zhdanov, R. J., and Zhenodarova, S. M. (1975). *Synthesis,* 222.

6 Oligosaccharide Synthesis on Polymer Supports

I.	Introduction	105
II.	Basic Principles of Polymer-Supported Oligosaccharide Synthesis	107
III.	Polymer Supports for the Synthesis of Oligosaccharides	108
IV.	Linkage of the First Sugar Molecule to the Polymer Support and Product Removal	110
V.	Protecting Groups in Oligosaccharide Synthesis Employing Polymer Supports	111
	A. Acyl-Type Protecting Groups	111
	B. Ether, Glycoside, and Acetal Groups	111
VI.	Mechanism and Steric Control in Successive Coupling of Monosaccharide Residues on a Polymer Support	112
VII.	Miscellaneous Applications of Polymers in the Carbohydrate Field	113
VIII.	Monitoring of Solid-Phase Oligosaccharide Synthesis	114
IX.	Advantages, Limitations, and Future Scope of the Use of Polymer Supports in Polysaccharide Synthesis	114
	References	116

I. INTRODUCTION

In the living organism, polysaccharides act as structural units as well as reserve sources of energy. Specific polysaccharides are known to have biological functions as well, e.g., heparin plays an important role in blood clotting and immunopolysaccharides are concerned with the control of antibody production. Sugar residues are also important constituents of such bioregulators as, nucleotides, glycoproteins, and glycolipids.

Many structural polysaccharides are homopolymers containing only one type of monosaccharide unit, linked in an ordered fashion. Some plant polysaccharides are copolymers containing several different monosaccharides. In some polysaccharides one type of sugar residue occurs as a graft on the main linear polysaccharide chain made of different sugar monomers (Whistler, 1973).

Although not many oligosaccharides are known to occur free in nature

(Guthrie and Honeyman, 1968), oligosaccharides of strictly defined constitution are of considerable interest as substrates for the study of enzyme mechanisms and of structures. Current methods of preparing oligosaccharides consist of either sequential synthesis in homogeneous solution or fractionation of partially hydrolyzed natural polysaccharides. Enzymatic hydrolysis can be specifically directed to cleave the graft units, leaving the main polysaccharide chain intact. Alternatively, enzymatic or chemical methods of hydrolysis can be used to cleave the linear chain at intervals, leaving the graft units intact. The oligosaccharides obtained during such hydrolyses consist of mixtures of di-, tri-, and higher oligosaccharides having unusual structures not found free in nature. In general, methods based on sequential synthesis of oligosaccharide in a homogeneous phase or on controlled hydrolysis of natural polysaccharides are not versatile enough to provide such unusual oligosaccharides.

Some of the naturally occurring carbohydrates are also found as glycosides with nonsugar components, whereas some others contain less common functional groups, such as deoxy-sugars, uronic acids, and sulfuric or phosphoric acid esters.

It is thus apparent that oligosaccharide syntheses and syntheses of sugar derivatives are of fundamental importance for biological and chemical applications in many fields of contemporary interest, such as immunology, enzymology, and pharmacology.

Carbohydrate chemists, employing conventional solution methods, have observed several difficulties in oligosaccharide synthesis, among which selective protection and deprotection of the groups and steric control of the coupling reaction are of great importance. In some cases, it is difficult or impossible to synthesize an oligosaccharide of a desired structure by the conventional methods. In many more cases the yields are low and the isolation of the products becomes a major challenge.

Just as in the case of peptide and nucleotide chemistry, there appears to be good scope for using solid-phase methods to circumvent some of the difficulties in oligosaccharide synthesis. Applications of polymer-mediated syntheses to oligosaccharides began at a much later stage compared with the peptide and oligonucleotides, but they now seem to have picked up momentum. The pioneer work in the field of polymer-support oligosaccharide synthesis was done by Frechet and Schuerch (1971), Guthrie *et al.* (1971), Excoffier *et al.* (1972), Belorizky *et al.* (1972), Zehavi and Patchornik (1973), and Pfaffli *et al.* (1972). Frechet and Pelle (1975) and Seymour and Frechet (1976) have carried out syntheses of specific monosaccharide derivatives using polymer supports and have used polymeric reagents for the protection of specific groups in the monosaccharides.

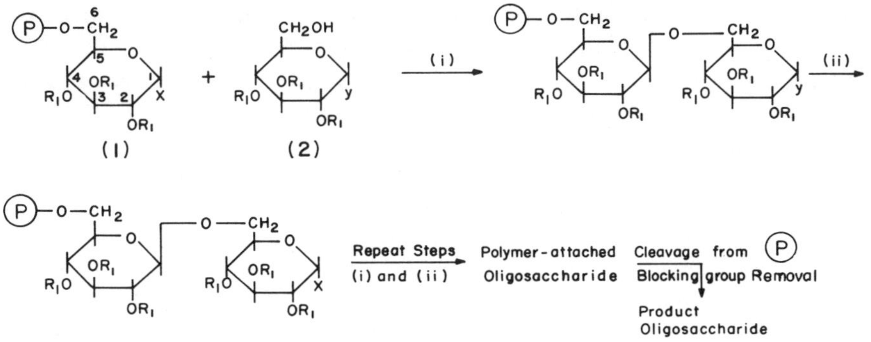

II. Basic Principles of Polymer-Supported Oligosaccharide Synthesis

[Scheme 6-1 diagram]

(P) = Polymer, R_1 = Protecting group, X = Reactive group at C-1,
Y = Unreactive group at C-1.

Scheme 6-1

II. BASIC PRINCIPLES OF POLYMER-SUPPORTED OLIGOSACCHARIDE SYNTHESIS

Two possible approaches have been suggested for polymer-mediated oligosaccharide syntheses (Schemes 6-1 and 6-2). Synthesis of a disaccharide is essentially a glycoside formation. The first unit (the glycosidic component sugar) (1) is linked to the polymer by an alcoholic hydroxyl group (such as C-6 or C-3) through an ester or an ether linkage, leaving the C-1 group free (or reversibly blocked). A suitably blocked second saccharide unit (2) having one free hydroxyl group is then coupled to the polymer-bound sugar. The C-1 hydroxyl of the polymer-bound sugar has

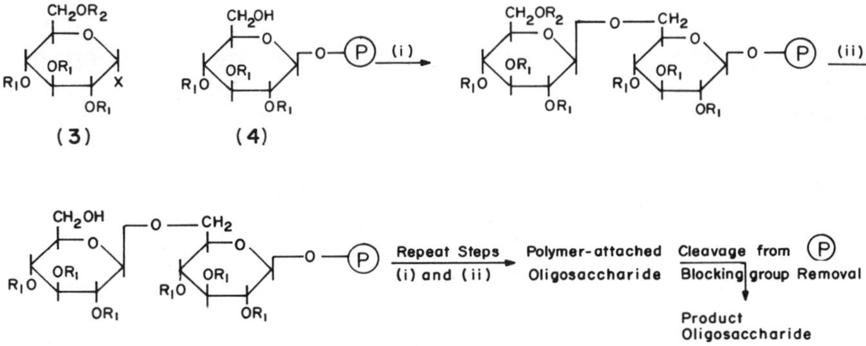

[Scheme 6-2 diagram]

R_1 = Blocking group, R_2 = Blocking group, removable in presence of R_1, X = Reactive group at C-1, (P)=Polymer

Scheme 6-2

to be activated in this process. For further lengthening of the chain, the C-1 group on the second unit is then converted into its reactive form and the process repeated to build up tri-, tetra-, and higher oligosaccharides (Scheme 6-1). The polymer-bound oligosaccharide product is finally cleaved from the polymer and unblocked to give the product.

Alternatively, the first sugar unit (4) is linked to the polymer via the anomeric center C-1 (unreactive polymer glycoside). The polymer-bound first sugar has only one free hydroxyl (generally at C-6) which is then coupled to the reactive C-1 (X) of the second sugar unit (3) (Scheme 6-2). 3 has one hydroxyl group (C-6) blocked by a group (R_2) that can be removed in preference to the hydroxyl groups blocked by R_1. Chain lengthening is carried by deblocking R_3 and repeating the coupling sequence.

The coupling procedures employed in polymer-mediated synthesis have been adopted from the solution methods of synthesis and will be discussed later. When the desired sequence of the oligosaccharide has been built up on the polymer support, its isolation has to be carried out by cleaving from the polymer and final deblocking of the groups. The stereospecificity and yield of the final products have been comparable or better than in the solution method.

III. POLYMER SUPPORTS FOR THE SYNTHESIS OF OLIGOSACCHARIDES

All the cases of polymer-supported syntheses in the field of carbohydrates reported so far have employed a styrene-based polymer. Polymers employed in polysaccharide synthesis are described in Table 6-1.

Both polymer glycosides and polymer esters of the first sugar unit have been made. No problems due to incompatibility of polymer and reactants have been reported. Since highly protected sugars are employed in the synthesis, the normal hydrophilic nature of the sugars seems to be decreased, making them compatible with styrene-based polymers. This also seems to permit the use of weakly polar solvents such as benzene and methylene chloride in oligosaccharide synthesis.

Guthrie and co-workers (1971, 1973) have described the use of a copolymer containing preformed monosaccharide units as esters of vinyl benzoate or vinylbenzene sulphonate and styrene. This is the only example of a polymer-supported synthesis wherein the substrate is first linked to the monomer and then polymerized to give the polymer-bound substrate. High and uniform loading of the polymer has been possible in this way, leaving no unreacted sites on the polymer. These noncross-linked polymers are soluble in solvents such as benzene, toluene, acetone,

TABLE 6-1

Polymer Supports Used in Oligosaccharide Synthesis

Polymer support and the active site of bonding	Type of polymer-saccharide bond and its cleavage	Method of coupling successive saccharide units	Synthesis achieved	References
Polystyrene, incorporating an allyl alcohol group	Polymer–saccharide glycoside. Ozonolysis of the alkenyl bond	C-1 Bromide (Koenigs-Knorr method)	Isomaltose	Frechet and Schuzerch, 1971, 1972a,b
Polystyrene, (soluble), incorporating an acid chloride group	Polymer–saccharide (C-6) ester[a]. Alkaline hydrolysis (NaOMe–Dioxane)	C-1 Bromide via cis-1,2-orthoester	—	Guthrie et al., 1971, 1973
Polystyrene (soluble), incorporating an arylsulphonyl chloride group	Polymer–saccharide (C-6) sulphonate ester[a] O-Alkyl fission of sulphonate ester AcOK–DMF		β-D-Gentiobiose octaacetate	Guthrie et al., 1971, 1973
Polystyrene (insoluble), incorporating an acid chloride group	Polymer-2-acetamidosugar (-3-O) ester. Ammonolysis, NH$_3$–MeOH toluene or alkaline hydrolysis	C-1 Bromide via cis-1,2oxazoline	2-Acetamido-6-O-(2-acetamido-3,4,6-tri-O-acetyl-2-deoxy- 6- D-glucopyranose) - 2-deoxy- α-D-glucopyronoside	Belorizky et al., 1972; Excoffier et al., 1972
Polystyrene incorporating an o-nitrobenzyl alcohol group (insoluble)	Polymer–saccharide glycoside. Photolysis of polymer O-nitro-benzyl glycoside at 320 nm		Isomaltose	Zehavi and Patchornik, 1973
Polystyrene (insoluble), incorporating chloromethyl or mercaptomethyl groups	Polymer–saccharide thioglycoside formed via the thio sugar	C-1-Chloride propanol–KOH	Isomaltose	Chiu and Anderson, 1976

[a] Polymer was prepared from a preformed monomer, incorporating sugar ester. The linear polymer is soluble in many organic solvents.

chloroform, and dioxane. They combine the advantage of homogeneous medium with polymer-supported synthesis.

Zehavi and Patchornik (1973) employed a light-sensitive solid-support incorporating o-nitrobenzylglycoside groups, from which the sugar may be cleaved by irradiation at 320 nm. Such a polymer has also been employed for polypeptide synthesis (Rich and Gurwara, 1973, 1975). The light sensitivity of the equivalent low-molecular-weight o-nitrobenzylglycosides had been established earlier (Zehavi *et al.*, 1972; and Zehavi and Patchornik, 1972).

Polymers incorporating aldehyde and phenyl boronic acid functions have been employed as polymer supports for the selective derivatization of sugars. These polymers also serve to block certain groups in the sugar molecules whose participation in the derivatization reaction is not desired (Hanessian *et al.*, 1974; Frechet and Pelle, 1975; Seymour and Frechet, 1976).

IV. LINKAGE OF THE FIRST SUGAR MOLECULE TO THE POLYMER SUPPORT AND PRODUCT REMOVAL

The very first report of a polymer-supported oligosaccharide synthesis (Frechet and Schuerch, 1971) was based on linking the first sugar to the polymer by a glycosidic linkage. It was realized that the intersugar glycoside bonds and the polymer glycoside bond of the first sugar residue would break simultaneously in any attempt to cleave the polymer–glycoside bond. Hence the polymer had an alkenol group (**5**) incorporated into it. This group could be cleaved oxidatively (by ozonolysis) rather than by hydrolysis. Oxidative cleavage of the polymer–glycoside (**6**) gave 2-ethanal glycoside (**7**) rather than the free oligosaccharide (Scheme 6-3).

In other cases a polymer–sugar ester bond was formed (Guthrie *et al.*,

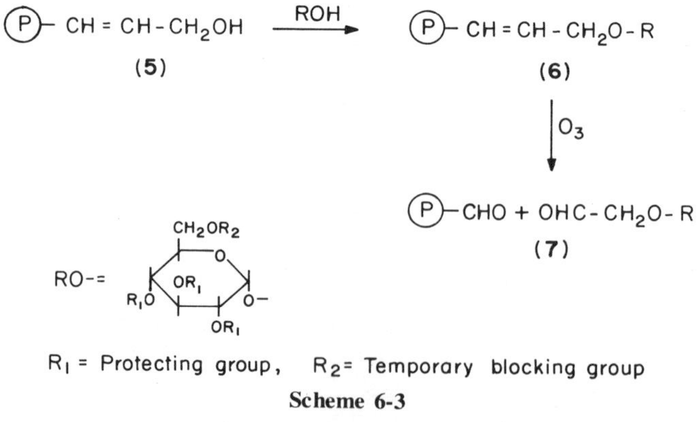

$$\text{(P)}-CH=CH-CH_2OH \xrightarrow{ROH} \text{(P)}-CH=CH-CH_2O-R$$

(5)   (6)

$$\downarrow O_3$$

$$\text{(P)}-CHO + OHC-CH_2O-R$$

(7)

R_1 = Protecting group, R_2 = Temporary blocking group

Scheme 6-3

1971; Guthrie *et al.*, 1973; and Excoffier *et al.*, 1972). While Guthrie and co-workers employed a copolymer of preformed monomer of *p*-vinyl benzoate of protected sugar, Excoffier made the polymer–sugar ester from a preformed polymer acid chloride and the protected sugar.

Alkaline saponification of polymer-bound sugar ester did not proceed very smoothly, but when polymer–sulfonate sugar esters were employed, the cleavage involving detosylation and O-alkyl fission of the sulfonate ester (KOAc–DMF) proceeded smoothly (Guthrie *et al.*, 1973), producing O-acetates. From this it may be concluded that the OAc group is not affected under these conditions.

V. PROTECTING GROUPS IN OLIGOSACCHARIDE SYNTHESIS EMPLOYING POLYMER SUPPORTS

The groups that normally need protection in oligosaccharide synthesis are the alcoholic and the glycoside (anomeric) hydroxyl. For amino sugars, the amino function also needs to be protected.

A. Acyl-Type Protecting Groups

Acyl groups (acetyl and benzoyl) have frequently been used for protection of hydroxyl groups in sugars. While the O-acetyl group is readily deblocked by alkaline saponification, ammonolysis, or methanolysis, the benzoyl group is cleaved only by prolonged alkaline saponification. Methanolysis to cleave the O-acetyl group also cleaves the polymer–sugar ester bond, but O-benzoyl groups are not cleaved under these conditions.

p-Nitrobenzoyl and certain other substituted benzoyl groups (R_2 in Scheme 6-2) have been used for temporary blocking of such hydroxyl groups. These must however, be cleaved (EtONa–dioxane) for subsequent chain elongation. During glycoside formation (or solvolysis) from the 6-O-acylpyranosyl 1-bromide, the 6-O-acyl groups were shown to participate in the solvolysis reaction and hence to determine the conformation (α or β) at the anomeric center at C-1 in the product (Frechet and Schuerch, 1972a,b; Ishikawa and Fletcher, 1969) and so to influence the outcome of the reaction.

B. Ether, Glycoside, and Acetal Groups

Apart from the acyl function, the protection of hydroxyl groups has also been carried out by etherification. The benzyl ether group has been the group of choice, because it can be cleaved either by acidolysis or by catalytic hydrogenation. Protection of the hydroxyl group by methylation is not much favored, but methyl glycoside formation has frequently been used for reversible blocking of the anomeric hydroxyl. The glycoside may also be formed with the polymer hydroxyl group. In this case it acts as a protecting group in addition to the polymer–sugar linkage (Frechet and Schuerch, 1971, 1972a; Zehavi and Patchornik, 1973). Acetal formation for protection of the hydroxy group has also been empolyed. Acetal groups are selectively cleaved by TFA, leaving the acyl groups intact.

VI. MECHANISM AND STERIC CONTROL IN SUCCESSIVE COUPLING OF MONOSACCHARIDE RESIDUES ON A POLYMER SUPPORT

Activation of the C-1 group of one sugar molecule is essential for the formation of glycosidic linkage between two sugar residues. The most widely used method of C-1 activation employs a C-1 halide (generally bromide, Koenigs and Knorr, 1901). This can react with the alcohol function in the second monosaccharide unit to give either an α- or β-glycoside, or a mixture of both. The steric control of the configuration at the anomeric center depends upon such factors as the nature of the substituents blocking the hydroxyl groups on the glucopyranosyl bromide molecules and the participation of these groups in the reaction. The reaction, which proceeds via S_N1 mechanism, should involve complete inversion of configuration at C-1 (α-glucopyranosyl bromide yielding β-glycoside and vice versa). Participation of other groups can, however, change the steric configuration at C-1, i.e., glycoside formation can proceed either without inversion of configuration at C-1 or may result in anomerization (Helferich and Zinner, 1962; Frechet and Schuerch, 1972a).

A more stereospecific course of glycoside formation by reaction of the polymer-bound C-1 bromide with the second monosaccharide hydroxyl component involves the initial formation of a cis-1:2-orthoester. This undergoes an acid-catalyzed stereospecific isomerization (Kochetkov *et al.*, 1967) to the corresponding glycoside (disaccharide).

With 2-acetamido sugars the activation of C-1 groups has been carried out using a C-1 chloride or via formation of a cis-1,2-oxazoline of the glycoside sugar molecule (Belorizky *et al.*, 1972; Excoffier *et al.*, 1972). This then reacts with the second sugar (hydroxy component) and a β-glycoside specifically formed. The latter two methods are suitable only for forming β-glycosides, whereas both α- and β-glycosides are obtained using C-1 halide activation. This is because orthoesters and oxazolines are formed only when C-1 and C-2 hydroxyls are cis, i.e., when C-1 is alpha. In general, the synthesis of α-glycosides is difficult by solution methods.

VII. MISCELLANEOUS APPLICATIONS OF POLYMERS IN THE CARBOHYDRATE FIELD

In addition to the synthesis of oligosaccharides on polymer supports, polymeric reagents have been used for the synthesis of sugar derivatives. Partially substituted derivatives of sugars are obtainable by solution methods. Thus, a pair of cis-hydroxyl groups in a sugar molecule is readily blocked by acetal (or ketal) formation or by formation of a phenyl boronic acid ester. The remaining hydroxyl group can be acylated or etherified, followed by deblocking (removal of protecting groups) to give a partially acylated or etherified sugar.

Partially derivatized sugars have been prepared using polymers as protecting groups. A polymer incorporating an aldehyde group was obtained by oxidation of chloromethylated polystyrene (Frechet and Schuerch, 1971) and its polymeric acetal was prepared by reacting the aldehyde with methyl-α-D-glycopyranoside, resulting in blocking of C-4 and C-6 hydroxyls. The polymer-bound sugar was then selectively benzoylated (PhCOCl in pyridine) or benzylated ($C_6H_5CH_2Cl$, THF, NaH) to give 2,3-disubstituted methyl-α-D-glucopyranosides in good yield. The derivatized sugar was cleaved from the polymer by hydrolysis, using TFA–dioxane (1:3) at room temperature. The same polymer has been used as a support for both sugar and nucleoside functionalization (Hanessian et al., 1974).

Polystyrylboronic acid has similarly been used for the protection of hydroxyl groups in polyols. It has been claimed that the esters of the polymeric boronic acid are more stable to moisture and that the polymeric by-product is re-usable (Seymour and Frechet, 1976). Both the aldehyde polymer and boronic acid polymer have been used for simultaneous protection of C-4-O and C-6-O in derivatization reactions. A polymer incorporating a trityl group has also been used for protection of the C-6-O group only (Frechet and Nuyens, 1976).

Eschenfelder et al. (1975) have described the use of the silver salt of a copolymer of maleic acid and 1,4-bis(vinyloxy)butane for the preparation of the glycosides from α-acetobromoglucose. The polymer, which acts as an acid binging reagent, gives a high yield of the corresponding β-acetoglucoside and maintains a high stereospecificity. This further development of the Koenigs–Knorr synthesis has the advantage that removal of γ-hydroxy or 1,4-dicarboxylic acids in a separate step is not necessary. Some by-products, such as 1-O-acylglucose, also remain linked to the polymer and are easily separated.

Oligonucleotide synthesis on polymer support generally involves the linking of carbohydrate moieties (ribose or deoxyribose) to the polymer (usually as the polymer trityl ether). Reactions of such polymer-bound sugar residues are discussed in Chapter 5.

Polymer-bound sugars have also been used as a chiral group in asymmetric synthesis (Kawana and Emoto, 1972, 1974).

There is one report (Benson, 1975) in which the proposed mechanism of the reaction of a sugar derivative (Morgan-Elson assay for 2-acylamido-2-deoxy-D-hexose sugars) has been tested using the sugar derivative bound to a polymer support. This approach will be discussed in Chapter 11.

VIII. MONITORING OF SOLID-PHASE OLIGOSACCHARIDE SYNTHESIS

Guthrie *et al.* (1971) developed a method for determining the incorporation of a sugar containing a preformed monomer in the soluble copolymer that is based on measurement of the optical rotation. Specific rotation was used in calculation of the weight fraction of units of optically active monomers in the copolymer, assuming a linear proportionality of both quantities. The polymers of optically active monomers, having asymmetric groups in the side chains at a distance further away than that of the β-position to the main chain, should have practically the same rotation as the low-molecular-weight model compound or the parent monomer.

In most of the other cases where insoluble polymers were employed, monitoring of the solid-phase reactions was carried out by hydrolytic delinking of the sugar components from the polymer and determining their concentration by optical rotation. Unreacted sugar in the reaction mixture can also be determined from optical rotation.

In many cases the product identification was done by tlc comparison with authentic samples, optical rotation, and glc of trimethylsilyl derivatives. Enzymatic digestion of the product with appropriate enzymes has been used to determine the configuration of the anomeric center.

IX. ADVANTAGES, LIMITATIONS, AND FUTURE SCOPE OF THE USE OF POLYMER SUPPORTS IN POLYSACCHARIDE SYNTHESIS

Compared to solid-phase polypeptide synthesis, there has been far less activity in polymer-supported oligosaccharide synthesis. Unlike the solid-phase synthesis of oligonucleotides (Letsinger and Mahadevan, 1965), which closely followed the first reported solid-phase polypeptide

IX. Advantages, Limitations, and Future Scope

synthesis (Merrifield, 1963), the first report of solid-phase oligosaccharide synthesis (Frechet and Schuerch, 1971) was made some eight years after Merrifield's first report. Oligosaccharide synthesis using polymer supports has yet to become an acceptable alternative to the solution method.

Solid-phase syntheses of oligosaccharides have the usual advantages of ease of product isolation and purification. However, no estimate of the prospects of such methods has been discussed. Certain limitations are apparent and are now discussed.

1. The type of polymer–first sugar residue bonds (i.e., ester or glycoside bonds) are also present in the product oligosaccharide. Hence, there is no clean method of unlinking the finished product from the polymer support without affecting internal bonds in the oligosaccharides. The only exception is the method of Zehavi and Patchornik (1973) which employs a light-sensitive solid support.

2. Most of the naturally occurring oligosaccharides have 1,4-glycosidic linkages. In case of the model compounds so far synthesized by solid-phase methods, 1,6-glycosidic bonds have always been made. It appears that the polymer support increases the selectivity of reaction for the less hindered 6-O position of a hexopyranose over that at the 4-O position.

The synthesis of an oligosaccharide with 1,4-glycosidic linkage using a polymer support has yet to be reported.

3. The ultimate yields of the products in solid-phase oligosaccharides have been low in spite of good loading capacities and high coupling yields. The losses have been attributed to the final unlinking of the product from the polymer as well as to incomplete deblocking of the protecting groups.

4. In spite of highly stereospecific coupling reactions, the outcome of coupling is not completely specific and anomeric products are formed. In general the stereospecificity of a reaction is somewhat lowered by use of polymer-supported reactants. It is perhaps for this reason that no synthesis of a disaccharide with an α-glycoside bond has been made.

5. In addition, some common difficulties, such as those observed in solid-phase peptide syntheses, are also known. These include the lack of suitable monitoring methods for each reaction step and problems arising out of incomplete reactions, which give rise to error sequences.

The binding of the first sugar residue to polymers derivatized to contain a trityl group may appear as a convenient alternative in oligosaccharide synthesis. Both binding and cleavage of sugars to such a polymer should be a simple operation. However, newer methods of activation of anomeric centers not requiring acidic conditions are required.

It is easy to predict that greater activity in the field of polymer-supported oligosaccharide synthesis will be necessary before it is established as a routine alternative to the solution method. Even then, the more prolonged and complex reactions necessary for the formation of glycosidic linkages will certainly hinder the automation of the process.

In contrast to oligosaccharide synthesis on polymer supports, the use of polymers as protecting groups in carbohydrate chemistry appears to be a more promising proposition and should be given attention as a convenient alternative to the solution method.

REFERENCES

Belorizky, N., Excoffier, G., Gagnaire, D., Utille, J. D., Vignon, M., and Vottero, P. (1972). *Bull. Soc. Chim. France* 4749.
Benson, R. L. (1975). *J. Org. Chem.* **40**, 1647.
Chiu, S.-H. L., and Anderson, L. (1976). *Carbohyd. Res.* **50**, 227.
Eschenfelder, V., Brossmer, R., and Wachler, M. (1975). *Angew Chem. Int. Ed.* **14**, 715.
Excoffier, G., Gagnaire, D., Utille, J. P. and Vignon, M. (1972). *Tetrahedron Lett.*, 5065.
Frechet, J. M. J., and Nuyens, L. J. (1976). *Can. J. Chem.* **54**, 926.
Frechet, J. M. J., and Pelle, G. (1975). *J. Chem. Soc. Chem. Commun.*, 225.
Frechet, J. M., and Schuerch, C. (1971). *J. Am. Chem. Soc.* **93**, 492.
Frechet, J. M., and Schuerch, C. (1972a). *Carbohyd. Res.* **22**, 399.
Frechet, J. M., and Schuerch, C. (1972b). *J. Am. Chem. Soc.* **94**, 604.
Guthrie, R. D., and Honeyman, J. (1968). "An Introduction to the Chemistry of Carbohydrates," 3rd ed. Oxford Univ. Press (Clarendon), London and New York.
Guthrie, R. D., Jenkin, A. D., and Stehlicek, J. (1971). *J. Chem. Soc.* C, 2690.
Guthrie, R. D., Jenkin, A. D., and Stehlicek, J. (1973). *J. Chem. Soc.* C, 2414.
Hanessian, S., Ogawa, T., Guidon, Y., Kamennot, S. L., and Roy, R. (1974). *Carbohyd. Res.* **38**, C15.
Helfferich, B., and Zinner, J. (1962). *Chem. Ber.* **95**, 2604.
Ishikawa, T., and Fletcher Jr., H. G. (1969). *J. Org. Chem.* **34**, 563.
Kawana, M., and Emoto, S. (1972). *Tetrahedron Lett.*, 4855.
Kawana, M., and Emoto, S. (1974). *Bull. Chem. Soc. Jpn.* **47**, 160.
Kochetkov, N. K., Khorlin, A. J., and Bochkov, A. F. (1967). *Tetrahedron* **23**, 693.
Koenigs, W., and Knorr, E. (1901). *Chem. Ber.* **34**, 957.
Letsinger, R. L., and Mahadevan, V. (1965). *J. Am. Chem. Soc.* **87**, 3526.
Merrifield, R. B. (1963). *J. Am. Chem. Soc.* **85**, 2149.
Pfaffli, P. J., Hixson, S. H., and Anderson, L. (1972). *Carbohyd. Res.* **23**, 195.
Rich, D. H., and Gurwara, S. K. (1973). *J. Chem. Soc. Chem. Commun.*, 610.
Rich, D. H., and Gurwara, S. K. (1975). *J. Am. Chem. Soc.* **97**, 1575.
Seymour, E., and Frechet, J. M. J. (1976). *Tetrahedron Lett.*, 1149.
Whistler, R. L. (1973). "Industrial Gums," 2nd ed. Academic Press, New York.
Zehavi, U., and Patchornik, A. (1973). *J. Am. Chem. Soc.* **95**, 5673.
Zehavi, U., and Patchornik, A. (1972). *J. Org. Chem.* **37**, 2285.
Zehavi, U., Amit, B., and Patchornik, A. (1972). *J. Org. Chem.* **37**, 2281.

7 Peptide Synthesis Using Polymeric Active Esters

    I. Introduction 117
   II. Principles of Peptide Synthesis Using Polymeric Active
       Esters 117
  III. Polymeric Active Esters Used for Peptide Synthesis . 118
  IV. Synthesis of Cyclic Peptides Using Polymeric Active
       Esters 119
   V. Scope and Limitations of the Polymeric Active Ester
       Method for Peptide Synthesis 122
           References 123

I. INTRODUCTION

A technique of peptide synthesis different from Merrifield's method has been introduced (Fridkin *et al.*, 1965a, 1965b, 1966; Wieland and Birr, 1966a, 1966b). This method is based on the use of polymer-supported amino acid active esters. Wunsch (1971) refers to this technique as "polymeric reagent synthesis." The method is claimed to be free from certain limitations of the classical solid-phase peptide synthesis; e.g., it offers the advantage of isolation and purification of the intermediate peptides, and the synthesis does not go "unchecked." However, the method has certain limitations. These will be discussed later.

II. PRINCIPLES OF PEPTIDE SYNTHESIS USING POLYMERIC ACTIVE ESTERS

In peptide synthesis it is necessary to activate the carboxyl group (COOH-group) during the reaction involving amide bond formation. Earlier methods of COOH-activation were based on the formation of acid chlorides, anhydrides (either mixed or symmetrical carboxyanhydrides, as well as mixed anhydrides of inorganic acids such as sulfuric or phosphorous acids), and azides (Schroder and Lubke, 1965). It was soon realized that the use of certain active esters (esters with weakly acidic hydroxyl groups) offered many advantages over the earlier methods of COOH-activation. In addition, coupling agents [e.g., carbodiimides, carbonyldiimidazole, and N-ethoxycarbonyl-2-ethoxy-1,2-dihydroquinoline (EEDQ)] have been found useful for *in situ* COOH-activation. In general, the active esters used in peptide synthesis are isolated in a pure state and then used in the coupling reaction whereas active acyl derivatives, such as those formed with DCC and EEDQ, are formed *in situ* and are not isolated.

In a polymeric active ester, the acyl component, A, is bound covalently via an active ester link to the polymeric carrier, Ⓟ. Such reagents (Ⓟ—A) readily react with the nucleophile, B, (generally the amino component in peptide synthesis) to give the product, A—B [Eq. (1)].

$$\text{Ⓟ—A + B} \longrightarrow \text{Ⓟ + A—B} \tag{1}$$

The reaction rates of the nucleophile with an N-hydroxysuccinimide ester resin have been shown to be influenced by the size and polarizability of the side chain on the amino acid nucleophile and by the solvent used in the reaction (Gut and Davidovich, 1976). In addition to carboxyl group blocked, single amino acids, which are usually used as nucleophiles, unblocked peptides have been used (Andreev *et al.*, 1977). In order to obtain high yields of the peptide, A—B, an excess of Ⓟ—A, which can be removed from the reaction mixture by filtration or centrifugation, is used. For further elongation of the peptide chain, successive coupling reactions with other polymeric esters may be carried out until the peptide of the desired sequence is obtained. Peptide synthesis using a polymeric active ester is illustrated in a typical example using a polymeric nitrophenyl (PNP) ester (Scheme 7-1). Patchornik and Kraus (1975) have classified these reagents as polymeric group-transfer reagents.

A common advantage of such reagents is found in the ease of separation of products from the spent reagent. Increased selectivity of the reagent due to steric hindrance has been observed, and in most cases the spent reagent may also be regenerated at low cost and re-used.

III. POLYMERIC ACTIVE ESTERS USED FOR PEPTIDE SYNTHESIS

Polymers have been functionalized to incorporate almost all types of low-molecular-weight active ester groups originally used in peptide synthesis. In Table 7-1 are listed a large number of polymeric active esters that have been used for peptide synthesis. For most of them, the polymer

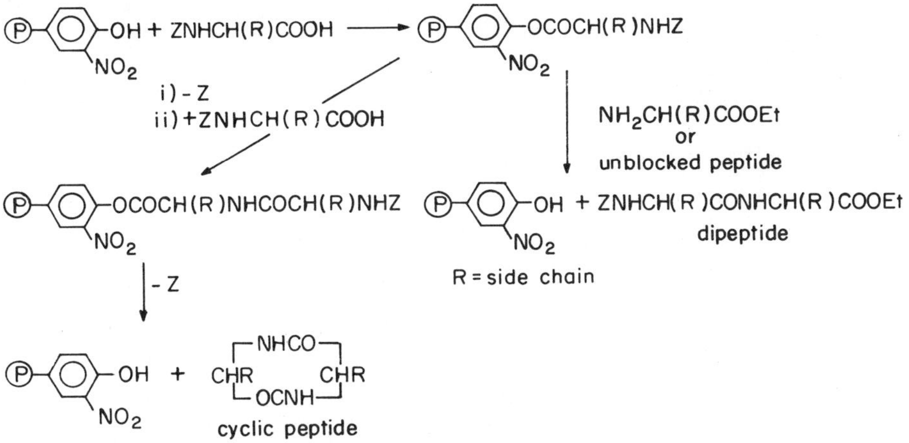

IV. Synthesis of Cyclic Peptides Using Polymeric Active Esters

Scheme 7-1

support used is polystyrene but in one case, polylysine has also been employed. Some of the polymeric active esters were prepared from functionalized monomers whereas in other cases, the preformed resin was functionalized after polymerization. A typical example of the functionalization of polystyrene with a 1-OH-benzotriazole group is illustrated in Scheme 7-2.

IV. SYNTHESIS OF CYCLIC PEPTIDES USING POLYMERIC ACTIVE ESTERS

When an N-blocked peptide is converted into its polymeric active ester by coupling it to a suitable functionalized polymer such as poly(4-

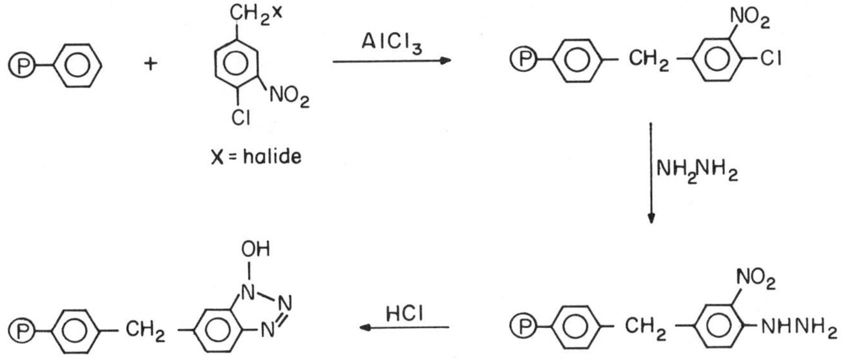

Scheme 7-2

TABLE 7-1

Polymeric Active Esters for Peptide Synthesis

Polymeric reagents	Remarks	References
Co[poly-(4-hydroxy-3-nitrostyrene) styrene]	Acylation of amines, synthesis of linear and cyclic peptides, including glutathione and bradykinin	Fridkin et al., 1965a; Sokolovsky et al., 1964; Wilchek et al., 1965; Williams, 1972
Copolystyrene incorporating (4-hydroxy-3-nitrophenoxy) methyl groups	Peptide synthesis	Fridkin et al., 1971
p-Nitrophenol formaldehyde condensation polymer	Simple peptide synthesis	Skylarov et al., 1966
(4-hydroxy-3-nitro)benzylated polystyrene	Nonpowdery polymer easy to filter; can be used in a column, used for synthesis of linear tetrapeptides, and peptide substrate for porcine elastase	Kalir et al., 1974; Fridkin et al., 1975; Atlas et al., 1975
Copolystyrene incorporating (3-hydroxy-4-nitrocarboxy phenyl)methyl groups	Tetrapeptide synthesis	Panse and Laufer, 1970
N^ϵ-(3-nitrotyrosylated) poly-DL-lysine	Simple linear peptides	Fridkin et al., 1965a
poly(3-nitro-4-mercaptostyrene)	Peptide synthesis	Skylarov et al., 1966
p,p'-dihydroxydiphenyl sulfone–formaldehyde condensation polymer	Linear peptide synthesis	Wieland and Birr, 1966a
Merrifield resin incorporating p-hydroxyphenylsulfone	Peptide synthesis	Marshall and Liener, 1970
Copolystyrene incorporating (4-hydroxy-thiophenoxy)methyl group	Synthesis of cyclic peptides, cyclization effected by peracid oxidation to sulfone followed by N-deblocking	Flanigan and Marshall, 1970
4-(methylthio)phenol formaldehyde	Synthesis of cyclic peptides, cyclization effected by peracid oxidation to sulfone followed by N-deblocking	Flanigan and Marshall, 1970

Polymer/Resin	Application	References
Poly-(5-vinyl-3-methyl-8-hydroxy)quinoline	Simple peptide synthesis	Manecke and Haake, 1968
4-hydroxyazobenzene-4'-sulfonate 4'-sulfonate bound to Dowex-1 resin	Simple peptide	Wieland and Birr, 1967
Merrifield resin incorporating 4-aminomethyl-2,2-diphenylethanediol	"Safety catch" activation of the polymer ester by TFA dehydration of 1,2-diol to alkenyl ester	Wieland, 1972
Copolyethylene-N-hydroxymaleimide	Linear peptide synthesis, use of DCC coupling avoided by use of O-sulfonate ester	Laufer et al., 1968; Fridkin et al., 1972; Sahni et al., 1977; Narita et al., 1972
Copolystyrene–N-hydroxymaleimide ("Macronet polymers")	Linear peptide synthesis	Rogozhin et al., 1973; Tsiryapkin et al., 1975; Andreev et al., 1976; Andreev et al., 1977
Co(polystyrene–N-hydroxymaleimide), DVB crosslinked. (prepared from the N-hydroxymaleimide amino acid ester)	Linear peptide synthesis and oligocaproic acids	Akiyama et al., 1976a,b
Co[polystyrene–N-(4-hydroxy-3-nitrophenyl)maleimide], linear and crosslinked.	Linear peptide synthesis	Teramoto et al., 1977
Polystyrene incorporating 1-hydroxybenzotriazole (P)—CH₂—[benzotriazole structure with OH(—COR)]	Synthesis of thyrotropin releasing hormone as well as several oligopeptides in an inhomogeneous two polymer system	Kalir et al., 1975; Jung et al., 1975; Heusel et al., 1977
Polystyrene incorporating 1,3-dimethyl-4-nitroso-5-aminopyrazole (P)—CH₂OCON—[pyrazole structure]—HO(—RCO)	Coupling followed by visual change of color (orange color of the active polymer changed to green color on exhaustion)	Guarneri et al., 1972

hydroxy-3-nitrostyrene), the carboxyl group of the peptide is activated, and at the same time each peptide molecule is isolated from the others because of immobilization on the polymer matrix. If the polymer-bound peptide is now N-deblocked, an intramolecular reaction is favored and cyclic amides are obtained in high yields (Fridkin et al., 1965a) (Scheme 7-1). A slightly different approach to cyclic peptide synthesis has been made by Flanigan and Marshall (1970) who have used a "dual-function support." This method differs from the previous one in that the peptide synthesis is carried out on the polymer support. The resin–peptide ester bond is then activated by a polymer-modification reaction, in this case by sulfide to sulfone oxidation. Upon N-deblocking, cyclization, accompanied by cleavage of the peptide from the support, takes place.

Even though cyclic peptides are formed in reasonable yields, suggesting site isolation in the polymer matrix, Crowley and Rapoport (1976) maintain that the cyclization is kinetically favored—because of a favorable balance of entropic and enthalpic relationships—rather than being directly attributable to site isolation.

V. SCOPE AND LIMITATIONS OF THE POLYMERIC ACTIVE ESTER METHOD FOR PEPTIDE SYNTHESIS

There is no doubt that, initially, the polymeric esters were used with the aim of illustrating that the peptide bond could be formed with their help. Until 1967 no synthetic applications of this method to naturally occurring peptides had been reported (Wunsch, 1971). However, the subsequent preparation of several linear and cyclic peptides—including the sequences of thyrotropin-releasing hormone (Kalir et al., 1975), LH-RH (Fridkin et al., 1977), and bradykinin (Fridkin et al., 1968)—in pure and biologically active form using polymeric active esters has proved the practical potentiality of the method.

A major drawback in the Merrifield (1963) method has been the "unchecked" synthesis of the peptide chain, since it can be cleaved from the resin only at the end of the sequence of reactions. This has frequently resulted in failure sequences and truncated sequences, as well as doubling of sequences because of interpolymer acylation reactions. In the polymeric active ester method, the peptide formed remains in solution. Thus it can be isolated, purified, and its purity checked before use in the subsequent coupling reaction. Therefore, it appears that the use of polymeric esters may provide a suitable and time-saving method for the synthesis of medium-sized, protected peptides, i.e., those containing

15–20 amino acids, that can subsequently be used for fragment condensation of larger peptides.

A problem that can be visualized in the synthesis of large peptides (containing 50 or more amino acids) by this method is the difficulty of acylating one polymer (the large peptide segment already synthesized) by another polymer (the active ester). Problems of peptide–polymer solubility and steric hindrance due to the polymer–polymer interaction may arise that may not permit the reaction to take place at all. Thus, in this respect Merrifield's method would seem to have an edge. However, recent observations by Fridkin and co-workers (1972) have indicated that peptide synthesis using polymeric esters does not preclude reaction with large peptides. It was shown by the reaction of insulin with Z-Ala-co(polyethylene-N-hydroxymaleimide) that the alanination of insulin was quantitative. No single strategy may be wholly suitable for the synthesis of large natural peptides, and, therefore, sometimes a combination of methods may work best (Wunsch, 1971).

REFERENCES

Akiyama, M., Narita, M., Seida, K., and Osawa, T. (1976a). *J. Poly Sci. Poly Chem.* **14**, 2173.
Akiyama, M., Shimizu, K., and Narita, M. (1976b). *Tetrahedron Lett.,* 1015.
Andreev, S. M., Tsiryapkin, V. A., Davidovich, Yu. A., Samoilova, N. A., and Rogozhin, S. V. (1976). *Bioorg. Chem.* **2**, 725.
Andreev, S. M., Tsiryapkin, V. A., Samoilova, N. A., Mironova, N. V., Davidovich, Yu. A., and Rogozhin, S. V. (1977). *Synthesis,* 303.
Atlas, D., Kalir, R., and Patchornik, A. (1975). *FEBS Lett.* **58**, 179.
Crowley, J. I., and Rapoport, H. (1976). *Acc. Chem. Res.* **9**, 135.
Davidovich, Yu. A., and Gut, V. (1976). *Collect. Czech Chem. Commun.* **41**, 1805.
Flanigan, E., and Marshall, G. R. (1970). *Tetrahedron Lett.,* 2403.
Fridkin, M., Patchornik, A., and Katchalski, E. (1965a). *J. Am. Chem. Soc.* **87**, 4646.
Fridkin, M., Patchornik, A., and Katchalski, E. (1965b). *Israel J. Chem.* **3**, 69.
Fridkin, M., Patchornik, A., and Katchalski, E. (1966). *J. Am. Chem. Soc.* **88**, 3164.
Fridkin, M., Patchornik, A., and Katchalski, E. (1968). *J. Am. Chem. Soc.* **90**, 2953.
Fridkin, M., Patchornik, A., and Katchalski, E. (1971). *Pept. Proc. Eur. Symp., 10th, 1969,* p. 166. North-Holland Publ., Amsterdam.
Fridkin, M., Patchornik, A., and Katchalski, E. (1972). *Biochemistry* **11**, 466.
Fridkin, M., Kalir, R., Warshowsky, A., and Patchornik, A. (1975). *Pept. Proc. Am. Pept. Symp., 4th, 1975,* p. 395. Ann Arbor Press, Ann Arbor, Michigan.
Fridkin, M., Hazum, E., Kalir, R., Rotman, M., and Koch, Y. (1977). *J. Solid-Phase Biochem.* **2**, 175.
Guarneri, M., Ferroni, R., Giori, P., and Benssi, C. A. (1972). *Pept. Proc. Am. Pept. Symp., 3rd, 1972,* p. 213. Ann Arbor Press, Ann Arbor, Michigan.
Gut, V., and Davidovich, Yu. A. (1976). *Collect. Czech. Chem. Commun.* **41**, 780.

Heusel, G., Bovermann, G., Göhring, W., and Jung, G. (1977). *Angew. Chem. Int. Ed.* **16,** 642.
Jung, G., Bovermann, G., Göhring W. and Heusel, G. (1975). *Pept. Proc. Am. Pept. Symp. 4th,* Ann Arbor Press, Ann Arbor, Michigan.
Kalir, R., Fridkin, M., and Patchornik, A. (1974). *Eur. J. Biochem.* **42,** 151.
Kalir, R., Warshawsky, A., Fridkin, M., and Patchornik, A., (1975). *Eur. J. Biochem.* **59,** 55.
Laufer, D. A., Chapman, T. M., Marlborough, D. I., Vaidya, V. M., and Blout, E. R. (1968). *J. Am. Chem. Soc.* **90,** 2696.
Manecke, G., and Haake, E. (1968). *Naturwissenschaften.* **55,** 343.
Marshall, G. R., and Liener, I. E. (1970). *J. Org. Chem.* **36,** 867.
Merrifield, R. B. (1963). *J. Am. Chem. Soc.* **85,** 2149.
Narita, M., Teramato, T., and Okawara, M. (1972). *Bull. Chem. Soc. Jpn.* **45,** 3149.
Panse, G. T., and Laufer, D. A. (1970). *Tetrahedron Lett.,* 4181.
Patchornik, A., and Kraus, C. M. (1975). *Pure and Appl. Chem.* **43,** 503.
Rogozhin, S. V., Davidovich, Yu. A., Andrev, S. M., and Yurtanov, A. I. (1973). *Dokl. Akad. Nauk. SSSR Ser. Khim.* **211,** 1356; **212,** 108.
Sahni, M. K., Jain, J. C., Narang, C. K., and Mathur, N. K. (1977). *Ind. J. Chem.* **15,** 481.
Schröder, E., and Lübke, K. (1965). "The Peptides," Vols. 1 and 2. Academic Press, New York.
Sokolovsky, M., Wilchek, M., and Patchornik, A. (1964). *J. Am. Chem. Soc.* **86,** 1202.
Skylarov, L. Yu., Grobunov, V. I., and Shchukine, L. A. (1966). *J. Gen. Chem. USSR* **36,** 2217.
Teramoto, T., Narita, M., and Okawara, M. (1977). *J. Polym. Sci. Poly Chem. Ed.* **15,** 1369.
Tsiryapkin, V. A., Andreev, S. M., Davidovich, Yu. A., Samoilova, N. S., Rogozhin, S. V., and Belikov, V. M. (1975). *Dokl. Akad. Nauk. SSSR Ser. Khim.* **223,** 1156.
Wieland, Th., and Birr, C. (1966a). *Angew. Chem.* **78,** 303.
Wieland, Th., and Birr, C. (1966b). *Angew. Chem. Int. Ed.* **5,** 310.
Wieland, Th., and Birr, C. (1967). *Chimica* (Switz) **21,** 582.
Wieland, Th., Birr, C., and Fleckenstein, P. (1972). *Annalen,* **756,** 14.
Wilchek, M., Sarid, S., and Patchornik, A. (1965). *Biochim. Biophys. Acta* **104,** 616.
Williams, R. E. (1972). *J. Polym. Sci.* A-1 **10,** 2123.
Wunsch, E. (1971). *Angew Chem. Int. Ed.* **10,** 786.

8 Solid-Phase Sequencing of Peptides and Proteins

I.	Introduction	125
II.	Solid-Phase Edman Degradation Using Polymeric Reagents	126
III.	Solid-Phase Degradation Employing Polymer-Bound Peptides	127
IV.	Other Polymer Supports for Solid-Phase Sequencing	129
V.	Attachment of the Peptide to the Polymer Support	129
VI.	Automation in Solid-Phase Sequencing	133
VII.	Solid-Phase Sequencing of Peptides from the Carboxyl Terminus	133
VIII.	Scope and Limitations of Solid-Phase Sequencing Methods	134
	References	136

I. INTRODUCTION

Undoubtedly, solid-phase methods have found their most important applications in the field of protein chemistry, especially for peptide synthesis. Another very important and successful application in this area is the sequencing of peptides and proteins by the solid-phase Edman degradation method. It can be said that the method is even better established than solid-phase peptide synthesis and that the uncertainties observed in solid-phase peptide synthesis do not exist in the solid-phase Edman degradation. Reviews of work in this area are available (Laursen, 1975; Laursen and Horn, 1977; Mross and Doolittle, 1977).

Of the two well-recognized methods for the sequencing of proteins (Stark, 1970), the older one [Sanger's (1945) 2,4-dinitrophenylation method] is not suited for sequential degradation. In the Edman degradation method (Edman, 1950; Ilse and Edman, 1963), one amino acid at a time is cleaved from the NH_2-terminus of the peptide to give the corresponding phenylthiohydantoin (PTH) or anilinothiazolone.

In the case of proteins or large peptides, the protein or the peptide itself acts as the polymer support from which amino acids are removed, one at a time, via cleavage as thiohydantoins. For smaller peptides such as those obtained by tryptic digestion of proteins, difficulties have been encountered in the solution methods of Edman degradation. This is because the increasingly smaller peptide fragments of low molecular weight tend to become intractable, and their separation from other low-molecular-weight substances becomes increasingly difficult. Although, various chromatographic and liquid extraction methods (Blackburn, 1970) were developed to separate and identify the degradation products (i.e., the thiohydantoins) from the residual peptides, much larger amounts of proteins (1.0–2.0 μmole) were required for such solution methods; even then, after 20 degradation cycles or more, the reaction mixture became intractable and highly contaminated with by-products. The use of carriers such as succinylated poly-L-ornithine, parvalbumin, apocytochrome c, and polybrene in an attempt to overcome this difficulty in liquid-phase automated peptide sequencing has been described (Silver and Hood, 1974; Rochat *et al.*, 1976; Bonicel *et al.*, 1977; Hunkapiller and Hood, 1978).

It was expected that, just as in solid-phase peptide synthesis, the Edman degradation would also be improved by anchoring either the reagent, i.e., the thiocyanate, or the peptide itself on a solid support. This has actually been found to be so.

II. SOLID-PHASE EDMAN DEGRADATION USING POLYMERIC REAGENTS

In an early attempt, an insoluble Edman reagent based on polystyrene was prepared and investigated (Stark, 1965; Manecke and Gunzel, 1968; Dowling and Stark, 1969). Two alternative routes are available for introducing the isothiocyanate group in the polystyrene polymer (Scheme 8-1).

The hydrophilic nature of the polymeric reagent was increased to make it compatible with the peptide by reacting about 60% of the functional groups (NCS) with glucosaminol. Although several steps of peptide degradation were achieved with the reagent polymer, the yields were generally low and larger amounts of peptides were required, by comparison with the dansyl-Edman procedure (Gray, 1967). The cleaved amino acids remained linked to the polymer as thiohydantoins, which, though easily separable from the residual peptide were difficult to characterize.

The major limitation of Stark's method was that the insoluble polymeric Edman's reagent kept the cleaved amino acid bound to it as a thiohydantoin, while the residual peptide remained in solution. It was therefore necessary to apply a subtractive sequencing method by first carrying out

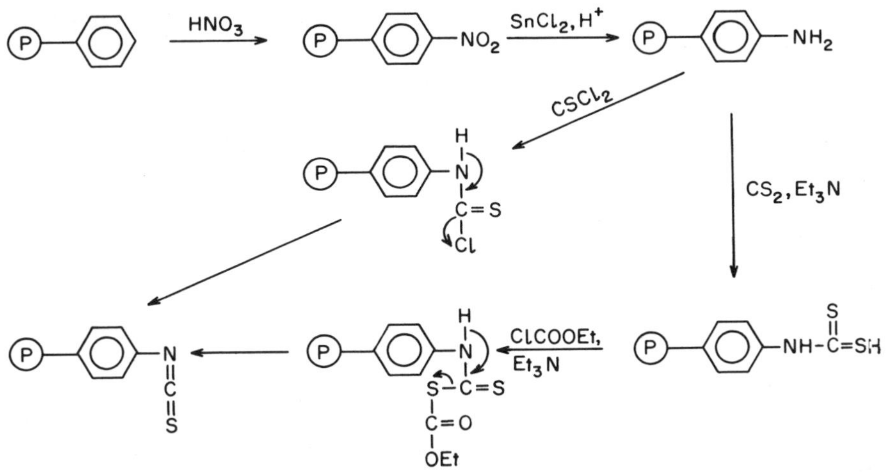

III. Solid-Phase Degradation Employing Polymer-Bound Peptides

Scheme 8-1

amino acid analyses on the residual peptide or cleaving the polymeric thiohydantoin and then identifying the bound amino acid. Most of these difficulties are overcome by the following method wherein the peptide is bound to the polymer and low-molecular-weight reagents are used.

III. SOLID-PHASE DEGRADATION EMPLOYING POLYMER-BOUND PEPTIDES

An alternative approach to solid-phase Edman degradation consists of binding the peptide to a resin support through its COOH-terminus and carrying out the degradation from the NH_2-terminus using the usual Edman reagents. This approach has been reported by Laursen (1966), Edman and Begg (1967), Schellenberger *et al.* (1967), and Dijkstra *et al.* (1967).

Laursen (1966, 1971) attached the peptide by means of its α-carboxyl group to a solid support and carried out the usual Edman degradation at the NH_2-terminus using phenyl or methyl isothiocyanate. The resin, which was functionalized according to Scheme 8-2, was prepared from chloromethylated co(polystyrene–DVB) (1%). The addition of ring amino groups to the polystyrene matrix results in increased polarity and swelling of the resin, which in turn facilitates the degradation.

In the procedure used by Laursen (1966), the NH_2-terminus of the peptide was blocked with a *t*Boc group and the COOH-terminus covalently linked to the preswollen resin via an amide bond by activation with

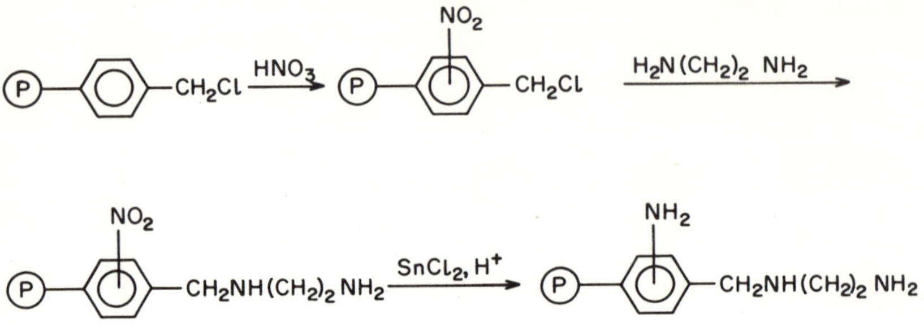

Scheme 8-2

carbonyldiimidazole. The degradation of the polymer-bound peptide was initiated in the usual Edman way. The terminal amino group of the polymer-bound peptide was first exposed by deblocking of *t*Boc groups with trifluoroacetic acid (TFA). Subsequently, the resin-bound peptide was reacted with phenyl or methyl isothiocyanate and the cleavage of the NH_2-terminal amino acid was carried in the usual manner (Scheme 8-3). Since anhydrous conditions were maintained in the resin column, a thiazolinone was initially formed by cleavage of the NH_2-terminal peptide, which was subsequently separated from the column and isomerized in aqueous acid to the corresponding phenyl- or methylthiohydantoin. The thiohydantoins were finally identified by tlc.

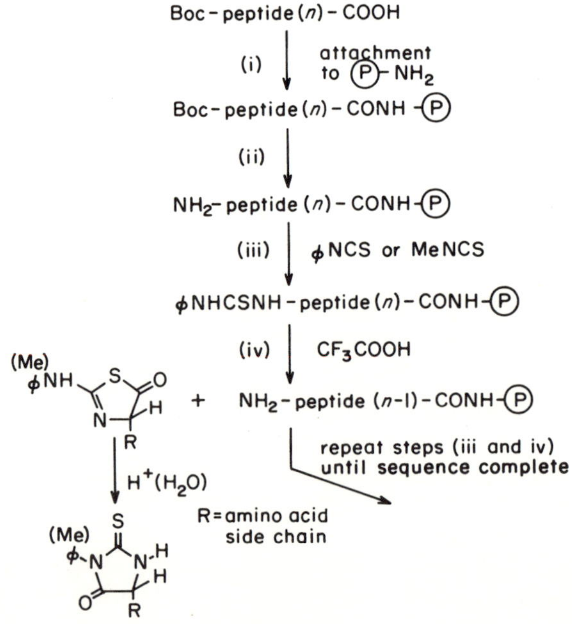

Scheme 8-3

With the development of the liquid-phase or solid-phase sequencer, the development of methods for identifying the released PTH amino acids at very low concentration also assumes importance, and glc methods have been employed along with automated sequencers. In glc, silylated PTHs are employed. High-pressure liquid chromatography has also been used for the identification of PTHs from automatic sequencers.

IV. OTHER POLYMER SUPPORTS FOR SOLID-PHASE SEQUENCING

Since Laursen's first report of solid-phase sequencing using a polymer-bound peptide, many other polymeric supports have been developed for this purpose. These are listed in Table 8-1. These polymeric supports are of three types: styrene-based polymers, silylated aminoalkyl glass, and polyamides.

Several of the polystyrene-based supports aim at increasing the size of the spacer-arm, thus making the attached peptide more easily accessible to the reagents. For example, several modified supports using a modified macroreticular polystyrene as a base have been reported (Inman *et al.*, 1977). Amino groups for peptide attachment were located at the end of poly(ethylene glycol) units that were directly attached to the polystyrene core.

Aminoalkyl glass, prepared by reacting controlled pore glass beads with 3-aminopropyltriethoxysilane, has also been used as a solid support (Wachter *et al.*, 1975). Although this support is of low capacity compared with the styrene-based polymers, it has proved useful for the attachment of larger molecules because the polymer does not swell with solvents and because the reactive groups are mainly confined to the bead surface. Another silylated aminoalkyl-controlled pore glass support (Wachter *et al.*, 1975) employs an even larger spacer-arm, i.e., an N^β-aminoethyl-(3-aminopropyl) group (the glass is marketed by Pierce Chemical Co.). This is said to be particularly suitable for peptides having more than 40 amino acid residues. In addition to the amino group-containing glasses, a recent report (Laursen and Horn, 1977) has demonstrated the utility of a carboxyl-containing glass as a support.

V. ATTACHMENT OF THE PEPTIDE TO THE POLYMER SUPPORT

Laursen (1966, 1971) used carbonyldiimidazole for carboxyl activation and coupling. Subsequently, the use of a soluble carbodiimide, *N*-ethyl-

TABLE 8-1

Polymer Supports Used in Edman Solid-Phase Sequencing

Polymer support	Mode of attachment of the peptide	References
Aminoethylenediamine- polystyrene resin	Terminal COOH via carbonyldiimidazole coupling	Laursen 1966, 1971; cf., Machleidt and Wachter, 1977
Ethylenediamine- polystyrene resin		Laursen, 1966, 1971
Macroreticular polystyrene incorporating 2-aminoethylthiomethyl groups	p-Phenylene diisocyanate coupling of lysine or ornithine	Inman et al., 1977
Triethylenetetramine- polystyrene resin	Terminal COOH, via carbonyldiimidazole coupling, also useful for homoserine lactone activated peptides	Previero et al., 1973; Schellenberger et al., 1972; Bridgen, 1975; Horn and Laursen, 1973; Machleidt and Wachter, 1977
Aminopolystyrene resin	p-Phenylene diisothiocyanate coupling of ϵ-lysine or ϵ-ornithine	Laursen et al., 1972; Machleidt and Wachter, 1977
Macroreticular polystyrene incorporating poly(ethylene glycol) groups and terminal amines	p-Phenylene diisocyanate coupling of lysine or ornithine	Inman et al., 1977
Corning 3-aminopropyl glass	Suitable for peptides containing 40 or more amino acid residues; suitable for coupling to ϵ-NH$_2$-lysine	Wachter et al., 1973; Machleidt et al., 1973; Wachter et al., 1975; Machleidt and Wachter, 1977
Corning carboxymethyl glass	Via the carboxy glass N-hydroxysuccinimide esters (lysine or ornithine)	Laursen, 1977
Corning 3-aminopropyl glass derivatized with p-aminobenzoic acid	Via diazotization and reaction with tyrosine and histidine	Chang et al., 1977
Polamide resin, incorporating β-alanine group β-alanylhexamethylenediamine polydimethylacrylamide resin	p-Phenylene diisothiocyanate coupling of ϵ-NH$_2$-lysine or ϵ-NH$_2$-ornithine	Atherton et al., 1976

V. Attachment of the Peptide to the Polymer Support

N'-(3-dimethylaminoisopropyl)carbodiimide, for this purpose has been reported (Previero et al., 1973); and more recently, the use of trifluroacetyl mixed anhydrides was suggested (Lee and Riordan, 1978).

In later work, Laursen's group (Laursen et al., 1972; Laursen, 1977) used a different approach for coupling of the peptide to the resin support. An amine–polystyrene resin was employed to which the peptide was coupled via its ϵ-amino groups (the lysine residues). p-Phenylene diisothiocyanate (DITC) was used as a bifunctional coupling reagent. In the case of arginine, deguanidation to ornithine was effected by hydrazine, and the latter was similarly coupled via its δ-amino group (Scheme 8-4). Simultaneously, the amino group of the NH_2-terminal amino acid was also linked, and this was cleaved by acid to the corresponding resin–thiazolinone, thereby exposing the amino group of the adjacent amino acid. Up to 100 nmoles of the peptide could be attached to 35 mg of the resin with a coupling yield of 80–100%. The amino acid sequence of a ribosomal protein was largely determined by this method, obviously proving the method's value. This method is suitable for attachment of peptides obtained by tryptic digestion of proteins that contain an NH_2-terminal lysine or arginine. A limitation of the method may be observed with proteins having no lysine residues. In such a case, any arginine residues in the peptide would need to be deguanidated with hydrazine, a

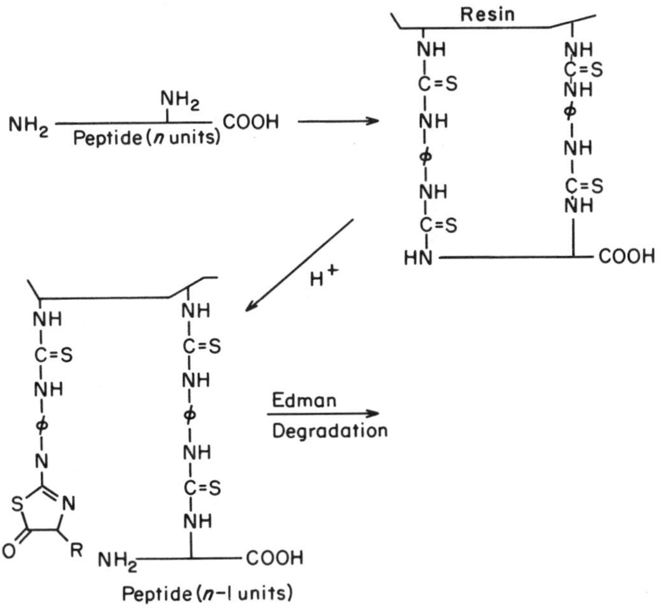

Scheme 8-4

reaction that may be accompanied by such side reactions as internal peptide bond cleavages.

Besides tryptic digestion, certain chemical methods are frequently used for fragmentation of proteins before the sequencing can be done. Highly selective cleavage of methionine-containing peptides by cyanogen bromide gives peptides with homoserine at the COOH-terminus. Cyanogen bromide-degraded peptide fragments are usually much larger than peptides obtained by tryptic digestion. Horn and Laursen (1973) have developed a simple method for attaching the peptides containing COOH-terminal homoserine to the TETA–resin. The peptide–resin link was established via reaction of homoserine lactone with TETA. The unreacted amino groups on the resin were then blocked with $CH_3N{=}C{=}S$ (Laursen, 1971). Poor solubility of the peptides was circumvented in some cases by adding a small amount of water in the coupling medium.

Silanized glass beads have been used as solid supports. Peptides were coupled to the beads by using either carbodiimide or *p*-phenylene diisothiocyanate. The coupled peptides have been satisfactorily degraded, with a repetitive yield of about 92% (Previero *et al.*, 1973; Schellenberger *et al.*, 1971, 1972; Wachter *et al.*, 1973). As an example, cytochrome *c* from *Canadida krusei* was attached to this support and the degradation of over 35 residues, with a repetitive yield of 95%, was achieved. This clearly demonstrates the usefulness of such supports for longer molecules (Machleidt *et al.*, 1973). The phenyl diisothiocyanate method has also been used for attachment of peptides to polyacrylamide supports (Atherton *et al.*, 1976).

Yet another bifunctional reagent (Herbrink *et al.*, 1975) for the attachment of peptides to support resin-containing amino groups has been reported. The new reagent, *N*-(*p*-isothiocyanatobenzoyl)-D-homoserine lactone (IBHL), has been used for binding the peptide via the lysine ϵ-amino group. During the binding reaction, the first amino acid at the NH_2-terminus is cleaved as a substituted phenylthiohydantoin, while the rest of the peptide is attached to the aminated glass through the ϵ-amino lysine group. The bound peptide is then degraded by the usual Edman method. The method has the limitation of being applicable only to those peptides which contain lysine residues (c.f. the DITC method of coupling), and the sequential analysis cannot continue beyond the amino acid next to the lysine in the bound peptide.

The IBHL reagent, now commercially available (Pierce Chem. Co.), is much more expensive than DITC. Recently, a prefunctionalized glass support, isothiocyanatophenylthiocarbamylaminopropyl glass (DITC–glass) (Machleidt *et al.*, 1975), has been marketed.

In addition, chemical methods based on reaction with group-specific reagents have been tried. Attachment of tyrosine- and histidine-containing peptides has been investigated (Chang et al., 1977). Attachment via methionine and tryptophan groups should also be possible (Shechter et al., 1977).

VI. AUTOMATION IN SOLID-PHASE SEQUENCING

An important aspect of solid-phase sequencing is that the process has permitted improvements in the already existing liquid-phase automatic sequencers (Edman and Begg, 1967). The liquid-phase automatic sequencers are based on a reaction vessel (spinning cup) in which the peptide is held by a charged carrier such as succinylated polyornithine in the form of an even, stable film. An ideal carrier should be completely inert to the sequencing reactions and should prevent extractive losses during the solvent wash step. In most of the liquid-phase methods, an exponential decrease in the yield has been observed because of noncovalent bonding of the peptides to the carriers. In addition, losses increase with the decreasing size of the peptides.

Covalent binding of peptides/proteins to insoluble solid polymers can permit many improvements in automatic sequencing. The most important is that all reactions can be carried in a column. Thus, solid-phase Edman degradation has resulted in significant improvements in the construction of automatic peptide sequencers.

Laursen and Bonner (1970) and Laursen (1971) first described the construction of an automatic solid-phase sequencer. An improved version of Laursen's automatic solid-phase automatic sequencer is now commercially available (Sequemat Inc., Watertown, Massachusetts).

In the automated process, the cleavage cycle is repeated on the peptide bound to the resin in the column, and during each cycle the cleavage product is washed off the column. Ultimately, the cleavage products are identified by tlc. Sensitivity in the detection and identification of the thiohydantoins is greatly increased by employing radioactive phenyl isothiocyanate (McKean et al., 1974).

For further details about the construction and operation of automatic protein sequencers, the reader is referred to a recent publication by Perham (1975).

VII. SOLID-PHASE SEQUENCING OF PEPTIDES FROM THE CARBOXYL TERMINUS

Recent reports (Williams and Kassell, 1975; Kassell et al., 1977; Darbre, 1977) describe a solid-phase approach for sequencing from the COOH-terminus. The peptide is bound, via an amide bond involving the NH_2-terminus of the peptide, to a porous glass support functionalized to contain hydroxysuccinimide groups. A side reaction, which may result in the binding of the ε-amino group of lysine, is prevented by carrying out one cycle of Edman degradation, and the ε-amino group of lysine is blocked by formation of a phenyl thiocarbamyl derivative.

The peptide bound to the solid support is then degraded from the COOH-terminus by the Stark procedure (1968), which involves reaction with acetic anhydride and ammonium thiocyanate. In presence of an acid, the COOH-terminal amino acid is cleaved as a thiohydantoin or iminohydantoin, leaving the remaining peptide attached to the solid support. Successive degradation steps are then carried out on the bound peptide.

The cleaved thiohydantoin or iminohydantoin can either be identified as such or hydrolyzed to the parent amino acid, which is then identified. Alternatively, the residual peptide from a portion of the support material can be cleaved and its amino acid analysis carried out so that the cleaved amino acid can be determined by subtractive methods.

Presently, the method has only been applied to certain small synthetic model peptides, and it is claimed that 1 μmole of the peptide is enough for carrying out several cycles of degradations. However, the method is yet to be tested as suitable for sequencing peptides obtained by tryptic digestion or by chemical cleavage of natural proteins.

Aside from the problem of obtaining reproducibility with commercially available supports (Kassell *et al.*, 1977), the method has a major drawback (cf. the limitation of Stark's original solution method). It does not work with certain amino acids, e.g., aspartic acid and proline, although solutions to these problems have been suggested (Kassell *et al.*, 1977). At present, it is difficult to predict if solid-phase sequencing from the COOH-terminus will become an acceptable alternative to methods using sequencing from the NH_2-terminus. It certainly demonstrates that solid phase chemists have adapted yet another reaction to solid-phase methods.

VIII. SCOPE AND LIMITATIONS OF SOLID-PHASE SEQUENCING METHODS

Solid-phase sequencing has been one of the latest and most successful applications of reactions on polymer supports. Automation is one of the important advantages of solid-phase sequencing. Also, it has the advantage of handling large peptides or even proteins. It is certainly a method that no one involved in large-scale primary structure work can afford to neglect.

VIII. Scope and Limitations of Solid-Phase Sequencing Methods

There is no doubt that primary structure determinations can be done manually. However, it is certain that most future work will be concerned with proteins, which, due to the difficulties in isolation, will not be available in large enough quantities to handle manually. There is no doubt that, in such cases, one must resort to an automatic sequencer with high sensitivity.

Many limitations of the solid-phase sequencing methods of Laursen (1971) have been observed. Some of these have subsequently been overcome in several of its modifications.

1. Problems arising from the linking of β- or γ-carboxylic groups to the resin in the case of aspartic or glutamic acid residues: During the linking of the N-blocked peptides to the resin support, the γ-carboxylic acid group in glutamic acid residue is also linked to the resin by an amide bond. Hence, during the formation of phenylthiohydantoins, the cleavage product remains linked to the resin. In case of the β-carboxyl group of aspartic acid, there is a possibility of formation of an imide on treatment with carbonyldiimidazole, and this blocks further degradation (Ondetti et al., 1968).

2. In the coupling reaction of the N-blocked peptide to the resin using carbonyldiimidazole, an attachment yield of 70% for insulin B-chain (30 amino acids) and 95% for A-chain (21 amino acids) was obtained. The average loading of peptide onto resin was about 0.2 μmole/50 mg of resin. It is desirable that the attachment yield be increased to nearly 100%. In one modification (Mross and Doolittle, 1970), the attachment yields for decapeptide were increased to 98% using a water-soluble carbodiimide in dioxane–water mixtures. Other coupling reagents such as DCC have been examined, but cross-linking was found to increase in such cases.

3. Although no problem has been observed in solid-phase degradation of serine, threonine phenylthiohydantoin is partially converted into dehydrothreonine (Laursen, 1971).

4. The average yield of phenylthiohydantoin with most of the amino acids amounted to 89%. With this yield, no problem was observed up to 18 cycles of degradation; but in further cycles, the contamination due to impurities increased to such an extent that interference was observed. It would be desirable, even necessary, to increase the thiohydantoin yield to about 95% in order to carry out 30 or more cycles of degradation successfully.

On the average, tryptic peptides from protein digests contain only about 10 amino acids. This size peptide can be sequenced easily by the solid-phase method, provided they do not contain aspartic acid residues. Sequencing of arginine and tryptophan peptides was not attempted in the original Laursen's method (1971). Only glutamic acid and serine, out of the remaining 18 common amino acids, may cause difficulty when present together in the same peptide, since under the conditions employed, it is not possible to detect their thiohydantoins.

REFERENCES

Atherton, E., Bridgen, J., and Sheppard, R. C. (1976). *FEBS Lett.* **64,** 173.
Blackburn, S. (1970). "Protein Sequence Determination." Dekker, New York.
Bonicel, J., Bruschi, M., Couchoud, P., and Bovier-Lapierre, G. (1977). *Biochimie* **59,** 111.
Bridgen, J. (1975). *In* "Solid-Phase Methods in Protein Sequence Analysis" (R. A. Laursen, ed.), p. 11. Pierce Chemical Inc., Rockford, Illinois.
Chang, J. Y., Creaser, E. H., and Hughes, G. J. (1977). *FEBS Lett.* **84,** 187.
Darbre, A. (1977). *Methods in Enzymol.* **42,** 357.
Dijkstra, A., Billict, H. A., Van Doninck, A. H., Van Velthuysen, H., Matt, L., and Beyerman, H. C. (1967). *Rec. Trav. Chim.* **86,** 65.
Dowling, L. M., and Stark, G. R. (1969). *Biochemistry,* **8,** 4728.
Edman, P. (1950). *Acta Chim. Scand.* **4,** 277, 283.
Edman, P., and Begg, G. (1967). *Eur. J. Biochem.* **1,** 80.
Gray, W. R. (1967). *Methods in Enzymol.* **11,** 469.
Herbrink, P., Tessler, G. I., and Lamberts, J. J. M. (1975). *FEBS Lett.* **60,** 313.
Horn, M. J., and Laursen, R. A. (1973). *FEBS Lett.* **36,** 285.
Hunkappiller, M. W., and Hood, L. E. (1978). *Biochemistry* **17,** 2124.
Ilse, D., and Edman, P. (1963). *Aust. J. Chem.* **16,** 411.
Inman, J. K., Garrett, C. D., and Appella, E. (1977). *In* "Solid-Phase Methods in Protein Sequence Analysis" (A. Previero and M.-A. Coletti-Previero, eds.), p. 81. North-Holland Publ., Amsterdam.
Kassell, B., Krishnamurti, C., and Friedman, H. L. (1977). *In* "Solid-Phase Methods in Protein Sequence Analysis" (A. Previero and M.-A. Coletti-Previero, eds.), p. 39. North-Holland Publ., Amsterdam.
Laursen, R. A. (1966). *J. Am. Chem. Soc.* **88,** 5344.
Laursen, R. A. (1971). *Eur. J. Biochem.* **20,** 89.
Laursen, R. A. (1975). *In* "Solid-Phase Methods in Protein Sequence Analysis" (R. A. Laursen, ed.). Pierce Chem. Co., Inc.,Rockford, Illinois.
Laursen, R. A. (1977). *Methods in Enzymol.* **42,** 277.
Laursen, R. A., and Bonner, A. G. (1970). *Fed. Proc.* **29,** 727.
Laursen, R. A., and Horn, M. J. (1977). *In* "Advanced Methods in Protein Sequence Determination" (S. B. Needleman, ed.), p. 21. Springer-Verlag, Berlin and New York.
Laursen, R. A., Horn, M. J., and Bonner, A. G. (1972). *FEBS Lett.* **21, 67.**
Lee, H. M., and Riordan, J. F. (1978). *Anal. Biochem.* **89,** 136.
Machleidt, W., and Wachter, E. (1977). *Methods in Enzymol.* **42,** 263.
Machleidt, W., Wachter, E., Scheulen, M., and Otto, J. (1973). *FEBS Lett.* **37,** 217.
Machleidt, W., Hofner, H., and Wachter, E. (1975). *In* "Solid-Phase Methods in Protein Sequence Analysis" (R. A. Laursen, ed.), Pierce Chem. Co., Inc., Rockford, Illinois.
Manecke, G., and Gunzel, G. (1968). *Naturwissenschaften.* **55,** 84.
McKean, D. J., Peters, E. H., Waldby, J. I., and Smithies, O. (1974). *Biochemistry,* **13,** 3048.
Mross, G., and Doolittle, R. (1970). Cited by R. A. Laursen (1971). *Eur. J. Biochem.* **20,** 89.

References

Mross, G., and Doolittle, R. (1977). *In* "Advanced Methods in Protein Sequence Determination" (S. B. Needleman, ed.), p. 1. Springer-Verlag, Berlin and New York.
Ondetti, M. A., Deer, A., Sheehan, J. T., Pluscec, J., and Kocy, O. (1968). *Biochemistry,* **7,** 4069.
Perham, R. N. ed. (1975). "Instrumentation in Amino Acid Sequence Analysis." Academic Press, New York.
Previero, A., Derancourt, J., Colletti-Previero, M. A., and Laursen, R. A. (1973). *FEBS Lett.* **33,** 135.
Rochat, H., Bechis, G., Kopeyan, C., Gregoire, J., and Rietschoten, J. V. (1976). *FEBS Lett.* **64,** 404.
Sanger, F. (1945). *Biochemistry,* **39,** 507.
Schecter, Y., Rubinstein, M., and Patchornik, A. (1977). *Biochemistry* **16,** 1424.
Schellenberger, A., Jeschkeit, J., Henkel, R., and Lehmann, H. (1967). *Z. Chem.* **7,** 192.
Schellenberger, A., Jeschkeit, J., and Larsen, R. A. (1971). *Eur. J. Biochem.* **20,** 89.
Schellenberger, A., Jeschkeit, J., Graubaum, H., Mech, C., and Sternkopf, G. (1972). *Z. Chem.* **12,** 62.
Silver, J., and Hood, L. (1974). *Anal. Biochem.* **60,** 285.
Stark, G. R. (1965). *Fed. Proc.* **24,** 225.
Stark, G. R. (1968). *Biochemistry* **7,** 1796.
Stark, G. R. (1970). *Adv. in Protein Chem.* **24,** 261.
Wachter, E., Machleidt, W., Hofner, H., and Otto, J. (1973). *FEBS Lett.* **35,** 97.
Wachter, E., Hofner, H., and Machleidt, W. (1975). *In* "Solid-Phase Methods in Protein Sequence Analysis" (R. A. Laursen, ed.), p. 31. Pierce Chem. Co. Inc., Rockford, Illinois.
Williams, M. J., and Kassell, B. (1975). *FEBS Lett.* **54,** 353.

9
Polymeric Supports in General Organic Chemistry

I.	Introduction	138
II.	Alkylation and Acylation of Esters Using Functionalized Carriers	139
III.	Dieckmann Cyclization of Polymer-Bound Esters	143
IV.	Cyclization of Large Ring Compounds on Polymeric Supports	145
V.	Monofunctionalization of Polymer-Bound Compounds	146
	A. Diols	146
	B. Diacids	149
	C. Dialdehydes	149
	D. Diamines	151
	E. Dihydroxy Aromatic Compounds	151
	F. Dithiols	151
VI.	Synthesis of Threaded Macrocyclic Systems (Hooplanes)	152
VII.	Photochemical Applications	153
	References	153

I. INTRODUCTION

As a result of the impetus provided by Merrifield's work with peptides and subsequent work with nucleotides and saccharides, many reports of the use of polymers as supports in general organic chemistry have appeared. In addition to the usual advantages associated with the use of polymeric supports, they offer the possibility of isolating molecules from each other on the polymer matrix. The isolation of one molecule from another on the matrix (the site-isolation or hyperentropic factor) opens the possibility that intramolecular reactions would be the preferred reactions where intramolecular alkylation or acylations were being done.

In using polymers as supports for molecules, a link between the polymer matrix and the supported molecule must be established. This link can serve not only to anchor the molecule but also to block one of the reactive groups in the supported molecule. If the participation of the reactive group in the subsequent reactions is not desired, the link serves to protect the group. Alternatively, other groups in the supported molecule can be blocked and the molecule freed from the support in a partially blocked state (specific site blocking). When difunctional compounds are used (e.g., diacids or dialcohols), monofunctionalization of the compound is possible. The use of large excesses of the starting difunctional compounds should lead to a substantial amount of monofunctionalization. Nevertheless, large quantities of the starting material are recovered, indicating dual site reaction. Attempts have been made to minimize the amount of dual site reaction, and a review of this aspect of the work has appeared (Leznoff, 1978).

One of the more unusual applications of polymer-supported molecules involves the synthesis of catenated molecules. This will be discussed later.

II. ALKYLATION AND ACYLATION OF ESTERS USING FUNCTIONALIZED CARRIERS

Methods of C-alkylation and C-acylation of compounds containing reactive methylene groups are well-known synthetic reactions. To evaluate the advantages of carrying out these reactions with the help of a polymer support, let us first consider the side reactions that can arise in solution methods (Cope *et al.*, 1957; Hudson and Hauser, 1940). (i) When an ester having more than one α-H is acylated, there is always a possibility of formation of diacylated products. (ii) Since the ester can also act as an acylating agent, self-condensation is another possible side reaction. Self-condensation is particularly pronounced in ethylphenyl acetate. In fact, dibenzyl ketone is synthesized by this reaction.

If the molecules of an enolizable ester are held on a polymer by covalent linking at sufficiently low levels of loading, they should be essentially isolated from molecular species of their own kind. Provided that the polymer is rigid and does not undergo conformational change to bring the anchored molecules very close to each other, self-condensation can be prevented. Even the possibility of diacylation will be greatly reduced by immobilizing one of the reactive species, because the hyperentropic efficacy (high dilution effect) of the reaction reduces statistical chances of the same methylene group's being acylated more than once.

Patchornik and Kraus (1970), Kraus and Patchornik (1971a,b), and Szabo *et al.* (1977) employed a polymeric benzylic ester for immobilizing an acid, whereas Camps *et al.* (1971) used the esters of 2-hydroxyethylpolystyrene. The latter polymer was prepared by copolymerization of p-vinylphenylethyl acetate and styrene (1:7) (Tanimoto and Oda, 1962) or by functionalization of preformed styrene using the reaction sequence bromination, lithiation, and alkylation with ethylene oxide. Camps et al. (1971) preferred a nonbenzylic polymeric ester for linking the acids. The reason for the choice of a nonbenzylic ester is not clear. The only imaginable advantage in using the 2-hydroxyethyl group is that it may act as spacer-arm, reducing the steric hindrance due to the polymer backbone. In the benzylation of polymer-bound isobutyric acid (Ⓟ—CH_2O—CO—$CHMe_2$) (Camps et al., 1971), the yield of benzyldimethylacetic acid was only 20% as compared with 80% isobutyric acid recovered. In contrast, high yields (45–85%) of monoalkylated products were obtained by Kraus and Patchornik (1971a) in the alkylation of a polymeric benzylic ester.

Acylation of polymer-bound phenylacetic acid and acetic acid according to Patchornik and Kraus (1970) is shown in Scheme 9-1. The reaction gave only one kind of ketone (20–43%) and the unreacted acids. In an analogous solution reaction, many 2,4-dinitrophenylhydrazine-positive compounds (ketones) were produced. In the p-nitrobenzoylation of a highly loaded (1.5 mmol/gm) polymeric-acetate, the expected 4'-nitroacetophenone was formed in 4% yield along with p-nitrobenzoylacetone (10%). The latter product was apparently formed from the ester self-condensation product Ⓟ—$CH_2CO_2CH_2COCH_3$. Thus, interresin reaction can become significant at high loading. However, a reexamination of these reactions (Kraus and Patchornik, 1974) indicated that 40% self-condensation took place in case of enolizable polymeric esters. Solution reaction under identical conditions provided 67% self-condensation; hence the authors concluded that short intervals between addition of base and electrophile are required to minimize self-condensation. Self-condensation can also be minimized by employing low reaction temperature and "popcorn" and macroporous resins. In fact, such intrapolymer transacylation reactions have also been observed in

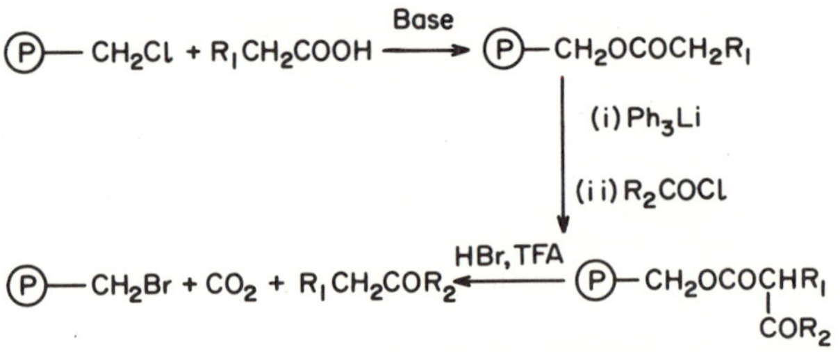

Ⓟ—CH_2Cl + R_1CH_2COOH $\xrightarrow{Base}$ Ⓟ—$CH_2OCOCH_2R_1$

$\downarrow$ (i) Ph_3Li

$\downarrow$ (ii) R_2COCl

Ⓟ—CH_2Br + CO_2 + $R_1CH_2COR_2$ $\xleftarrow{HBr, TFA}$ Ⓟ—$CH_2OCOCHR_1$ | COR_2

Scheme 9-1

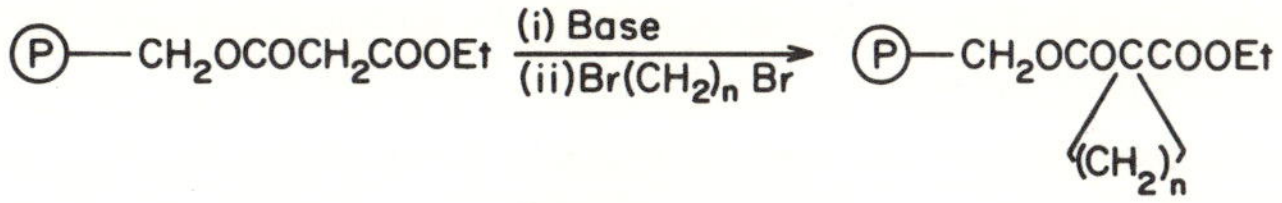

II. Alkylation and Acylation of Esters Using Functionalized Carriers

$$\text{\textcircled{P}}-CH_2OCOCH_2COOEt \xrightarrow[\text{(ii) Br(CH}_2)_n\text{Br}]{\text{(i) Base}} \text{\textcircled{P}}-CH_2OCOC(COOEt)(CH_2)_n$$

Scheme 9-2

solid-phase peptide synthesis (Beyerman *et al.*, 1972), resulting in the chain-doubling phenomenon.

Patchornik and Kraus (1970) also mentioned their studies on the formation of carbocyclic compounds by α,α-dialkylation of polymer-bound malonic ester by reaction with α,ω-dibromoalkanes (Scheme 9-2). When such reactions are carried out with malonic esters, simultaneous formation of α,ω-alkanedioic acids $HO_2C(CH_2)_{n+2}COOH$ is inevitable.

Alkylation of the primary amino groups of resin-bound amino acids has been reported (Szabo *et al.*, 1977). Mono- and dialkylations using benzyl chloride, methyl iodide, *n*-butyl iodide, and *p*-nitrobenzyl bromide were studied. It was shown that intersite interaction could occur, depending on the level of loading on the polymer.

Alkylation of β-diketones using polymers as supports for the intermediate enolate anion has also been reported (Gelbard and Colonna, 1977). Reaction of several cyclohexyl β-diketones with Amberlite IRA-900 formed a resin-linked β-diketonate that could be readily alkylated. Similarly, it was shown that alkylation of phenoxide anions can be readily effected when the anions are supported via an ionic bond with the resin (Gelbard and Colonna, 1977). Alkylations of β-diketones were also shown to occur if a fluoride-substituted, strongly basic resin (Amberlyst A26, A27, or Dowex MSA-1) is used (Miller *et al.*, 1978). In this case, the presence of the fluoride ion was necessary before reaction would proceed. Considering the report on the use of Amberlite IRA-900—a very similar resin—the necessity for a fluoride ion is puzzling. In the same publication, the O-alkylation of phenols, the sulfenylation of β-diketones, and the Michael addition of a thiol to an α,β-unsaturated ketone were also investigated.

Although the polymeric material was not being used strictly as a molecular support, the generation of polymer-supported enolate anions was also inferred when ketones were reacted with C_8K (potassium–graphite) and high-surface sodium or potassium on carbon–graphite or alumina (Hart *et al.*, 1977). Monoalkylation in these reactions is favored but is also accompanied by about 10% ketone reduction.

In the reactions discussed, effective site-isolation of the functionalized groups does not take place with low-cross-linked (2%) polymers (see Chapter 3, Section IV). Intermolecular (strictly speaking, intrapolymeric) reaction becomes significant when the loading of the polymer is high. This has been demonstrated by Kraus and Patchornik (1971b) by using a benzylic polymer esterified (low-loading) with an enolizable (phenylacetic, β-phenylpropionic, hexanoic, or octanoic) acid as well as with a nonenolizable (benzoic or chlorobenzoic) acid. On reacting the mixed polymeric ester with a base (Ph_3CLi), inter-condensation of the esters took place. After ester hydrolysis and decarboxylation, mixed ketones were obtained in yields much higher than those obtainable in solution methods (35–90% vs 20–42%). In another reaction, a polymeric enolizable ester was found to be indefinitely stable in the presence of polymeric trityllithium (base), indicating that interpolymeric reaction does not take place. However, upon addition of a low-molecular-weight enolizable ketone, the reaction takes place readily. The ketone is first enolized by the polymeric base and is then acylated by the second polymeric reagent to give the β-diketone (Patchornik and Kraus, 1976). Thus, because of the proximity of the moieties bound to the same polymer, their forced combination can be induced. The intrapolymeric nature of the reaction was established by reacting two batches of the polymeric ester (one enolizable and the other nonenolizable) under similar conditions. Work-up of the reaction mixture gave no ketone, indicating that interresin reactions did not occur.

The claim that site-separation is achieved in the reactions discussed earlier has been questioned (Crowley and Rapoport, 1976; Neckers, 1978). According to these authors, the Claissen condensation of two molecules of an ester resulting in self-condensation cannot be expected to compete with acylation using acid chlorides or acid anhydrides (more powerful acylating agents). Hence, the formation of mixed ketones as the major product cannot be attributed mainly to site-separation. Furthermore, if site-separation did take place by polymer binding, then why were self-condensation products formed in other reactions? The evidence for effective site-separation is contradictory because self-condensation has taken place in some cases and not in others. This difference is better explained in terms of a competitive reaction that takes place in one case and not in the other.

It has been repeatedly suggested (Crowley and Rapoport, 1976) that reactions used for testing hyperentropic utility must be selected very carefully. A normally rapid reaction that effectively competes with side reactions, even at high concentration, cannot be used to test the ability of the resin to isolate the attached moieties. Even when the proper choice of a reaction has been made, consideration must be given to the modification of chemical reactivity caused by resin binding and subsequent modification of side reactions.

III. DIECKMANN CYCLIZATION OF POLYMER-BOUND ESTERS

In the reports of Patchornik and Kraus (1970) and Camps *et al.* (1971), it was mentioned that Dieckmann cyclization of diesters of the type ⓟ—OCO(CH$_2$)$_n$CO$_2$R (n = 8 and 14) gave 9- and 15-membered rings, respectively, but no details were made available. Crowley and Rapoport (1970) described an interesting application of a unidirectional Dieckmann cyclization of polymer-bound benzyltriethylcarbinyl pimelate for the synthesis of specifically radiolabeled cyclohexanone derivatives (Scheme 9-3). By using the highly hindered triethylcarbinyl moiety and a hindered base (potassium triethylcarbinylate), the Dieckmann cyclization of pimelate ester was exclusively effected in such a way that the benzylic ester carbonyl participated in the reaction. The resulting keto ester bore a ^{14}C-1 label. The participation of the benzyl ester carbonyl group in the cyclization reaction was demonstrated by a kinetic study of this reaction. Since the pimelic acid diester 1 bears the ^{14}C-1 radiolabel, the autocleaved keto ester 2 should bear the whole radiolabel at C-1 if the benzyl ester carbonyl participates in the cyclization. This was proved by measuring the radiolabel in the cyclohexanone obtained by hydrolysis after decarboxylation of the keto ester 2. The use of a polymeric benzyl ester in this reaction provided the advantage of separation by filtration, thereby allowing the easy measurement of radioactivity in the solution as well as in the polymeric by-product (Crowley and Rapoport, 1970). When a nonpolymeric benzyl ester was used, the separation of the two keto esters, 2 and 3, could not be achieved, even by chromatography. The results of radiolabel distribution on polymer 3 and the autocleaved product 2 are shown in Table 9-1. According to these workers, unidirectional cyclization is basically dependent upon the use of one ester that is much more hindered than the other; this advantage is observed in both cases, i.e., with benzylic ester 1b and polymeric benzyl ester 1a.

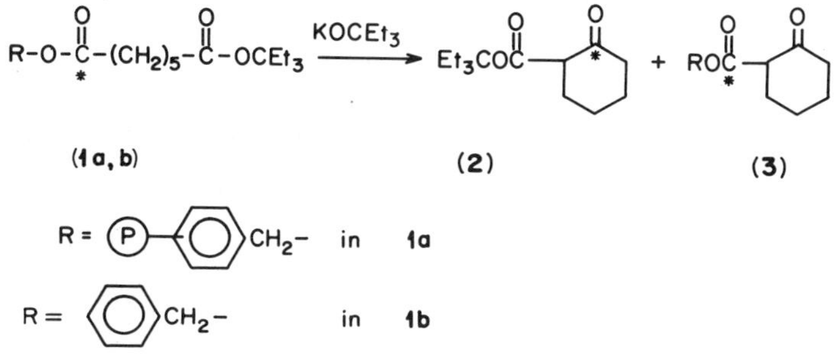

Scheme 9-3

TABLE 9-1

Yield and Radioactivity Label Distribution in Resin and Autocleaved Keto Esters

	Yield (%)	Keto label	Ester label
Autocleaved keto ester	46	99.7	0.3
Resin-retained keto ester	10	48	52

By carrying out the Dieckmann condensation using a polymeric benzyl ester, Crowley and Rapoport (1976) demonstrated the advantage of a purification effect in solid-phase synthesis in addition to hyperentropic efficacy. The Dieckmann cyclization of a nonpolymeric benzylic ester produces a mixture of two soluble keto esters whose separation by some conventional method becomes necessary (a rather difficult separation).

Unexpectedly, some ^{14}C radiolabel did appear on the ester carbonyl of resin-retained and autocleaved esters. The ratio of ^{14}C radiolabel of ester carbonyl and keto group could be determined by measuring the radioactivity of the carbon dioxide produced by keto ester hydrolysis and decarboxylation, as well as residual activity in the ketone. The ratio of ^{14}C label in ester carbonyl and keto groups is presented in Table 9-1.

The formation of ^{14}C-ester carbonyl label was explained by assuming that some transesterification took place (Crowley *et al.*, 1973) even when using the hindered triethylcarbinyl ester, resulting in scrambling of ^{14}C label (Scheme 9-4). Two paths for scrambling of ^{14}C label, shown in Scheme 9-4, involve the participation of polymeric benzyl alkoxide ion (Path A) and triethylcarbinyl alkoxide ion, followed by retransesterification by polymer alkoxide (Path B), respectively. From the ^{14}C-label distribution, it was concluded that Path A was the major route for transesterification. Furthermore, it was shown that, if the triethylcarbinyl ester

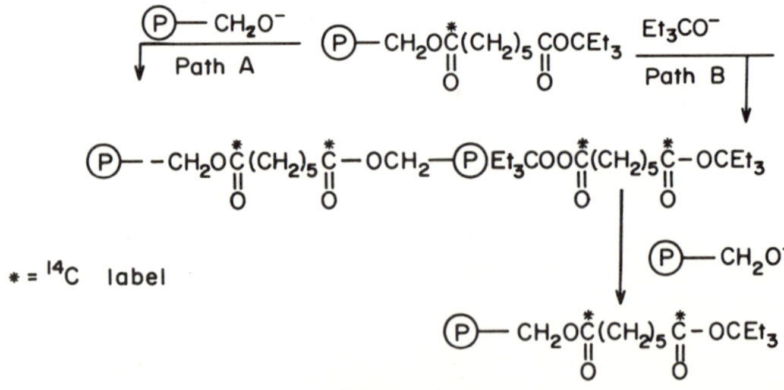

Scheme 9-4

group is substituted by an ethyl group, the transesterification becomes more competitive and produces extensive scrambling of ^{14}C label. These results clearly demonstrate that interresin reactions were suppressed not because of the isolation of reactants by immobilization (thereby effecting an environment of "high dilution"), but because of the more competitive nature of the cyclization reaction as compared with transacylation.

IV. CYCLIZATION OF LARGE RING COMPOUNDS ON POLYMERIC SUPPORTS

In an attempt to cyclize larger polymer-bound alkanedioic esters (4), both autocleaved (5) and resin-retained keto esters (6) were formed in poor yields, indicating nonselectivity for unidirectional ring closure (Crowley and Rapoport, 1976). It was also indicated that intraresin site–site reactions are rapid compared to cyclization because dimeric diketo esters (7) were formed as major products in the attempted cyclization of sebacyl resin (Scheme 9-5).

In an attempted cyclization of ω-cyanopelargonyl thiol resin ester (8) (Crowley et al., 1973), diketodinitrile (9) was isolated in addition to much smaller quantities of 2-cyanocyclononane (10) (Scheme 9-6). This again points to site–site condensation as the most straightforward explanation. That site-separation is not effective is also proved by studies of resin-bound transition metal complexes, where bridging of polymer-bound ligands via metal ion is known to take place (Collman and Reed, 1973; see Chapter 14). However, in some functionalized polymers, for example hydroxyl polymers (Patchornik and Kraus, 1970) and carboxylic acid polymers (Letsinger et al., 1964), non-hydrogen-bonded hydroxyls are clearly revealed by ir spectra of pure polymers. Swelling and solvation of the polymer could change the picture completely.

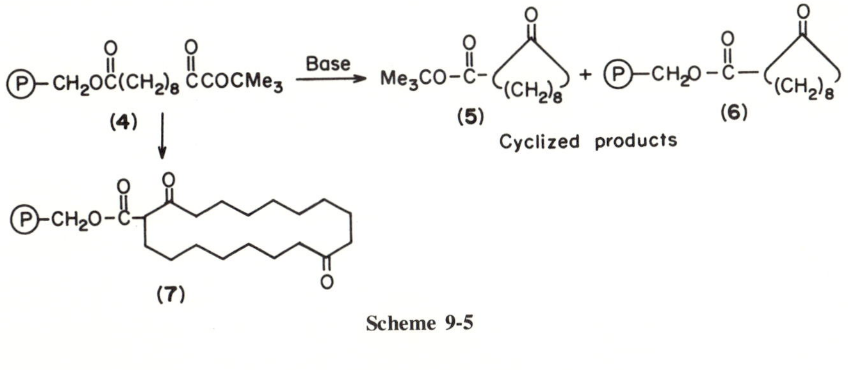

Scheme 9-5

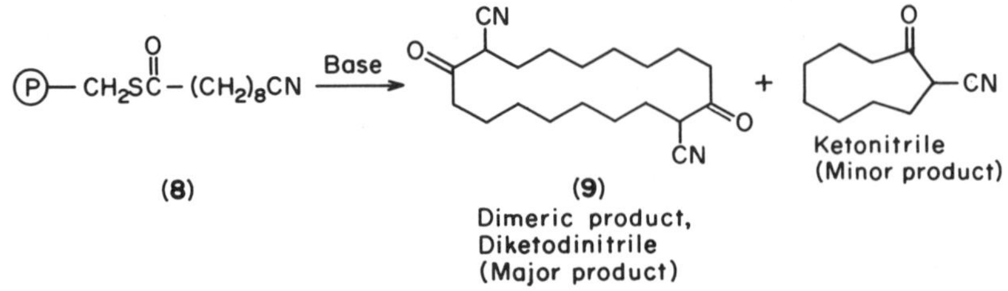

$$\text{(P)}-CH_2\overset{O}{\underset{}{S}}C-(CH_2)_8CN \xrightarrow{\text{Base}} \text{(9) Dimeric product, Diketodinitrile (Major product)} + \text{Ketonitrile (Minor product)}$$

(8)

Scheme 9-6

V. MONOFUNCTIONALIZATION OF POLYMER-BOUND COMPOUNDS

A. Diols

Just as the alkanedioic acids can be bound to the polymer by one of the carboxyl group by an ester bond, diols can be bound to a polymer (acid chloride) via one of the hydroxyl groups, leaving the other hydroxyl group free for derivatization (Wong and Leznoff, 1973; Leznoff and Wong, 1972; Fyles et al., 1978). There is strong evidence that "double binding" of the diols is a competitive reaction; and, during recovery of the monofunctionalized diols, large amounts of the free diols are also liberated from the polymer. These diols must have been formed by hydrolysis of the polymeric diester. However, site-separation is only partial (Wong and Leznoff, 1973; Leznoff and Wong, 1972). The monofunctionalization of polymer-bound diols according to Wong and Leznoff (1973) and Leznoff and Wong (1972) is represented in Scheme 9-7. A benzoyl chloride polymer has also been used (Fyles et al., 1978). These authors concluded that the swelling properties of the resin, by-product contamination, and

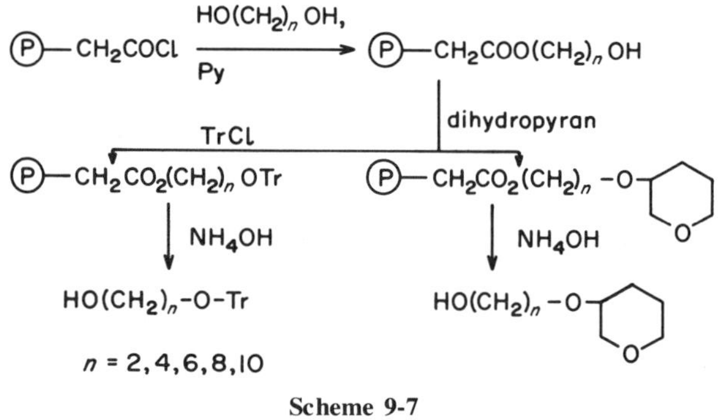

$n = 2, 4, 6, 8, 10$

Scheme 9-7

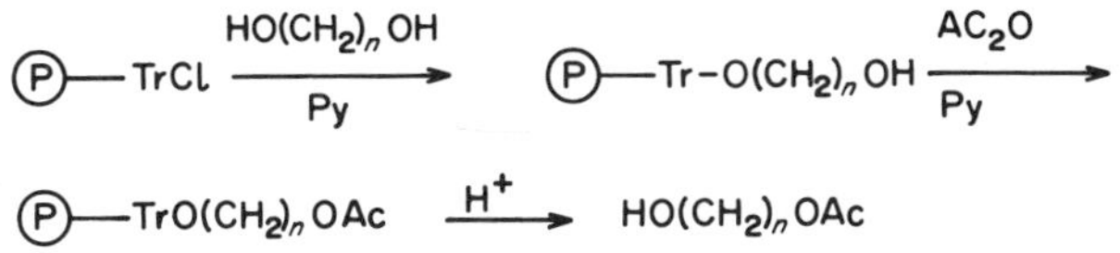

V. Monofunctionalization of Polymer-Bound Compounds

$$\text{(P)}-\text{TrCl} \xrightarrow[\text{Py}]{\text{HO(CH}_2)_n\text{OH}} \text{(P)}-\text{Tr}-\text{O(CH}_2)_n\text{OH} \xrightarrow[\text{Py}]{\text{Ac}_2\text{O}}$$

$$\text{(P)}-\text{TrO(CH}_2)_n\text{OAc} \xrightarrow{\text{H}^+} \text{HO(CH}_2)_n\text{OAc}$$

Scheme 9-8

poor recyclability along with the disadvantages mentioned earlier make this polymer a poor choice.

Another type of polymer, used for the monoprotection of hydroxyl functions in sugars and in diols, was functionalized to contain a trityl group (Fréchet and Hague, 1975; Fyles and Leznoff, 1976; Fréchet and Nuyens, 1976). Fréchet and Nuyens (1976) carried out several selective functionalizations (bromination and acylation) of polyhydroxy alcohols bound to the polymeric support (Scheme 9-8). In protection of hydroxyl group by etherification with polymeric trityl chloride, the selectivity for primary hydroxyl versus secondary and tertiary hydroxyl was observed. Similar selectivity is observed with low-molecular-weight trityl chloride. Since the 2% cross-linked polymer failed to provide significant hyperentropic efficacy (site-separation), Fyles and Leznoff (1976) attempted the use of a 20% cross-linked functionalized polymer, but such a polymer proved to be too fragile under stirring or vacuum-drying conditions.

Finally, a unique functionalization procedure was evolved that simultaneously increased cross-linking, particularly near the reaction site (Fyles and Leznoff, 1976). It was argued that if two or three phenyl groups of the original polymer were joined through a carbon atom to give the trityl function, the resulting polymer would be highly cross-linked; furthermore, the proximal phenyl groups would be functionalized into the same trityl group. This was achieved by reacting lithiated polymer two or even three times with the same substrate. Scheme 9-9 (routes a and b) shows the synthesis of the trityl group-containing polymer by reaction of the lithiated polymer with methyl benzoate and phosgene, respectively, thus quenching the proximal phenyllithium groups. The degree of functionalization was low (0.07–0.03 mmol/gm) because side reactions also produced benzophenone and methyl benzoate groups in the reaction. The resins produced according to this procedure are characterized by (i) low degree of functionalization, (ii) reduced proximity of the functional groups on the polymer backbone, (iii) reduced conformational flexibility due to high cross-linking of the polymer chain through triarylmethyl groups, and (iv) cross-linking located particularly on polymer strands near the site of functionalization which makes the translocation of such sites difficult.

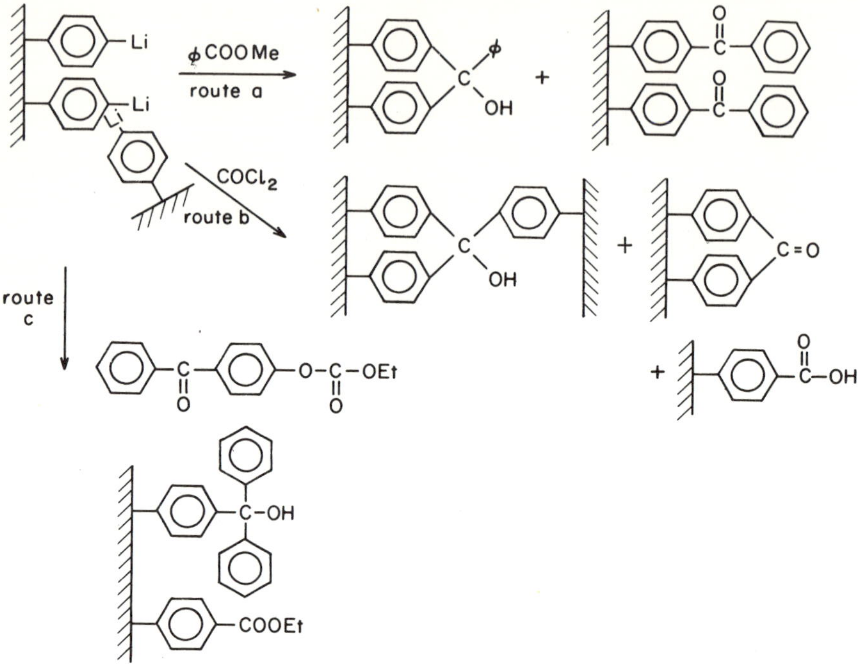

Scheme 9-9

When 1,10-decanediol was bound to such a polymer, the relative yield of monoacetate was 100% in one case, indicating that effective site-separation had taken place. This appears to be the most successful case of achieving hyperentropic efficacy using polymeric supports.

An extension of this work on microenvironment control has been made (Fyles *et al.*, 1978). Lithiated polystyrene was reacted with *p*-benzoylphenylethyl carbonate and the product resin hydrolyzed to give a resin containing proximate trityl alcohol and benzoic acid groups (Scheme 9-9, route c). "Double-binding" of diols to the trityl groups loaded on this resin and a similar resin, in which the groups were arranged in a random fashion, was found, indicating that complete site isolation was not achieved despite attempts to rigidify the local environment.

The synthesis of insect sex attractants using a polymeric support essentially involves the principle of monoprotection of diols by trityl ether. From the polymer-bound diols, the synthesis of compounds of general formula $AcO(CH_2)_n-CH=CH-(CH_2)_nCH_3$ has been achieved (Leznoff and Fyles, 1976; Leznoff *et al.*, 1977; Fyles *et al.*, 1978) via three different routes (Scheme 9-10).

V. Monofunctionalization of Polymer-Bound Compounds

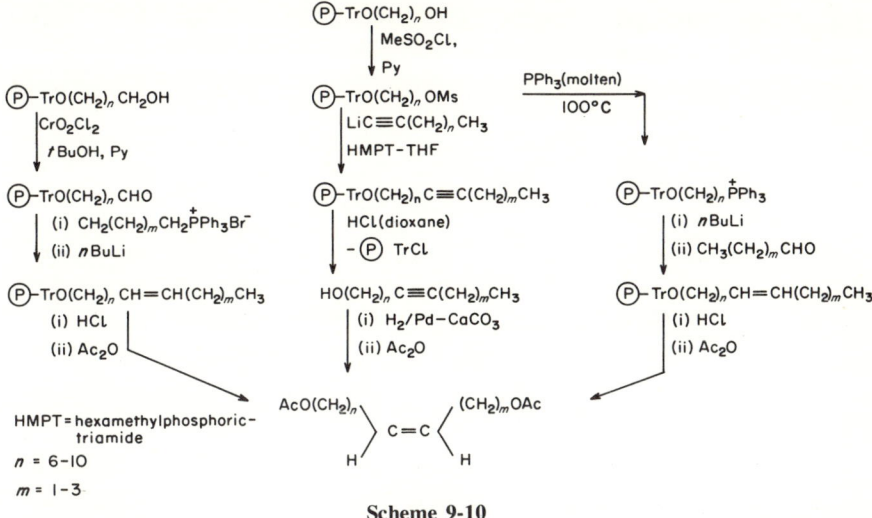

Scheme 9-10

HMPT = hexamethylphosphoric-triamide
$n = 6-10$
$m = 1-3$

B. Diacids

Leznoff and Goldwasser (1977) used a hydroxymethylated polymer for monoblocking of symmetrical diacid chlorides. The unblocked acid chloride group was then converted to an amide and the polymer-bound derivatives were cleaved from the polymer and characterized as monoamide monoesters (Scheme 9-11). The functionalized products were obtained in high yield and in pure state, indicating that there was no observable double binding of the acids on polymer support. It was stressed that proper choice of the reaction conditions can alter the ability of the polymer to undergo interresin reactions.

C. Dialdehydes

Besides diols and diacids, polymers have been used for protection of one of the aldehyde groups in symmetrical dialdehydes (Leznoff and

Scheme 9-11

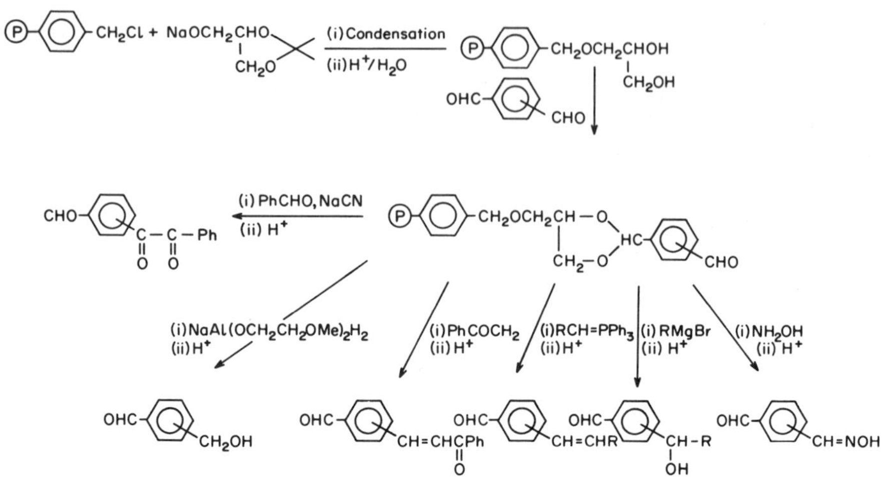

Scheme 9-12

Wong, 1973). Diol-containing resins used for the purpose of protection of aldehydes group via formation of the acetal were synthesized by Leznoff and Wong (1973) (Scheme 9-12). Although the resins readily formed acetals with m- and p-phthaldehyde, the highly hindered o-phthaldehyde and other ortho-substituted aromatic aldehydes as well as certain symmetrical aliphatic aldehydes failed to form acetals with this resin. In contrast, the 1,3-diol resin (Fig. 9-1) reacted with these less-reactive

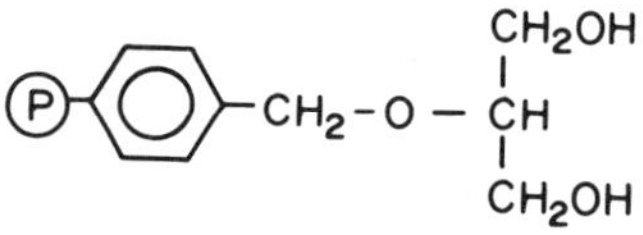

Fig. 9-1

aldehydes to give polymeric six-membered acetals (Fig. 9-2) (Leznoff and Greenberg, 1976).

Once the dialdehyde was anchored on one of these resins, the free aldehyde group could be made to undergo a number of reactions (Scheme

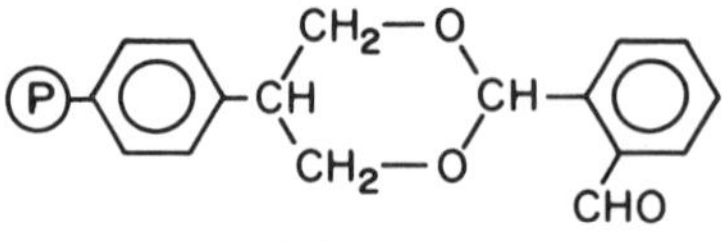

Fig. 9-2

9-12) such as oximation, Wittig reaction, crossed aldol condensation, Grignard reaction, and reduction with metal hydrides. Some of these monofunctionalized derivatives were prepared for the first time with the help of polymer supports, whereas certain others were only formed in low yield by solution methods even when an excess of the aldehydes was used. The resin support in these reactions could be recovered by acid cleavage of the resin-bound product and could be reused. A multistage synthesis of 4,4'-diformylstilbene was also performed using the polymer-protected dialdehyde (Wong et al., 1974), and the syntheses of some carotenoids have been reported (Leznoff and Sywanyk, 1977).

D. Diamines

Monofunctionalization of diamines has been reported (Leznoff and Dixit, 1977). Symmetrical diamines were bound to a benzyl chloride support that had been first reacted with p-nitrobenzyl carbonate. The aminocarbonates thus obtained were benzoylated and then treated with trifluoroacetic acid–trifluoroacetic anhydride in chloroform to yield the N-benzoyl-N'-trifluoroacetyl-1-ω-diaminoalkanes (50–80% yields).

E. Dihydroxy Aromatic Compounds

The preparation of monoethers of symmetrical dihydroxy aromatic compounds has been reported (Leznoff and Dixit, 1977). The dihydroxy aromatic compounds were bound to a benzoyl chloride support via a monoester bond, reacted with diazomethane, and cleaved from the resin by base hydrolysis. Yields ranged from 20 to 75%, depending on the compound. Catechol and 1,4-dihydroxynaphthalene, although attached to the polymer, did not yield any product. It was found that double-binding to the polymer was minimal and was not the cause of lowered yields.

F. Dithiols

In a set of experiments designed to test site-isolation within the polystyrene matrix, Farrall and Fréchet (1978) attempted to prepare the monoalkylated product of 1,4-butanedithiol. Large amounts of double coupling (dialkylation) were observed when the chloromethylated polystyrene used was reacted with excess sodium methoxide and 1,4-butanedithiol in DMF. Double coupling could be reduced substantially (to less than 5%) by employing a three-phase system consisting of $(n\text{Bu})_4\text{N}^+$ OH^- and 1,4-butanedithiol in a benzene–water solvent.

VI. SYNTHESIS OF THREADED MACROCYCLIC SYSTEMS (HOOPLANES)

The synthesis of threaded macrocyclic systems such as the hooplanes, catenanes, and knots presents great difficulties because of the special topology of these molecules (Wasserman, 1962). The normal synthetic routes to such molecules (Schill et al., 1972) have been long and arduous, particularly for the preparation of such substances in quantities greater than 1 mg.

Harrison and Harrison (1967) realized the possibility of utilizing an insoluble polymer support for the synthesis of such systems. Their strategy consisted of attaching the macrocyclic ring system to an insoluble polymer and then repeating the threading step several times (Scheme 9-13). Even if only a small percentage of macrocyclic rings were threaded, undesirable nonthreaded by-products could be removed simply by filtration and washing of the insoluble polymer supporting the macrocyclic ring. In order to secure higher yields—and for statistical reasons—the operation could even be automated. Ultimately, the threaded macrocyclic product (hooplane) could be cleaved from the polymer and separated from the unthreaded by-product. An overall yield of 6% of the threaded macrocycle was obtained in this way. This also demonstrates the potential of using insoluble polymers to retrieve minor components from a complex reaction mixture.

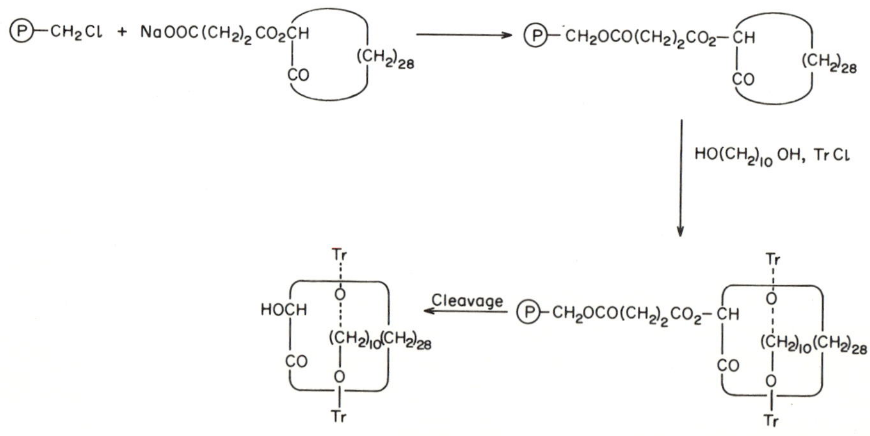

Scheme 9-13

VII. PHOTOCHEMICAL APPLICATIONS

Few examples of the use of polymers as supports for molecules being subjected to photolyzing reactions have been reported. In using organic-based polymers as supports the prime deterrent resides in the polymer carrier. If the wavelength of the photolyzing light is at all close to that of functional groups in the polymer, photochemical transformation and degradation of the polymer can occur. In addition, if one of the excited states of the molecule being subjected to the photolysis is sufficiently energetic it too can cause polymer chain degradation. Other aspects of photochemistry in organic polymers have been outlined (Farid *et al.*, 1979). Inorganic-based polymers suffer less from the problem of photochemically induced degradation and, indeed, one recent report describes the use of silica gel as a support for subjecting molecules to the Photo-Fries rearrangement (Avnir *et al.*, 1978).

REFERENCES

Avnir, D., de Mayo, P., and Ono, I. (1978). *J. Chem. Soc. Chem. Commun.*, 1109.
Beyerman, H. C., de Leer, E. W. B., and von Vossen, W. (1972). *J. Chem. Soc. Chem. Commun.*, 929.
Camps, F., Castells, M. J., Ferrando, M. J., and Font, J. (1971). *Tetrahedron Lett.*, 1713.
Collman, J. P., and Reed, C. A. (1973). *J. Am. Chem. Soc.* **29**, 2048.
Cope, A. C., Holmes, H. L., and House, H. O. (1957). *In* "Organic Reactions" (R. Adams, ed.), Vol. IX. Wiley, London.
Crowley, J. I., and Rapoport, H. (1970). *J. Am. Chem. Soc.* **92**, 6363.
Crowley, J. I., and Rapoport, H. (1976). *Acc. Chem. Res.* **2**, 135.
Crowley, J. I., Harvey III, T. B., and Rapoport, H. (1973). *J. Macromol. Sci. Chem.* A, **7**, 1117.
Farid, S., Martic, P. A., Daly, R. C., Thompson, D. R., Specht, D. P., Hartman, S. E., and Williams, J. L. R. (1979). *Pure Appl. Chem.* **51**, 241.
Farrall, M. J., and Fréchet, J. M. J. (1978). *J. Am. Chem. Soc.* **100**, 7998.
Fréchet, J. M. J., and Haque, K. E. (1975). *Tetrahedron Lett.*, 3055.
Fréchet, J. M. J., and Nuyens, M. J. (1976). *Can. J. Chem.* **54**, 926.
Fyles, T. M., and Leznoff, C. C. (1976). *Can. J. Chem.* **54**, 935.
Fyles, T. M., Leznoff, C. C., and Weatherston, J. (1978). *Can. J. Chem.* **56**, 1031.
Gelbard, G., and Colonna, S. (1977). *Synthesis*, 113.
Harrison, I. T., and Harrison, S. (1967). *J. Am. Chem. Soc.*, **89**, 5723.
Hart, H., Chen, B. C., and Peng, C. L. (1977). *Tetrahedron Lett.*, 3121.
Hudson, B. E., and Hauser, C. R. (1940). *J. Am. Chem. Soc.* **62**, 2457.
Kraus, M. A., and Patchornik, A. (1971a). *Israel J. Chem.* **9**, 269.
Kraus, M. A., and Patchornik, A. (1971b). *J. Am. Chem. Soc.* **93**, 7325.
Kraus, M. A., and Patchornik, A. (1974). *J. Polym. Sci. Polym. Symp.*, **47**, 11.

Letsinger, R. L., Kornet, M. J., Mahadevan, V., and Jerina, D. M. (1964). *J. Am. Chem. Soc.* **86,** 5163.
Leznoff, C. C. (1978). *Acct. Chem. Res.* **11,** 327.
Leznoff, C. C., and Dixit, D. M. (1977). *Can. J. Chem.* **55,** 3351.
Leznoff, C. C., and Fyles, T. M. (1976). *J. Chem. Soc. Chem. Commun.,* 251.
Leznoff, C. C., and Goldwasser, J. M. (1977). *Tetrahedron Lett.,* 1875.
Leznoff, C. C., and Greenberg, S. (1976). *Can. J. Chem.,* **54,** 3824.
Leznoff, C. C., and Sywanyk, W. (1977). *J. Org. Chem.* **42,** 3203.
Leznoff, C. C., and Wong, J. Y. (1972). *J. Org. Chem.* **50,** 2892.
Leznoff, C. C., and Wong, J. Y. (1973). *J. Org. Chem.* **51,** 3756.
Leznoff, C. C., Fyles, T. M., and Weatherston, J. (1977). *Can. J. Chem.* **55,** 1143.
Miller, J. M., So, K. H., and Clark, J. H. (1978). *J. Chem. Soc. Chem. Commun.,* 466.
Neckers, D. C. (1978). *Chem. Technol.* 108.
Patchornik, A., and Kraus, M. A. (1970). *J. Am. Chem. Soc.* **92,** 7587.
Patchornik, A., and Kraus, M. A. (1976). *Encycl. Polym. Sci. Technol.,* Supplement No. 1, 469–492.
Schill, G., Logemann, E., and Vetter, W. (1972). *Angew. Chem. Int. Ed.* **11,** 1089.
Szabo, L., Coppens, E., Clauder, O., and Pecher, J. (1977). *Bull. Soc. Chim. Belg.* **86,** 35.
Tanimoto, S., and Oda, R. (1962). *Kogyo Kagaku Zasshi,* **64,** 932 [C. A. 1962, **57,** 4854].
Wasserman, E., (1962). *Sci. Am.* **207,** 94.
Wong, J. Y., and Leznoff, C. C. (1973). *Can. J. Chem.,* **51,** 2452.
Wong, J. Y., Manning, C., and Leznoff, C. C. (1974). *Angew. Chem. Int. Ed.,* **13,** 666.

10 Polymer-Supported Asymmetric Synthesis and Resolution of Racemates Using Asymmetric Polymeric Materials

I.	Introduction.	155
II.	Asymmetric Syntheses on Polymeric Supports	156
III.	Resolution of Racemates Using Polymeric Materials .	158
	A. Conventional Methods	158
	B. Resolution of Amino Acids by Ligand-Exchange Chromatography	159
	C. Resolution of Racemates Using Asymmetric Templates	161
	References	162

I. INTRODUCTION

When a compound having an asymmetric center is synthesized from symmetrical molecules with the help of achiral reagents, the product is invariably a mixture of D- and L-isomers. In order to establish the identity of the synthetic product with naturally occurring substances, it is necessary to resolve the compound. Chemical methods for resolution are well-known.

An alternative to this is to carry out the synthesis in an asymmetric environment (created by use of asymmetric reagents, catalyst, etc.) (Valentine and Scott, 1978). It is then possible to prepare a specific isomer exclusively or at the very least in larger yield. Such a synthesis is referred to as an asymmetric synthesis.

Polymer-supported reagents have been used both for asymmetric synthesis and for chemical resolution of racemates.

II. ASYMMETRIC SYNTHESES ON POLYMERIC SUPPORTS

For the synthesis of an optically active organic compound from an achiral substrate, derivatization of the compound with a chiral reagent is necessary. This derivatization must also be reversible, so that after the derivatized substrate has been transformed into the optically active product, the reaction can be reversed to recover the asymmetric product. An alternate approach involves reacting an achiral substrate with a chiral reagent or catalyst. Both these approaches have been used for asymmetric synthesis [i.e., polymer supports used for reversible derivation (binding) of the substrate and chiral polymer supports used to bind an otherwise homogeneous catalyst].

The reduction of an α-keto acid, after derivatization to an ester with an optically active alcohol (L-menthol), is a well-known example of asymmetric synthesis. In an extension of this method, Kawana and Emoto (1972, 1974) synthesized a polymer-bound sugar (furanose) having only one free hydroxyl group (3) (Scheme 10-1) by using triphenylmethylcarbinol polymer (1) (Hayatsu and Khorana, 1966, 1967; Cramer *et al.*, 1966). The free hydroxyl group of the polymer-bound sugar was esterified with phenylglyoxyl chloride to give the corresponding polymeric ester (4). Grignard asymmetric reaction of the polymeric chiral ester with methylmagnesium iodide, followed by saponification, gave atrolactic acid (5) in

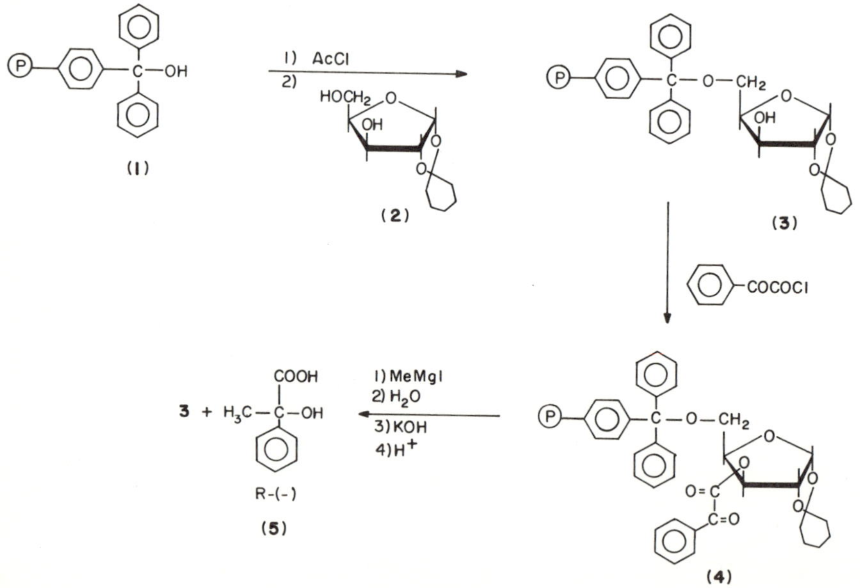

Scheme 10-1

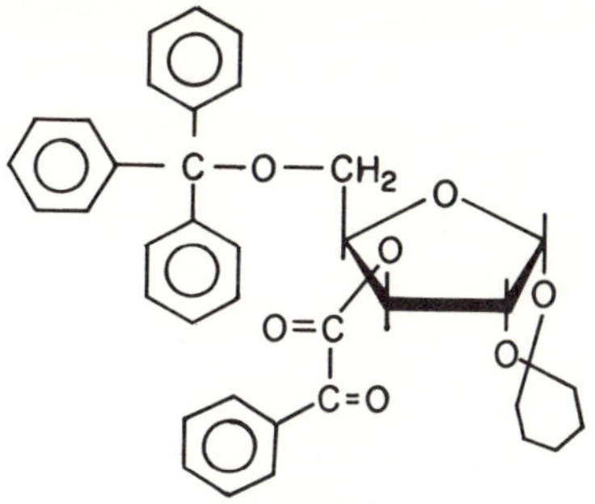

II. Asymmetric Syntheses on Polymeric Supports

Fig. 10-1

chemical and optical yields of 73% and 58%, respectively. The polymer could be reused.

To illustrate the advantage of the polymeric support in this example, the same synthesis was carried using the analogous nonpolymeric chiral ester (Figs. 10.1 and 10.2) and chemical and optical yields of 68% and 53%, respectively, were obtained.

According to these workers, the bulkiness of the substituent on C-4 of the furanose ring system played an important role in increasing the stereoselectivity of the Grignard reagent. In addition, the selectivity could have been further enhanced by the conformational rigidity around the sugar and acid groups in the cross-linked polymer.

In these cases, the polymer was used as an asymmetric support to induce the formation of optically pure product (cf. Worster *et al.*, 1979). Few reports of the use of polymer-bound asymmetric reagents seem to exist in the literature. In this application, the reagent is used either to promote the asymmetric coupling of two groups or to add a group to a compound in an asymmetric manner. By far the largest number of applications have been those in which the polymer-bound asymmetric centers act as catalysts. Asymmetric catalysts, based on either amino acids or cinchona alkaloids, have been used to catalyze the Michael reaction in an

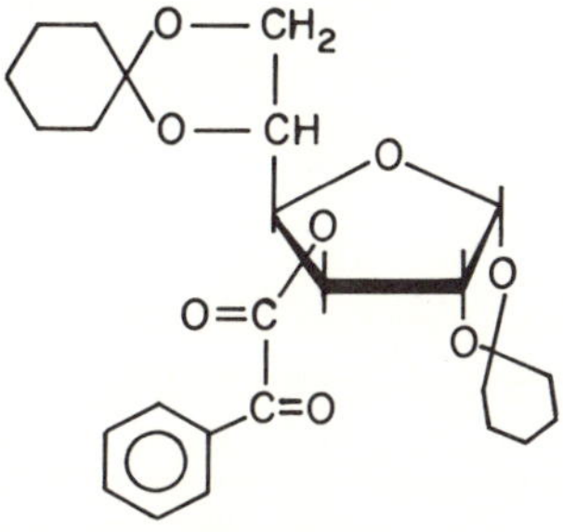

Fig. 10-2

$$\underset{CH_3}{\overset{O}{\underset{|}{>C=C-C}}}\overset{O}{\underset{\diagdown}{\diagdown}} \xrightarrow{RSH} RS-\underset{|}{\overset{|}{C}}-\underset{|}{\overset{H}{\underset{*}{C}}}-C\overset{O}{\underset{\diagdown}{\diagdown}}$$

Scheme 10-2

asymmetric fashion (Scheme 10-2) (Inoue et al., 1972; Ueyanagi and Inoue, 1977; Hermann and Wynberg, 1977, Kobayashi and Iwai, 1978). Polymeric asymmetric hydrogenation catalysts, usually based on a rhodium-phosphine complex, have been prepared. These will be alluded to in more detail in Chapter 14.

III. RESOLUTION OF RACEMATES USING POLYMERIC MATERIALS

A. Conventional Methods

The polymer-aided resolution of racemic mixtures of simple compounds has relied mainly on chromatographic methods (Buss and Vermeulen, 1968; Rogozhin and Davankov, 1968; Losse and Kuntze, 1970; Boyle, 1971). The polymers have been used either in their traditional roles to separate diastereoisomers formed by chemical derivatization with a chiral group or to provide an asymmetric surface upon which the enantiomers may be directly separated. A great deal of the work in this area has been aimed at the separation of DL-amino acids. The succeeding sections will illustrate the techniques and supports that have been used to separate these and similar compounds.

Chemical resolutions, based on the conversion of a racemic mixture into a set of diastereoisomers, are among the most frequently used methods of separating enantiomers (Rogozhin and Davankov, 1968; Rogozhin, 1971 and Boyle 1971). Various chromatographic methods have also been applied to the separation of diastereoisomers—the polymer being used merely as a support. Some of these chromatographic methods, such as gas chromatography, are mainly analytical in scope (Bonner, 1972; Iwase and Murai, 1974; Iwase, 1974, 1975; Maestas and Morrow, 1976; Konig et al., 1977). High-pressure liquid chromatography (Furukawa et al., 1975; Furukawa et al., 1977; Sousa et al., 1978), Sephadexes (Popova et al., 1973; Yoneda and Yoshizawa, 1976) and ion-exchange chromatography (Manning and Moore, 1968) have also been used in their traditional roles to separate diastereomeric mixtures.

Naturally asymmetric polymers have been used to separate racemates directly, without prior derivatization. For example, cellulose (Weichert,

1970; Handes et al., 1974), acetylated cellulose (Hesse and Hagel, 1975), starch (Hess et al., 1978), pectic acid (Popova and Kratchanov, 1972), alginic acid (Kratchanov et al., 1969), alginic acid–silica gel (A.M. El Din Awad and O.M. El Din Awad, 1974) and agarose-supported bovine serum albumin (Stewart and Doherty, 1973) have all been used to separate various amino acids and other compounds. One report of the preferential absorption of L-phenylalanine by kaolin (Jackson, 1971) has been refuted (Bonner and Flores, 1974).

In addition to the naturally asymmetric polymers, polymers specifically designed to incorporate asymmetric units have been used to separate racemates. The asymmetric units either originated from natural sources or were designed to provide a host–guest environment.

In general, the asymmetric units originating from natural sources have been amino acids (which have been attached to polystyrenes), polysiloxanes, Sephadexes, and polyacrylamides. The resulting polymers have been used in gas chromatographic separations (Frank et al., 1978), liquid chromatographic separations (Suda et al., 1970; Baczuk et al., 1971; Blaschke, 1974; Blaschke and Schwanghart, 1976), and "ligand-exchange" chromatographic separations.

B. Resolution of Amino Acids by Ligand-Exchange Chromatography

For resolution of amino acids, the method of ligand-exchange chromatography has received by far the most emphasis. The utility of this chromatographic technique in areas other than the resolution of racemates has been reviewed (Davankov and Semechkin, 1977). Only results pertinent to the use of this chromatographic method in the resolution of racemates will be discussed here.

Complexation of a racemic mixture by multidentate metal ligands is a highly stereoselective process. The stereoselectivity arises from the dense packing of the ligands into the complex coordination sphere. The dense packing results from the fact that the chelating ligand has several binding sites participating in the formation of the asymmetric complex (Davankov and Mitchell, 1972; Davankov et al., 1971). Since there are diastereoisomeric possibilities inherent in the interaction of metal ions with α-amino acids, the idea has found useful applications in the resolution of racemates (Davankov and Semechkin, 1977).

The metal ions that have been used are those having unoccupied d-orbitals, thereby allowing asymmetric arrangements of ligands about the central metal core. Cu^{2+}, Fe^{3+}, Ni^{2+} and Zn^{2+} have been the most widely used of these ions.

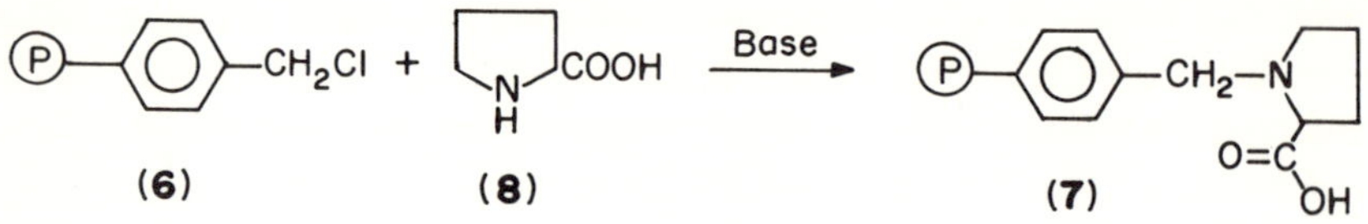

Scheme 10-3

The sorbents that have been used range from ones intentionally synthesized to contain multidentate ligands to ones with naturally asymmetric coordination sites. The most widely studied of the former are the micronet polystyrenes, containing proline and its derivatives (Davankov et al., 1973, 1974a,b; Semechkin et al., 1977; Davankov and Zolotarev, 1978a,b,c). The resins are prepared by reacting the chloromethylated material with proline or one of its derivatives (Scheme 10-3). Treatment of the resulting proline-containing polymer with solutions of metal ions, usually Cu^{2+}, results in the formation of a copper–proline chelate. This chelate is then reacted with a solution of the racemic amino acid to form the asymmetric complex (Fig. 10-3). The stability constants of the resin-bound copper–L-proline amino acid complex depends on the configuration of the amino acid bound in the second copper chelating site (for example, see Leach and Angelici, 1969; Snyder and Angelici, 1973). Increasing the concentration of ammonium ions in the external solution labilizes the complexes to different extents. Chromatographic separation of the amino acid enantiomers can be effected by passing a gradually increasing concentration of ammonium ions through a column containing the complexes. It has been found that many amino acid enantiomers may be separated by this procedure.

In addition to the proline polymers, polymers containing N-carboxymethyl-L-valine (Snyder et al., 1972), N-(β-hydroxyethyl)-D-propylenediaminotetraacetic acid (Humbel et al., 1970; Bernauer et al., 1971), L-histidine (Guetté et al., 1978), and N-benzyl-L-leucine (Tsuchida et al., 1976) have been used to separate racemates of amines and amino acids.

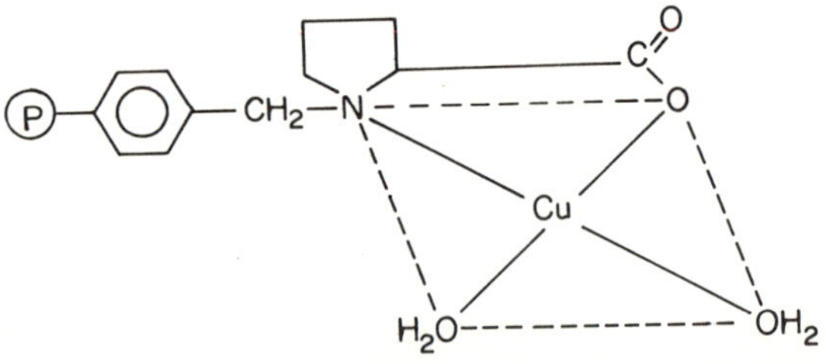

Fig. 10-3

III. Resolution of Racemates Using Polymeric Materials

The separation of kinetically inert octahedral complexes of Cr^{3+} and Co^{3+} on naturally asymmetric sorbents has been reviewed (Rogozhin and Davankov, 1968).

C. Resolution of Racemates Using Asymmetric Templates

Two recent developments have led to the preparation of polymers containing asymmetric templates. In a recent set of reports, the synthesis of an enzyme-like polymer was described (Wulff *et al.*, 1973; Wulff *et al.*, 1977a,b; Wulff, 1977; Wulff *et al.*, 1978). The polymer was prepared with functional groups juxtaposed in an exact, predetermined steric relationship by polymerizing monomers around an optically active template— either D-glyceric acid or 4-nitrophenyl-α-D-mannopyranoside-2,3,4,6-di-O-(4-vinylphenylboronate) (Scheme 10-4). The template was removed from the polmeric matrix leaving behind cavities that were asymmetric. The template molecules were found to bind to the resulting polymers more strongly than similarly shaped molecules.

This novel approach is based on resolution of a racemate by asymmetric cavities in the polymer rather than by the formation of diasteriomeric pairs of molecules. The asymmetry within the polymer is not permanent

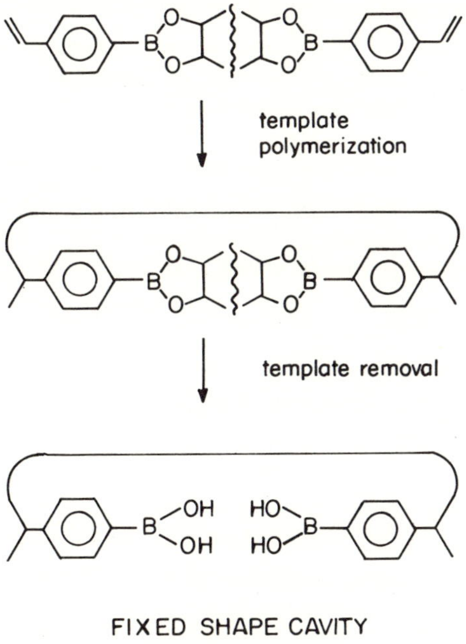

FIXED SHAPE CAVITY

Scheme 10-4

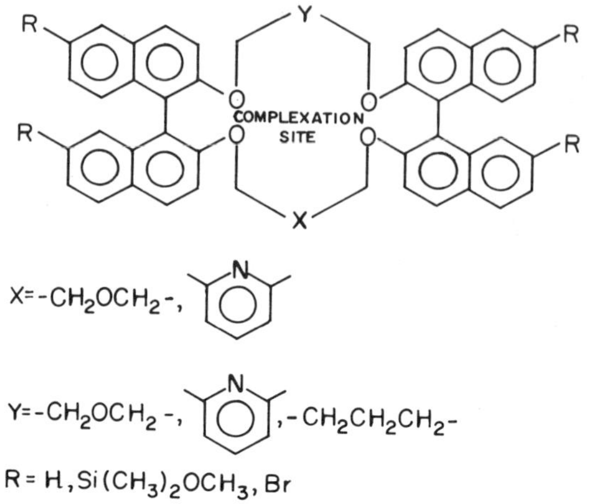

$X = -CH_2OCH_2-$, [pyridine]

$Y = -CH_2OCH_2-$, [pyridine], $-CH_2CH_2CH_2-$

$R = H, Si(CH_3)_2OCH_3, Br$

Fig. 10-4

however. Solvent-swollen polymers suffer the loss of the asymmetric cavities. This is the result of the internal flexibility that can be shown to occur in polymers (see Chapter 4). Recent experiments (Wulff et al., 1978) have shown that asymmetric areas formed within the polymer chains do not arise solely from the three dimensional structure generated by crosslinking the polymer.

The alternate system of template-containing polymers is that developed by Cram and associates. It relies on the preparation of a set of host–guest molecules (Fig. 10-4) that are subsequently attached to the polymer matrix (Sogah and Cram, 1976; Sousa et al., 1978). Polymer matrices that have been used include a macroreticular polystyrene and silica gel. High separation factors could be achieved for amino acid esters containing bulky side-chain substituents.

The host–guest system requires that the host system be designed to accommodate the guest (i.e., tailored to fit the guest). The best hosts require a three-point attachment (i.e., to be trilocular) and the degree of chiral recognition is influenced by bulk solvation factors, temperature, and counter-ion interactions. All these factors have to be taken into consideration when the design of a host is contemplated. The subtle influence of each on the interaction of the host and guest needs to be the subject of continuing work.

REFERENCES

Baczuk, R. J., Landrum, G. K., Dubois, D. J., and Dehm, H. C. (1971). J. Chromatogr. **60,** 351.

References

Bernauer, K., Jeanneret, M. F., and Vonderschmitt, D. (1971). *Helv. Chim. Acta.* **54**, 297.
Blaschke, G. (1974). *Chem. Ber.* **107**, 237.
Blaschke, G., and Schwanghart, A. D. (1976). *Chem. Ber.* **109**, 1967.
Bonner, W. A. (1972). *J. Chromatogr. Sci.* **10**, 159.
Bonner, W. A., and Flores, J. (1973). *Curr. Mod. Biol.* **5**, 103.
Boyle, P. D. (1971). *Q. Rev. Chem. Soc.* 326.
Buss, D. R., and Vermeulen, T. (1968). *Ind. Eng. Chem.* **60**, 12.
Cramer, F., Helbig, R., Hettler, H., Scheit, K. H., and Seliger, H. (1966).*Angew. Chem. Int. Ed.* **6**, 601.
Davankov, V. A., and Mitchell, P. R. (1972). *J. Chem. Soc.* (**C**) 1012.
Davankov, V. A., and Semechkin, A. V. (1977). *J. Chromatogr.* **141**, 313.
Davankov, V. A., and Zolotarev, Yu. A. (1978a). *J. Chromatogr.* **155**, 285.
Davankov, V. A., and Zolotarev, Yu. A. (1978b). *J. Chromatogr.* **155**, 295.
Davankov, V. A., and Zolotarev, Yu. A. (1978c). *J. Chromatogr.* **155**, 303.
Davankov, V. A., Rogozhin, S. V., and Kurgahov, A. A. (1971). *Izv. Akad. Nauk. Uzb. SSR, Ser. Khim.* 204.
Davankov, V. A., Rogozhin, S. V., Semechkin, A. V., and Sachkova, T. P. (1973). *J. Chromatogr.* **82**, 359.
Davankov, V. A., Rogozhin, S. V., Semechkin, A. V., Baranov, V. A., and Sannikova, G. S. (1974a). *J. Chromatogr.* **93**, 363.
Davankov, V. A., Rogozhin, S. V., and Semechkin, A. V. (1974b).*J. Chromatogr.* **91**, 493.
El Din Awad, A. M., and El Din Awad, O. M. (1974). *J. Chromatogr.* **93**, 393.
Frank, H., Nicholson, G. J., and Bayer, E. (1978). *Angew. Chem. Int. Ed.* **17**, 363.
Furukawa, H., Sakikibara, E., Kamei, A., and Ito, K. (1975). *Chem. Pharm. Bull. (Japan)* **23**, 1625.
Furukawa, H., Mori, Y., Takeuchi, Y., and Ito, K. (1977). *J. Chromatogr.* **136**, 428.
Guetté, J.-P., Guetté, M., Sepulchre, M.-O., and Blanchard, J.-M. (1978). *C. R. Acad. Sci. Paris* **287C**, 589.
Handes, L. V., Kido, R., and Schmaeler, M. (1974). *Prep. Biochem.* **4**, 47.
Hayatsu, H., and Khorana, H. G. (1966). *J. Am. Chem. Soc.* **88**, 3182.
Hayatsu, H., and Khorana, H. G. (1967). *J. Am. Chem. Soc.* **89**, 3880.
Hermann, K., and Wynberg, H. (1977). *Helv. Chim. Acta.* **60**, 2208.
Hess, H., Burger, G., and Musso, H. (1978). *Angew. Chem. Int. Ed.* **17**, 612.
Hesse, G., and Hagel, R. (1975). *Liebigs. Ann. Chem.,* 996.
Humbel, F., Vonderschmitt, D., and Bernauer, K. (1970). *Helv. Chim. Acta.* **53**, 1983.
Inoue, S., Ohashi, S., and Unno, Y. (1972). *Polymer J.* **3**, 611.
Iwase, H. (1974). *Chem. Pharm. Bull.* **22**, 1663, 2075.
Iwase, H. (1975). *Chem. Pharm. Bull.* **23**, 1604.
Iwase, H., and Murai, A. (1974). *Chem. Pharm. Bull.* **22**, 1455.
Jackson, T. A. (1971). *Experientia* **27**, 242.
Katchanov, C. G., Popova, M. I., Obretenov, T. Z., and Ivanov, N. (1969).*J. Chromatogr.* **43**, 66.
Kawana, M., and Emoto, S. (1972). *Tetrahedron Lett.,* 4855.
Kawana, M., and Emoto, S. (1974). *Bull. Chem. Soc. J.* **46**, 160.
Kobayashi, N., and Iwai, K. (1978). *J. Am. Chem. Soc.* **100**, 7071.
Konig, W. A., Rahn, W., and Eyem, J. (1977). *J. Chromatogr.* **133**, 141.
Leach, B. E., and Angelici, R. J. (1969). *J. Am. Chem. Soc.* **91**, 6296.
Losse, G., and Kuntze, K. (1970). *Z. Chem.* **10**, 22.
Maestas, P. D., and Morrow, C. S. (1976). *Tetrahedron Lett.,* 1047.
Manning, J. M., and Moore, S. (1968). *J. Biol. Chem.* **243**, 5591.
Popova, M. I., and Kratchanov, C. G. (1972). *J. Chromatogr.* **72**, 192.

Popova, M. I., Kratchanov, C. G., and Kuntcheva, M. J. (1973). *J. Chromatogr.* **87**, 581.
Rogozhin, S. V. (1971). *Vestn. Akad. Nauk Kaz SSR* **40**, 56.
Rogozhin, S. V., and Davankov, V. A. (1968). *Russ. Chem. Rev.* **37**, 565.
Semechkin, A. V., Rogozhin, S. U., and Davankov, V. A. (1977). *J. Chromatogr.* **131**, 65.
Snyder, R. V., and Angelici, R. J. (1973). *J. Inorg. Nucl. Chem.* **35**, 523.
Snyder, R. V., Angelici, R. J., and Meck, R. B. (1972). *J. Am. Chem. Soc.* **94**, 2660.
Sogah, G. D. Y., and Cram, D. J. (1976). *J. Am. Chem. Soc.* **98**, 3038.
Sousa, L. R., Sogah, G. D. Y., Hoffman, D. H., and Cram, D. J. (1978). *J. Am. Chem. Soc.* **100**, 4569.
Stewart, K. K., and Doherty, R. F. (1973). *Proc. Natl. Acad. Sci. U.S.A.* **70**, 2850.
Suda, H., Hosono, Y., Hosokawa, Y., and Seto, T. (1970). *J. Chem. Soc. J., Ind. Chem. Section,* **73**, 1250.
Tsuchia, E., Nishikawa, H., and Terada, E. (1976). *Eur. Polym. J.* **12**, 611.
Ueyanagi, K., and Inoue, S. (1977). *Makromol. Chem.* **178**, 235.
Valentine, D., Jr., and Scott, J. W. (1978). *Synthesis,* 329.
Weichert, R. (1970). *Ark. Kemi.* **31**, 517.
Worster, P. M., McArthur, C. R., and Leznoff, C. C. (1979). *Angew. Chem. Int. Ed.* **18**, 221.
Wulff, G. (1977). *Nachr. Chem. Tech.* **25**, 239.
Wulff, G., Sarhan, A., and Zabrocki, K. (1973). *Tetrahedron Lett.,* 4329.
Wulff, G., Vesper, W., Grobe-Einsler, R., and Sarhan, A. (1977a). *Makromol. Chem.* **178**, 2799.
Wulff, G., Grobe-Einsler, R., Vesper W., and Sarhan, A. (1977b). *Makromol. Chem.* **178**, 2817.
Wulff, G., Zabrocki, K., and Hohn, J. (1978). *Angew. Chem. Int. Ed.* **17**, 535.
Yoneda, H. and Yoshigawa, T. (1976). *Chem. Lett.,* 707.

11 Application of Polymeric Supports in Identifying Reaction Intermediates

I.	Introduction	165
II.	General Strategy Used for Trapping Reaction Intermediates Employing Polymeric Supports	166
	A. Cyclobutadiene Intermediates	166
	B. Benzyne Intermediate Trapping	167
	C. Reaction Intermediates in Transacylation Reactions	167
	D. Reaction Intermediates in Phosphorylation Reactions	169
	E. Enol Intermediates in C-Acylation Reactions	170
	F. Reaction Mechanisms of the Morgan-Elson Assay	171
	G. The Fries Rearrangement	172
	H. The Conant-Swan Reaction	172
	References	172

I. INTRODUCTION

The formation of certain short-lived intermediates during a reaction has generally been shown spectroscopically, but trapping of the intermediate provides more convincing proof of the reaction mechanism. Polymeric supports have found a novel application in the trapping of these intermediates, thereby throwing light on the reaction mechanism.

Complete site-separation does not occur when the functional groups are bound to resins having low cross-linking (Crowley and Rapoport, 1976) (see Chapter 3), and intrapolymeric reactions can become a reality when loading is high. In addition to this, it is observed that reactive functional groups bound on different resin beads do not react (Rebek and Gavina, 1974, 1975a,b; Rebek et al., 1975a,b). It appears that in a bis-heterogeneous mixture (reactants anchored on different polymers), the reactive sites are unable to come into proximity, i.e., close enough to bring about an intrapolymer chemical reaction.

II. GENERAL STRATEGY USED FOR TRAPPING REACTION INTERMEDIATES EMPLOYING POLYMERIC SUPPORTS

The general setup for trapping a reaction intermediate is shown in Scheme 11-1. The solid phase (I) (resin precursor) can be activated by a suitable reaction to give the low-molecular-weight reaction intermediate, which is soluble and can pass into the liquid phase (solvent). The liquid phase is also in contact with another solid phase (II) (resin trap) so that the reaction intermediate reacts at once with the second resin to give an adduct. The trapped intermediate can either be identified on the resin itself or after cleaving it from the resin. This "three-phase" approach is best illustrated by the following examples.

A. Cyclobutadiene Intermediates

Rebek and Gavina (1974, 1975a) devised an interesting trap for cyclobutadiene, consisting of the resin (2) functionalized with a maleimide group. Another resin (1), functionalized with a complex of tricarbonylcyclobutadiene–iron and o-phenanthroline, served as the precursor of the cyclobutadiene. Highly reactive cyclobutadiene was generated by oxidation of iron(II) in the resin complex by either ceric ion or pyridine oxide (Scheme 11-2). The two resins (1 and 2) were suspended in the same solvent. The maleimide resin, acting as a dienophile, reacted with the cyclobutadiene to give a Diels–Alder type adduct resin (3). The product, adduct resin 3, was separated from the reaction mixture by floatation and screening, and the adduct group was cleaved from the resin by methylamine to yield resin 4, N,N'-dimethylmaleimide (5) and its cyclobutadiene adduct (6). This approach is related to a suggestion made earlier in a Bio-Rad Bulletin (1973).

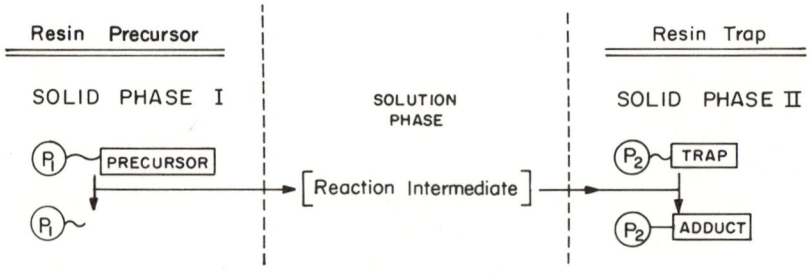

Scheme 11-1

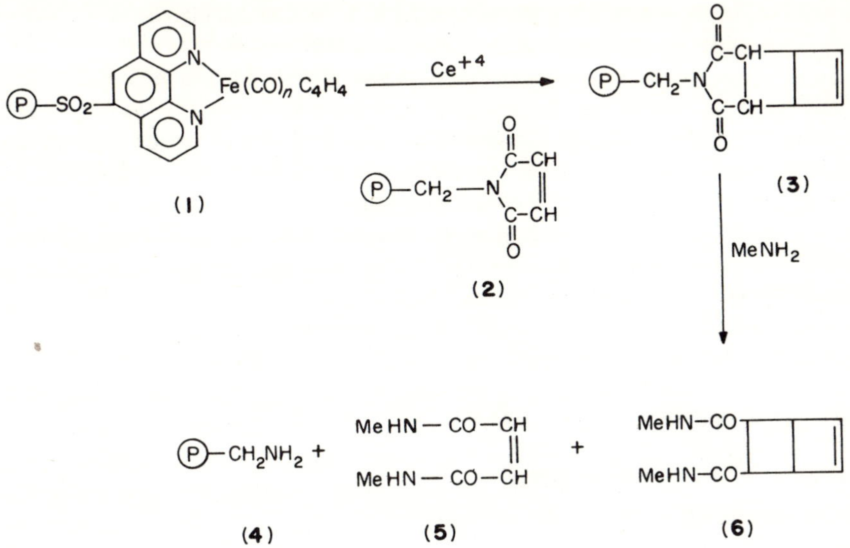

II. General Strategy Used for Trapping Reaction Intermediates

Scheme 11-2

B. Benzyne Intermediate Trapping

Investigations of polymer chain flexibility have been reported (Jayalekshamy and Mazur, 1976) (see Chapter 3). Generation of the benzyne intermediate (**8**, Scheme 11-3) from the benzotriazole (**7**) and titration with the aryne trapping reagent, tetracyclone, showed that the aryne was site-isolated for about one minute. Such studies would suggest that it might be possible to investigate the formation and the physicochemical properties of site-isolated, reactive intermediates using polymeric supports as a "matrix isolation" medium.

C. Reaction Intermediates in Transacylation Reactions

Another interesting application of a three-phase approach to identifying N-acylimidazole as the intermediate in the imidazole-catalyzed hydrolysis

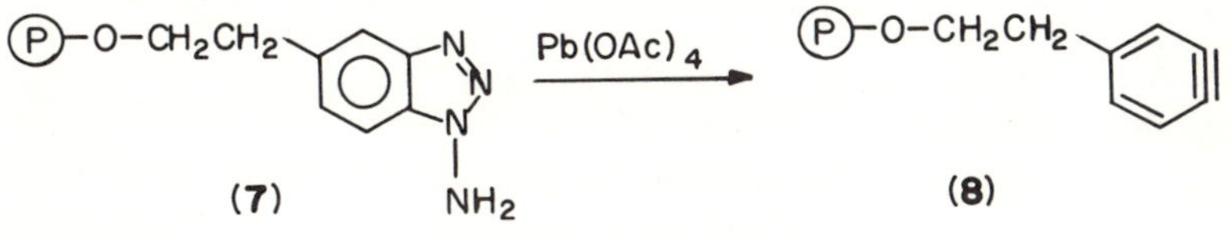

Scheme 11-3

of *o*-nitrophenyl esters has been made by Rebek *et al.* (1975). The formation of acylimidazoles as reaction intermediates has been observed spectrophotometrically (Holmquist and Bruice, 1969). The resin precursor in this case was an active ester (resin *o*-nitrophenyl ester) (**9a**) prepared according to the method of Kalir *et al.* (1974). In the presence of imidazole, which acted as a nucleophilic catalyst for ester hydrolysis, the acyl group was transferred to the imidazole to give *N*-acylimidazole. The second resin phase (a benzylamine resin) (**10**), prepared via Gabriel synthesis from a Merrifield resin (Weinshenker and Shen, 1972), acted as a trap for the acyl group (Scheme 11-4).

Nucleophilic catalysis of the acyl transfer reaction could be detected by using ^{14}C-labeled glycine and then by scintillation counting of the two resins. Other catalysts effective in the acyl transfer reaction have efficiencies in the order: imidazole > *N*-hydroxybenzotriazole > *N*-hydroxyphthalimide > *N*-hydroxysuccinimide.

Intermediates in elimination reactions could also be detected using this three-phase test on the carbamylation reaction (R = NHC_6H_5 in resin **9**). The reaction was catalyzed by triethylamine or a "proton sponge" [1,8-bis(dimethylamino)naphthalene]. The intermediate, phenyl isocyanate, was readily observed (infrared spectra) between the two resins and was trapped as the urea derivative of the resin (**11**).

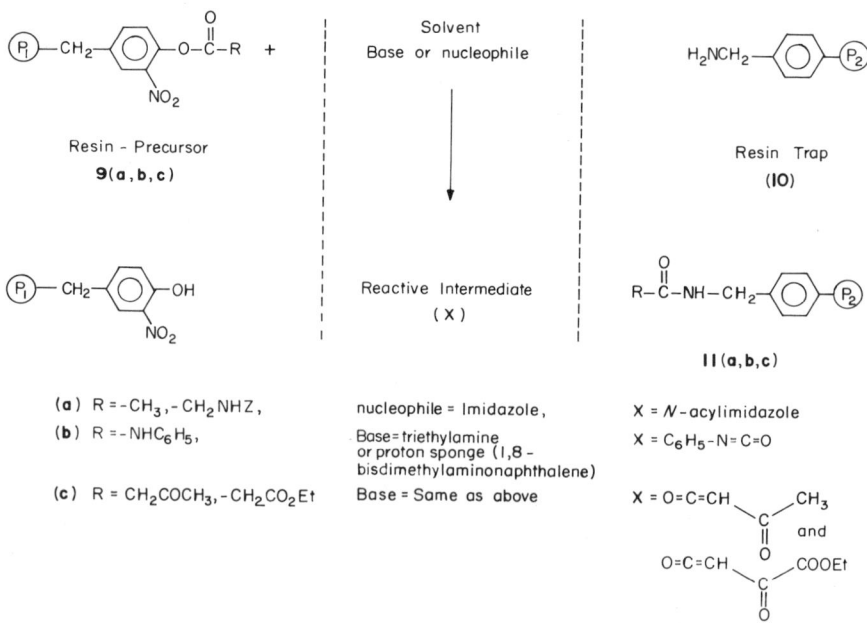

Scheme 11-4

Similarly, the acyl transfer from the resin β-keto ester (9) (R = CH_2CO_2Et and R = CH_2COCH_3) was supposed to take place via the intermediate formation of a ketene. Although the formation of ketene could not be demonstrated in this reaction, it could still be trapped on the amine–resin. The formation of a ketene as an acylating species was assumed by analogy to Bruice's E_1cB reaction (Holmquist and Bruice, 1969).

D. Reaction Intermediates in Phosphorylation Reactions

The mechanism of the phosphorylating species involved in dicyclohexylcarbodiimide (DDC)-promoted nucleic acid synthesis has been investigated by using a polymeric support (Blackburn et al., 1966). Polystyrenes containing hydroxyl groups (benzyl and phenethyl alcohols) were phosphorylated with DCC and β-cyanoethylphosphate, whereas the equivalent polymers containing the phosphate monoester (Ⓟ)–$(CH_2)_nOPO_3H$, $n = 1,2$) did not yield any diester with DCC and a soluble alcohol. The results suggested that, in contrast to the mesitylene sulfonyl chloride-promoted condensation, DDC-promoted condensation involved a trimetaphosphate intermediate.

Rebek and Gavina (1975a,b) and Rebek et al. (1978) also used the previous setup for detecting reaction intermediates in phosphorylation reactions. The precursor resin consisted of a phosphorylated

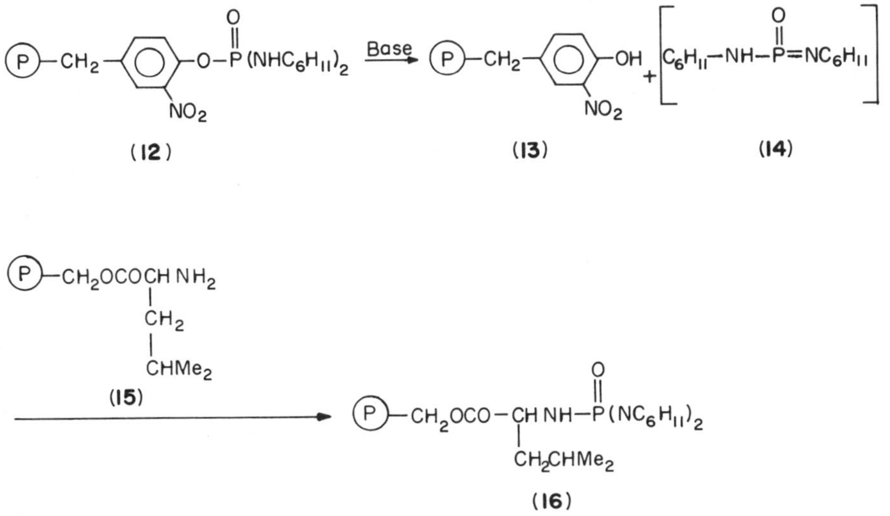

Ⓟ–CH_2–〈O〉(NO_2)–O–P(=O)($NHC_6H_{11})_2$ →(Base) Ⓟ–CH_2–〈O〉(NO_2)–OH + [C_6H_{11}–NH–P(=O)=NC_6H_{11}]

(12)   (13)   (14)

Ⓟ–$CH_2OCOCHNH_2$ | CH_2 | $CHMe_2$

(15)

⟶ Ⓟ–CH_2OCO–CHNH–P(=O)($NC_6H_{11})_2$ | CH_2CHMe_2

(16)

Scheme 11-5

o-nitrophenol polymer (12) that was prepared by reaction of N,N'-dicyclohexylphosphorodiamidic chloride with polymeric o-nitrophenol (13). The resin trap was an L-leucine ester attached to a Merrifield resin (15) in the usual manner. The reactive intermediate (14) was generated by adding a base (proton sponge) which N-phosphorylated the L-leucine resin. The product, N-leucylphosphorodiamide was finally identified by cleaving it from the resin trap (16) by transesterification (Scheme 11-5). This experiment, while illustrating the existence of a "reaction intermediate," did not allow its exact nature to be determined.

E. Enol Intermediates in C-Acylation Reactions

Striking effects of two highly reactive groups, rendered mutually inert by anchoring on polymers, were demonstrated by making a polymeric tritylithium. This was mixed in a solvent with a polymeric active ester and no observable reaction took place. However, upon addition of an enolizable ketone, reaction occurred immediately (Cohen et al., 1977). In this case, the soluble enolizable ketone was first enolized by the polymeric base and then acylated by the polymeric ester (Scheme 11-6). Here the formation of the enol as the reactive intermediate in the reaction was established. This experiment is basically different from those conducted by Rebek's group, i.e., the reactive intermediate was formed from a low-molecular-weight compound (ketone) by reaction with a polymeric base.

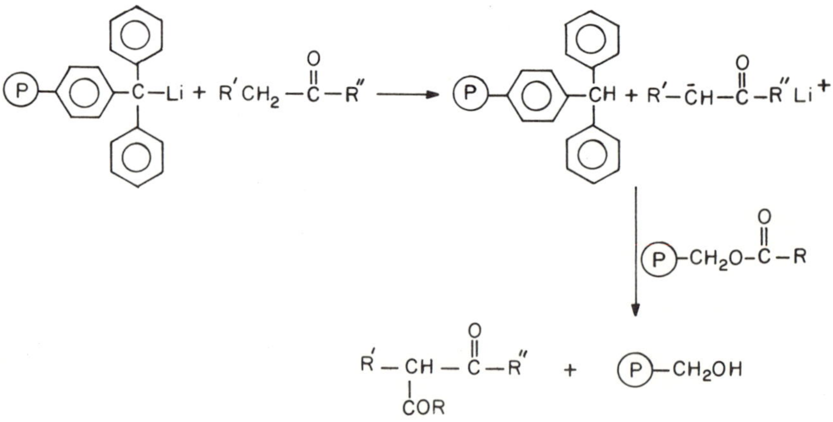

Scheme 11-6.

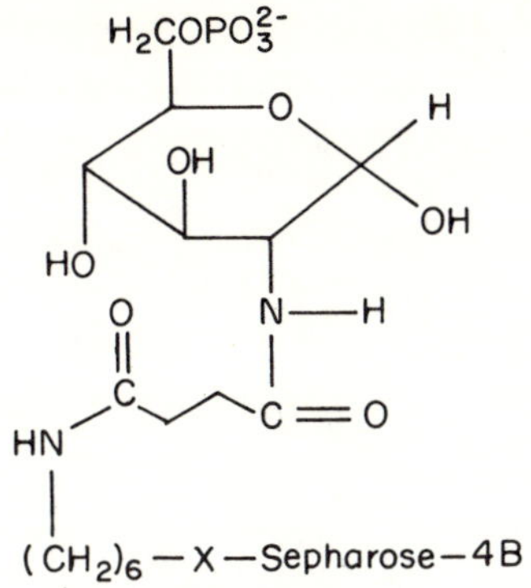

II. General Strategy Used for Trapping Reaction Intermediates 171

Fig. 11-1

F. Reaction Mechanisms of the Morgan-Elson Assay

A useful application of compounds bound to solid supports for testing the mechanism of a reaction has been reported by Benson (1975). During a sequence of reactions, the fate of a functional group or side chain is difficult to determine. In the case in point, the Morgan-Elson Assay (1934), a 2-acylamido-2-deoxy-D-hexose sugar was bound to succinylated Sepharose-4B (Fig. 11-1) or CPG-glass 7 (Fig. 11-2) and a modified Morgan-Elson reaction was carried out. The reaction involves an elimination, accompanied by the formation of a chromophore. After formation, the colored product could either remain linked to the polymer or could pass into the solution, depending upon whether the resin–compound linkage (aminoacyl) was cleaved during the reaction. It was conclusively proved, by carrying out the reaction on a polymer-bound sugar, that elimination of the aminoacyl group did not take place during the reaction.

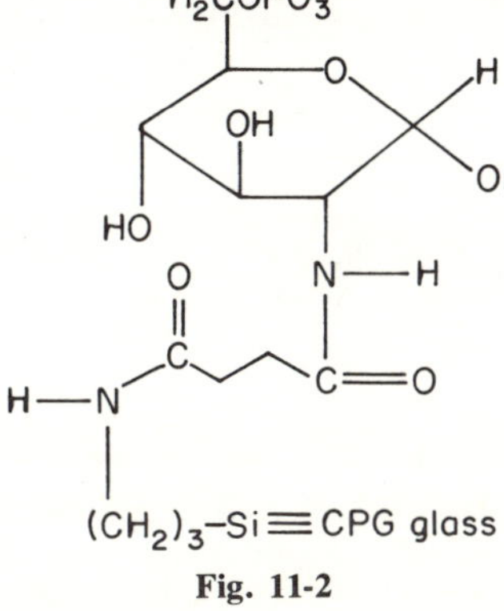

Fig. 11-2

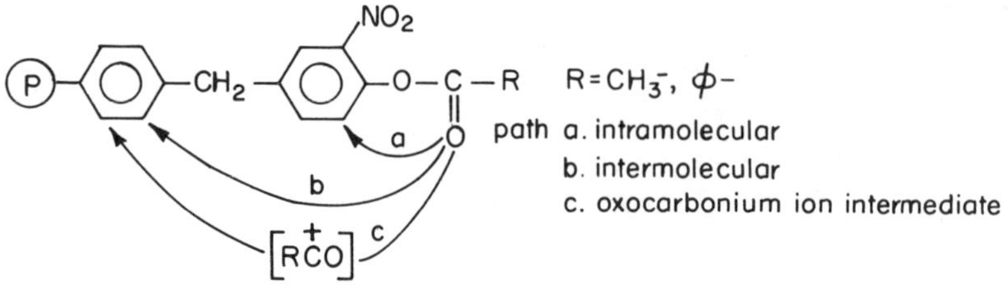

Scheme 11-7

G. The Fries Rearrangement

Acetoxy and benzyloxy esters of 4-hydroxy-3-nitrobenzylated polystyrene and 3-polystyrylmethyl-8-quinolinol were subjected to Fries rearrangement conditions (1.8 M AlCl$_3$ in nitrobenzene) (Warshawsky *et al.*, 1978) (Scheme 11-7). The intermolecular rearrangement pathway, or the oxocarbonium ion pathway, was favored because of the finding that the acyl group was transferred to the nonphenolic ring exclusively. Interpolymeric reactions, using two polymers which could be bead-size separated, implied that the oxocarbonium ion route was preferred.

H. The Conant-Swan Reaction

Polymers were used to test the intermediacy of a phosphorylated compound in phosphorylations observed during base-promoted decomposition of β-halo phosphonate (Conant-Swan reaction) (Rebek *et al.*, 1977, 1978). Polymer-bound β-chlorophosphonate was prepared and decomposed in the presence of an amine-bearing polymer of different bead size. It was shown that phosphorylation of the amine-bearing polymer occurred, indicating the intermediacy in solution of a phosphorylating species.

REFERENCES

Benson, R. L. (1975). *J. Org. Chem.* **40**, 1647.
Bio-Rad (1973). "Bio Beads for Organic Synthesis," Bio Rad Labs, Richmond, California.
Blackburn, G. M., Brown, M. J., and Harris, M. R. (1966). *J. Chem. Soc. Chem. Commu.*, 611.
Cohen, B. J., Kraus, M. A., and Patchornik, A. (1977). *J. Am. Chem. Soc.* **99**, 4165.
Crowley, J. I., and Rapoport, H. (1976). *Acc. Chem. Res.* **9**, 135.
Holmquist, B., and Bruice, T. C. (1969). *J. Am. Chem. Soc.* **91**, 2982, 2985, 2993, 3003.
Jayalekshamy, P., and Mazur, S. (1976). *J. Am. Chem. Soc.* **98**, 6710.
Kalir, R., Fridkin, M., and Patchornik, A. (1974). *Eur. J. Biochem.* **42**, 151.
Morgan, W. T. J., and Elson, C. L. A. (1934). *Biochem. J.* **28**, 988.

References

Rebek, J., and Gavina, F. (1974). *J. Am. Chem. Soc.* **96,** 7112.
Rebek, J., and Gavina, F. (1975a). *J. Am. Chem. Soc.* **97,** 1591.
Rebek, J., and Gavina, F. (1975b). *J. Am. Chem. Soc.* **97,** 3453.
Rebek, J., Brown, D., and Zimmerman, S. (1975). *J. Am. Chem. Soc.* **97,** 454.
Rebek, J., Gavina, F., and Navarro, C. (1977). *Tetrahedron Lett.,* 3021.
Rebek, J., Gavina, F., and Navarro, C. (1978). *J. Am. Chem. Soc.* **100,** 8113.
Weinshenker, N. M., and Shen, C. (1972). *Tetrahedron Lett.,* 3281.
Warshawsky, A., Kalir, R., and Patchornik, A. (1978). *J. Am. Chem. Soc.* **100,** 4544.

12 Polymer-Bound Reagents

I.	Introduction	174
II.	Polymeric Oxidizing Reagents	175
	A. Peracids	175
	B. Chromium-Containing Reagents	175
	C. Silver Carbonate—Celite	176
	D. A Polymeric Thioanisolyl Resin for Selective Oxidations and Homologation	177
	E. Miscellaneous Oxidizing Agents	179
III.	Polymeric Oxidation–Reduction Reagents	180
IV.	Polymeric Reducing Reagents	181
V.	Polymeric Group Transfer Reagents	183
	A. Halogenating Agents	184
	B. Acylating Agents	187
	C. Alkylating Agents	188
	D. Polymeric Nucleophiles	188
	E. Wittig and Ylid Reagents	189
	F. Tosyl Azides as Diazo Transfer Reagents	191
VI.	Polymeric Coupling Agents	191
	A. Carbodiimides	191
	B. Sulfonyl Chlorides	192
	C. Reagent Based on EEDQ	193
	D. Reagents Using Polymer-Supported Triphenylphosphine	193
VII.	Miscellaneous Reagents	194
	References	195

I. INTRODUCTION

Polymeric reagents suitable for nonsequential single-step reactions have been utilized by several research groups. Even though the utility of such reagents is limited to small-scale syntheses, the concept of immobilization of a chemical reagent in a polymeric form has been demonstrated. One of the values of such reagents, in common with solid-phase methods, is the simplification of separation procedures and, as has been demonstrated in many cases, the regeneration of reagents. Increased selectivity and other advantages have also been observed.

II. POLYMERIC OXIDIZING REAGENTS

Polymeric reagents have been used to carry out many reactions. In general the reagents may be classified as oxidizing reagents, redox reagents, reducing agents, and group transfer reagents.

II. POLYMERIC OXIDIZING REAGENTS

A. Peracids

One of the earliest reports of the use of an insoluble polymeric reagent relates to the peracids (Helfferich and Luten, 1964). The polymeric peracids were prepared from a copolymer containing both carboxylic and sulfonic acid groups. Alkenes were oxidized to α-glycols (sulfonic acid-catalyzed opening of initially formed epoxides to the glycol probably took place). The polymers were fragile and could be used only for a few oxidation–reduction cycles. These resins were called "oxygen transfer resins." Takagi (1967) reported the preparation of peracids based on polymethacrylic acid (Amberlite XE-89). Though effective in epoxidations, these resins were unstable (exploding on impact) and recycling resulted in a sharp decrease in polymer capacity.

Harrison and Hodge (1974), and Frechet and Haque (1975) reported the synthesis of polymeric peracids based on polyvinylbenzoic acid. The polymer peracid (4) can be synthesized from the corresponding polymer acid (1), acid chloride (2), or aldehyde (3) (Scheme 12-1). High yields (97%) of the peracid were obtained using methanesulfonic acid and 70% H_2O_2. Both a macroreticular and a swellable resin could be used. The reaction using the macroreticular resin was independent of solvent (i.e., swelling) whereas the reaction using the swellable resin was smooth only in solvents that caused the polymer to swell. Both types of polymers could be regenerated, but there was a variation in the oxidation capacity (iodometric titration) of the polymers upon recycling.

One aim of the workers in the field of solid-phase synthesis has been the development of reagents that can be used to carry out reactions in a column and that can be regenerated in a single step. To some extent this has been achieved in the epoxidation of the alkenes, opening up the possibility of automating the process.

B. Chromium-Containing Reagents

Polymer-supported chromic acid (Amberlyst A-26, $HCrO_4^-$-form) has been used to oxidize primary and secondary alcohols to carbonyl compounds (Cainelli et al., 1976a). The same reagent was used to prepare

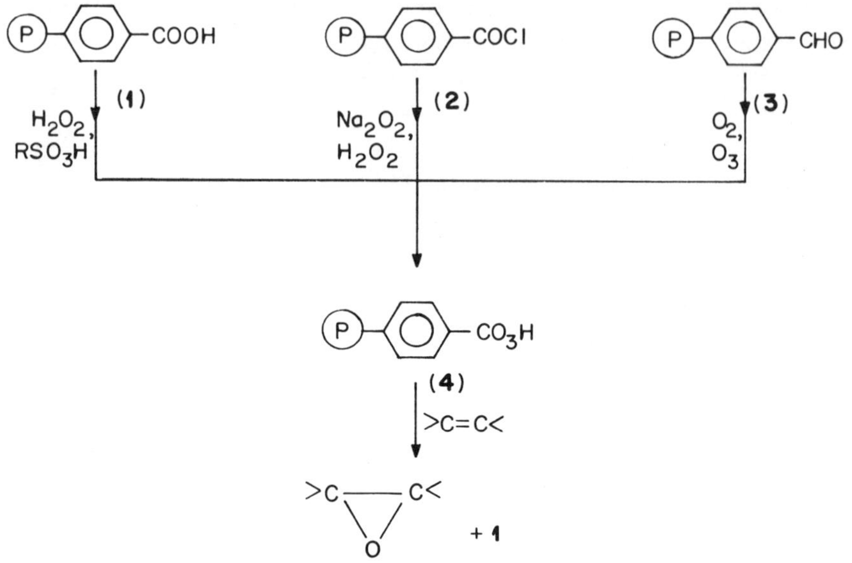

Scheme 12-1

aldehydes and ketones from allylic and benzylic halides (Cardillo *et al.*, 1976). The resin is now commercially available (Fluka and Polysciences).

Chemisorbed chromyl chloride (CrO_2Cl_2 on silica–alumina) has been used for the oxidation of primary and secondary alcohols in good to excellent yields (75–100%) (San Filippo, Jr., and Chern, 1977).

An insertion complex of chromium trioxide and graphite was also shown to have oxidizing properties (Lalancette *et al.*, 1972). Primary alcohols were oxidized to aldehydes whereas secondary alcohols remained untouched. Recent results have suggested that the oxidizing agent is a surface deposit of Cr_3O_8 (Ebert *et al.*, 1974). The complex is commercially available (Seloxcette, Alfa Division, Ventron Corp.).

Polyvinylpyridinium chlorochromate has been prepared by reacting cross-linked polyvinylpyridine with CrO_3 in acid solution (Frechet *et al.*, 1978). It has been shown to be similar in oxidizing properties to the polymer-supported chromic acid polymer previously mentioned (Cainelli *et al.*, 1976a). It could be readily regenerated.

C. Silver Carbonate—Celite

Silver carbonate supported on Celite (Fetizon's reagent) has found wide application in organic chemistry (Fieser and Fieser, 1977). It has been used mainly for the oxidation of alcohols and lactones for the oxidative

II. Polymeric Oxidizing Reagents

couplings of phenols, and for improved Konigs-Knorr coupling reactions of glycosides. Several studies since the initial report (Fetizon and Golfier, 1968) have dealt with the mechanistic aspects of the reaction (Fetizon et al., 1972; Kakis et al., 1974). The data support a mechanism (Scheme 12-2) where there is reversible absorption of the alcohol on the oxidant surface so that HCOH groups are coplanar. A subsequent irreversible homolytic shift of electrons results in the production of hydrogen ion, silver atoms, carbon dioxide, and water.

D. A Polymeric Thioanisolyl Resin for Selective Oxidations and Homologation

Polystyrene can be functionalized to incorporate a methyl sulfide group (Tanimoto et al., 1967; Farrall and Frechet, 1976). The resulting polymer (**5**) (Scheme 12-3), upon reaction with chlorine in the presence of triethylamine, forms an S-chlorosulfonium chloride resin (**6**). This chlorinated thioanisole polymer acts as a selective oxidant for alcohols (Crosby et al., 1975).

When α,ω-diols (**7**) were oxidized with the chlorinated thioanisole polymeric reagent, monoaldehydes (**8**) were formed in higher yields (50.2%) than the dialdehydes (**9**) (2.2%). This shows that isolation of polymeric functional groups has taken place. Since the degree of cross-linking was low (3%), the site-isolation was attributed to repulsion of similarly charged groups on the polymer.

Another application of this resin was made in an homologation reaction. This was an extension of Corey and Jautelat's (1968) reaction, originally

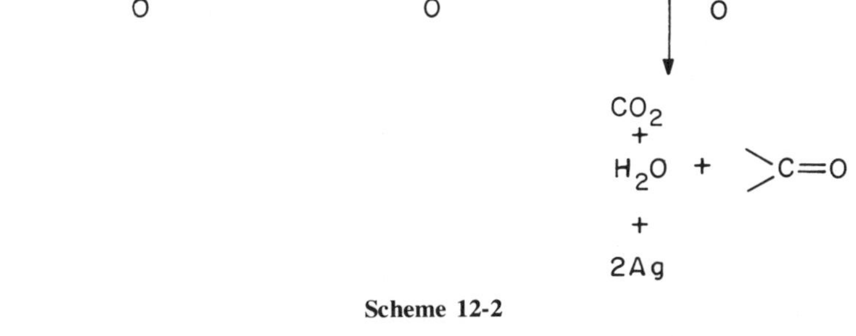

Scheme 12-2

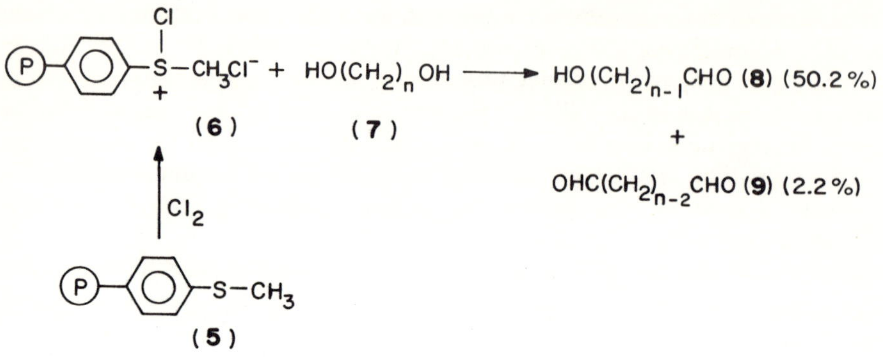

Scheme 12-3

developed by using the low-molecular-weight reagent, phenylthiomethyllithium. The corresponding polymeric reagent was prepared (Crosby and Kato, 1977) in an analogous way (Scheme 12-4). The polymeric anionic reagent was reacted with iodooctane at −70°C, and the cleavage of the higher homologs was carried out via quaternization with CH_3I, followed by cleavage with NaI (Scheme 12-4). It can be seen that the parent polymeric thioanisole reagent regenerates during the cleavage and may thus be reused.

In order to examine whether or not matrix isolation of the reactive functional groups had taken place in this case, the reaction was carried out on α,ω-diiodobutane. Surprisingly, 1,6-diiodohexane (a dual homolog) was formed in high yield (64%) as compared with only a 19% yield of 1,5-diiodopentane (the mono-homolog) indicating that the bispolymeric

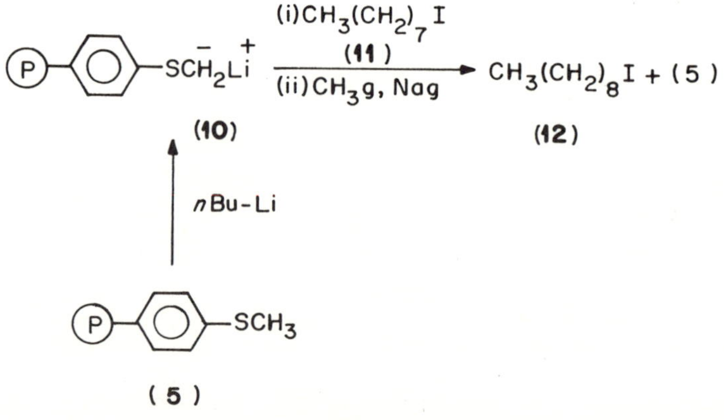

Scheme 12-4

reaction had taken place to a large extent. Thus, site-isolation was only partly achieved in the reaction.

With the polymeric chlorosulfonium resin and the thiomethyllithium resin, the experimental conditions for the reaction, e.g., the degree of cross-linking in the resin, solvents (which determine the swelling of the resin), concentration of functional groups, reaction time and temperature, were identical; yet the selectivities of the reagents for mono- and bis-reaction were completely reversed. This difference was thought to result from lithium counterion ionic clustering in the case of anionic resin and, thus, high concentrations of reactive groups. Similar behavior was also observed in the case of other polyelectrolytes (Oosawa, 1971; Flory, 1953; Patterson, 1971).

E. Miscellaneous Oxidizing Agents

N-Chloropolyamide derived from Nylon-6,6 has also been prepared and used for oxidation of several organic and inorganic substrates (Kaczmer, 1973). Presumably the reaction proceeds via formation of hypochlorite [Eq. (1)].

$$-CO(CH_2)_4CON(CH_2)_4-N- \xrightarrow{H_2O} -CO(CH_2)_4N(CH_2)_4N- + 2\ HOCl \quad (1)$$
$$\quad\quad\quad\ |\quad\quad\quad\quad\quad |\quad\quad\quad\quad\quad\quad\quad\quad\quad\quad |\quad\quad\quad\quad\ |$$
$$\quad\quad\quad\ Cl\quad\quad\quad\quad\ Cl\quad\quad\quad\quad\quad\quad\quad\quad\quad\quad H\quad\quad\quad\ H$$

Oxidation of aromatic sulfides with N-chloronylon-6,6 in the presence of optically active alcohols has been reported to give good yields of the sulfoxide and a limited induction of asymmetry (Sato *et al.*, 1977).

It is surprising that oxidations using polymeric NBS (N-bromosuccinimide) were not thoroughly investigated by earlier workers. It has been found that co(polyethylene-N-bromomaleimide) is a useful reagent for selective oxidation of alcohols (Sahni, 1977). Since allylic or benzylic brominations with NBS are carried in a heterophase and since both NBS and the by-product (succinimide) are insoluble in the solvent (CCl_4), product separation offers little problem. On the other hand, the oxidation reaction with NBS is carried in a homogeneous phase using polar solvents, e.g., pyridine or *tert*-butanol. In such a case, the Ⓟ-NBS has been found to offer an advantage in product separation since the excess of reagent and by-products can be removed by filtration. In the oxidation of secondary alcohols, the selectivity of Ⓟ-NBS was found to be comparable to that of NBS (Sahni, 1977).

Oxidation of thiols to disulfides by a 2-polyvinylpyridine/bromine complex has been reported (Christensen and Heacock, 1978). Thiols of various types (aromatic and aliphatic) were shown to form disulfides in good to excellent yields (58–96%).

Various oxidations using alumina or alumina-supported oxidizing agents have been reported. Alumina alone was shown to oxidize 2- and 3-nitrofluoren-9-ol and 2,4,7-trichlorofluoren-9-ol to the respective 9-ones (Pan et al., 1975). Fluoren-9-ol was unchanged upon passing a benzene solution through a column of alumina. Oppenauer-type oxidations of secondary alcohols and oxidation of β-hydroxy sulfides have been carried out with chloral on alumina (Posner et al., 1976; Posner and Chapdelaine, 1977). Alumina-supported sodium metaperiodate was used to oxidize various sulfides to the respective sulfoxides (Liu and Tong, 1978). Good to excellent yields were obtained with various aliphatic and aromatic sulfides by stirring a solution of the sulfide in 95% ethanol with the oxidant (Merck acidic Al_2O_3-90/sodium metaperiodate: 334/1 gm/mol).

Several other inorganic supports have been used for oxidizing agents. Dry ozonization of amines absorbed on silica gel has been reported to give moderate to good yields of the corresponding nitro compounds (Keinan and Mazur, 1977). Anilines react to give poor yields of the aromatic nitro compounds. Thallium(III) nitrate absorbed on K-10 montmorillonite clay has been shown oxidatively to rearrange alkyl aryl ketones, benzo-fused cycloakanones, and simple olefins (Taylor et al., 1976). Permanganate on Linde molecular sieves, silica gel, and montmorillonite clays have been used to oxidize alcohols to aldehydes and ketones (Regen and Koteel, 1977).

III. POLYMERIC OXIDATION–REDUCTION REAGENTS

A large number of redox polymers (electron-exchange polymers) have been reported. To a large extent, the use of these polymers has been limited to oxidation–reduction reactions of inorganic systems, such as Fe^{2+}–Fe^{3+}, Sn^{2+}–Sn^{4+}, Ce^{3+}–Ce^{4+}, and Ti^{3+}–Ti^{4+}. A comprehensive account of such redox polymers has been published by Cassidy and Kun (1965), Cassidy (1972), Lindsey (1974), and Manecke (1974).

Many of the redox polymers prepared so far incorporate the quinone or aminoquinone groups in the polymer matrix. These polymers have been synthesized by polymerization of vinyl derivatives of hydroquinones as well as by condensation of formaldehyde–hydroquinone.

Some of the recently reported quinone-based redox polymers (Manecke, 1974) have the structural units represented in Fig. 12-1. In these polymeric quinones, an attempt has been made to increase the hydrogen acceptor property of the quinone group by introducing electron-withdrawing substituents. Such substituents are known to increase the redox potential of the quinone–hydroquinone system (cf. 2,3-dichloro-

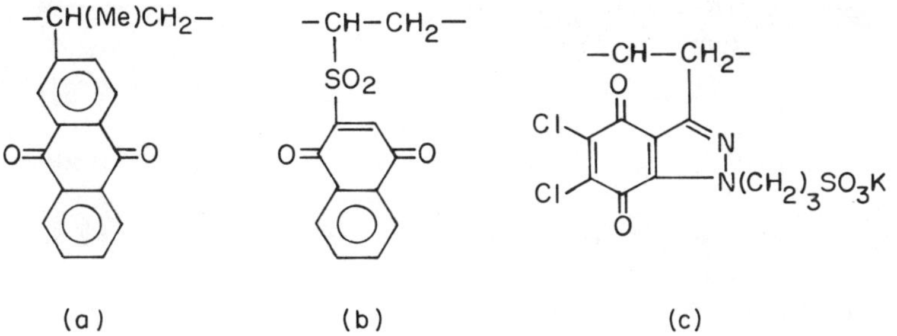

IV. Polymeric Reducing Reagents

Fig. 12-1

4,5-dicyanobenzoquinone), and such quinones have been used for the dehydrogenation and oxidation of organic compounds.

IV. POLYMERIC REDUCING REAGENTS

Reductions of organic compounds with polymer-supported reagents have included reports of the reduction of disulfides to thiols, of the use of alumina-supported materials for reduction of alkenes and ketones, and of the various reductions brought about by polymer-supported metal hydrides. The reductions of disulfides to thiols, mainly concerned with biologically related materials, will be discussed later (Chapter 15).

The use of alumina-supported materials has been reviewed (Posner, 1978). Alumina has been used to support various noble metals and the products used to hydrogenate alkenes (Sermon et al., 1974; Schwartz and Bathija, 1976). Cannizzaro and Meerwein-Pondorf-Verley reductions have been reported using either alumina or alumina impregnated with 2-propanol (Lamb et al., 1974; Posner, 1978; cf., Posner et al., 1977).

Potassium–graphite has been shown to reduce cyclic aliphatic ketones to the corresponding alcohols in good yields (Lalancette et al., 1972), and a commercial product is available (Graphimet K-29, Alfa Division, Ventron Corp.).

Various metal hydrides have been incorporated into polymeric supports. Weinshenker et al. (1975) synthesized an insoluble polymeric organotin dihydride (Scheme 12-5) which could be used for the reduction of carbonyl compounds. Active hydrogen in the polymeric reagent could be determined by reaction with an excess of iodooctane, followed by determination of the product by glc.

Selectivity in the reduction of carbonyl compounds by polymeric hydrides has been observed in the case of dicarbonyl compounds such as

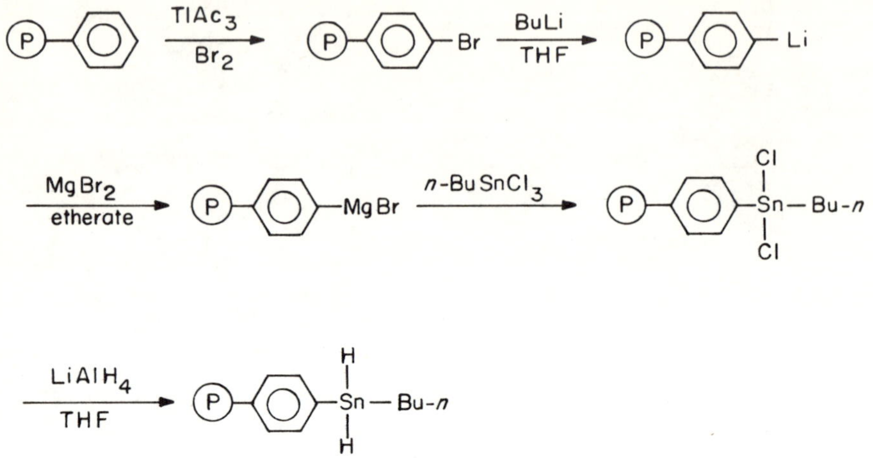

$$\text{(P)}\!-\!\!\bigcirc\;\xrightarrow[\text{Br}_2]{\text{TlAc}_3}\;\text{(P)}\!-\!\!\bigcirc\!-\!\text{Br}\;\xrightarrow[\text{THF}]{\text{BuLi}}\;\text{(P)}\!-\!\!\bigcirc\!-\!\text{Li}$$

$$\xrightarrow[\text{etherate}]{\text{MgBr}_2}\;\text{(P)}\!-\!\!\bigcirc\!-\!\text{MgBr}\;\xrightarrow{n\text{-BuSnCl}_3}\;\text{(P)}\!-\!\!\bigcirc\!-\!\underset{\underset{\text{Cl}}{|}}{\overset{\overset{\text{Cl}}{|}}{\text{Sn}}}\!-\!\text{Bu-}n$$

$$\xrightarrow[\text{THF}]{\text{LiAlH}_4}\;\text{(P)}\!-\!\!\bigcirc\!-\!\underset{\underset{\text{H}}{|}}{\overset{\overset{\text{H}}{|}}{\text{Sn}}}\!-\!\text{Bu-}n$$

Scheme 12-5

terephthaldehyde, which gave up to 86% of the monoalcohol. The formation of the monoalcohol was thought to be the result of the restricted accessibility of one aldehyde group whereas the other aldehyde group becomes reduced and remains bound to the polymer as tin alkoxide (Fig. 12-2).

The polymeric tin hydride also possessed a selectivity in reducing halides in the presence of other functional groups, e.g., α-bromoacetophenone was reduced to acetophenone. In this respect it is, therefore, superior to lithium aluminum hydride and the yields of reduction products are generally high.

The polymeric reagent was also claimed to be more stable and less odorous and toxic than the volatile tin hydride. In addition, it simplifies the work-up procedure for isolation of the product, and the spent polymeric by-product is regenerable.

A similar reagent has been prepared by absorbing tributyltin hydride onto silica gel (Fung *et al.*, 1978). This reagent has been used for the selective reduction of aldehydes.

Another polymeric hydride used for reduction of carbonyl compounds is a poly(4-vinylpyridine)borane synthesized by Hallensleben (1974).

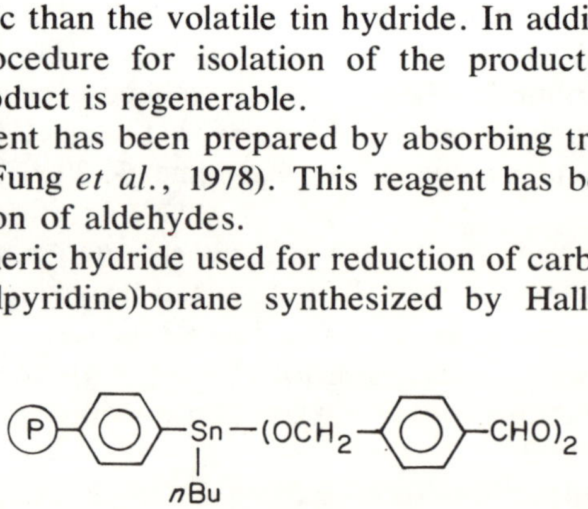

$$\text{(P)}\!-\!\!\bigcirc\!-\!\underset{\underset{n\text{Bu}}{|}}{\text{Sn}}\!-\!(\text{OCH}_2\!-\!\!\bigcirc\!-\!\text{CHO})_2$$

Fig. 12-2

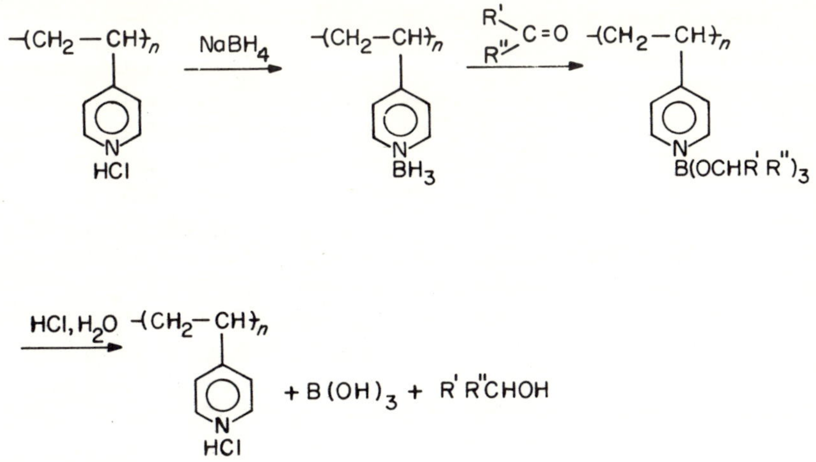

V. Polymeric Group Transfer Reagents

Scheme 12-6

Though the reduction yields with the polymeric borane were generally low, the polymeric hydride, based on macroporous resin, could be used in a column and could be regenerated. Its synthesis and use are represented in Scheme 12-6.

Sodium borohydride has been incorporated into alumina (Hodosan and Servan, 1969), cellulose matrices (Perrier and Benerito, 1976), and ion-exchange resins (Gibson and Bailey, 1977). The latter reagent was readily prepared using Amberlite IRA-400 and Amberlyst A-26. A more extensive survey of the properties of this material has been published (Cook et al., 1978; Cook, 1978). The borohydride resin (Resin–Borohydride) is a developmental product of Ventron Corporation, Beverly, Massachusetts.

Cyanoborohydride has been incorporated into Amberlyst A-26 and used in reductive aminations, reductive methylations, and for reducing pyridinium ions to their tetrahydro derivatives (Hutchins et al., 1978).

Reduction of amides to nitriles by polymer-supported phosphine oxide has been described (Relles and Schluenz, 1974; Harrison et al., 1977).

V. POLYMERIC GROUP TRANSFER REAGENTS

The following sections will examine the uses to which polymers have been put in exchanging and interchanging groups within and between molecules. The groups referred to are, needless to say, those which are larger than the hydrogen atoms and electrons that have been the subject of the preceding sections.

A. Halogenating Agents

Polymeric reagents have been used to introduce halogens into molecules either by nucleophilic displacement or by electrophilic addition. Anion exchange resins carrying fluorine, chlorine, bromine, and iodine anions have been used to exchange one halide for another in various alkyl halides (Cainelli *et al.*, 1976b). Fluorine transfer from Amberlyst A-26 was used to prepare compounds containing ^{18}F (Robinson, 1973, 1975; De Klein *et al.*, 1977). Various labeled alkyl and acyl fluorides were prepared by this route. Sulfonyl fluorides have been prepared by using a standard amine anion-exchange resin loaded with fluoride anions to exchange halogen anions with various sulfonyl chlorides (Borders, Jr. *et al.*, 1972). Addition of fluorine to phenyl-substituted alkenes by a polymeric aryliodine(III) difluoride has been reported (Zupan, 1977). In this case the polymer behaves as an electrophile and adds to the alkene forming a carbonium ion which is, in turn, attacked by a fluoride ion. Subsequent break-up of the polymer–molecule adduct leads to the formation of 1,1-difluoroalkanes.

The addition of chlorine to various polymer-supported chlorinating reagents has been the subject of several reports. Graphite–SbCl$_5$ ($C_{24}SbCl_5$) has been used to replace secondary bromo, iodo, and tosyl groups by chlorine in several aliphatic compounds (Bertin *et al.*, 1974; Kagan, 1976).

Recently, a polymeric trisubstituted phosphine dichloride (**14**) was prepared by the reaction of phosgene with polymeric trisubstituted phosphine oxide (**13**) (Scheme 12-7) (Relles and Schluenz, 1974). This polymeric phosphine dichloride (**14**) was used for the preparation of acid chlorides. The by-product polymer was also regenerable.

The polymeric triphosphine oxide (**13**) (Scheme 12-8) can be prepared by oxidation of the corresponding phosphine with peracetic acid. It could also be synthesized directly from chloromethylated copolystyrene by reaction with methyldiphenylphosphinite (Arbuzov reaction).

In addition to its use in the synthesis of acid chlorides, polymeric phosphine dichloride also converts alcohols to the corresponding chloride.

Another reagent, polymeric triphenylphosphine (Regen and Lee, 1975;

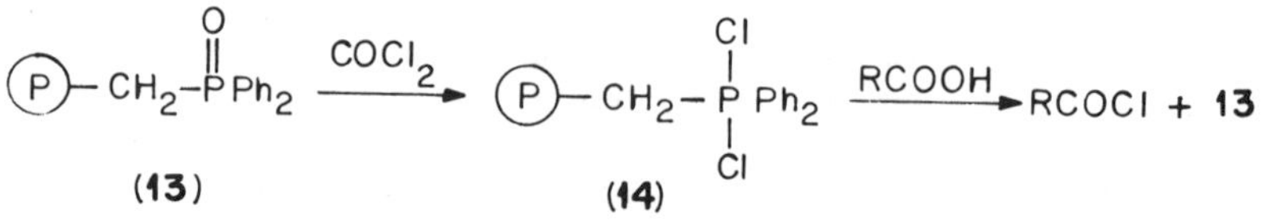

$$\text{(P)}-CH_2-\overset{O}{\overset{\|}{P}}Ph_2 \xrightarrow{COCl_2} \text{(P)}-CH_2-\overset{Cl}{\underset{Cl}{\overset{|}{P}}}Ph_2 \xrightarrow{RCOOH} RCOCl + \mathbf{13}$$

(**13**)   (**14**)

Scheme 12-7

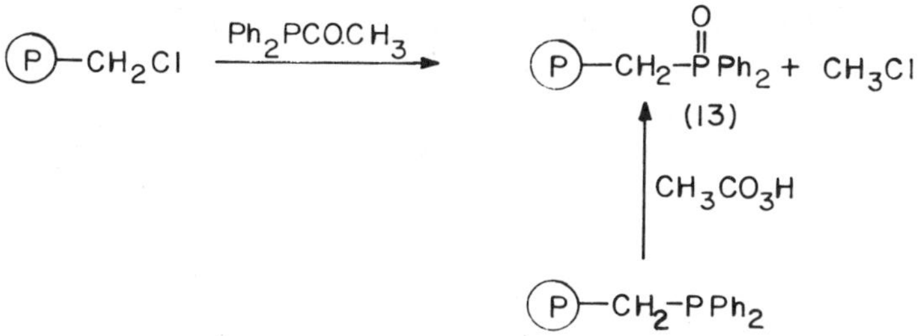

V. Polymeric Group Transfer Reagents

Scheme 12-8

$$\text{(P)}-CH_2Cl \xrightarrow{Ph_2PCOCH_3} \text{(P)}-CH_2-\overset{\overset{O}{\|}}{P}Ph_2 + CH_3Cl$$

$$(13)$$

$$\uparrow CH_3CO_3H$$

$$\text{(P)}-CH_2-PPh_2$$

Sherrington et al., 1977) reacted with alcohols in presence of CCl_4 (Scheme 12-9) to give the corresponding alkyl chlorides. The by-product, the polymeric phosphine oxide, could be reconverted into the reagent polymer by reduction with chlorosilane. The reaction has the advantage of preceding at neutral pH and is free from gaseous HCl formation, as in case of dichlorophosphine reagent. The yields of alkyl chlorides varied from 60 to 99%. The same polymer was also used to prepare acid chlorides (Hodge and Richardson, 1975). Acid chlorides were also prepared using poly(*p*-vinylbenzoyl chloride) in either a cross-linked or linear form (Hallensleben, 1973). Chlorinations were also claimed to be possible using poly(*p*-styryl iodide/chloride) (Hallensleben, 1972).

Polymeric analogs of NBS have been prepared. The parent polymers, polymaleimide (**18**) and co(polyethylene–maleimide) (**16**), are readily prepared by polymerization of the corresponding monomers (Scheme 12-10). Further functionalization, i.e., *N*-bromination of the polymer, is carried by the same sequence of reactions as with the low-molecular-weight analogs. (Yanagisawa et al., 1969a,b; Sahni, 1977).

The allylic bromination of cyclohexene was successfully done using co(polyethylene–*N*-bromomaleimide). When polymeric *N*-bromosuccinimide (⑨-NBS) was used for bromination of cumene, products other than those of benzylic bromination were also formed (Scheme 12-10) (Yaroslavsky et al., 1970a,b). The change in mechanism has been attributed to the polar environment provided by neighboring succinimide units in ⑨-NBS. Polymeric *N*-chloromaleimide, synthesized by Yaroslavsky and Katchalski (1972), on reaction with ethylbenzene, also gave products due to aromatic substitution.

Brominations of ketones and alkenes in good to excellent yields by

$$\text{(P)}-\text{C}_6\text{H}_4-PPh_2 + ROH + CCl_4 \longrightarrow \text{(P)}-\text{C}_6\text{H}_4-\overset{\overset{O}{\|}}{P}Ph_2 + RCl + CH_3Cl$$

Scheme 12-9

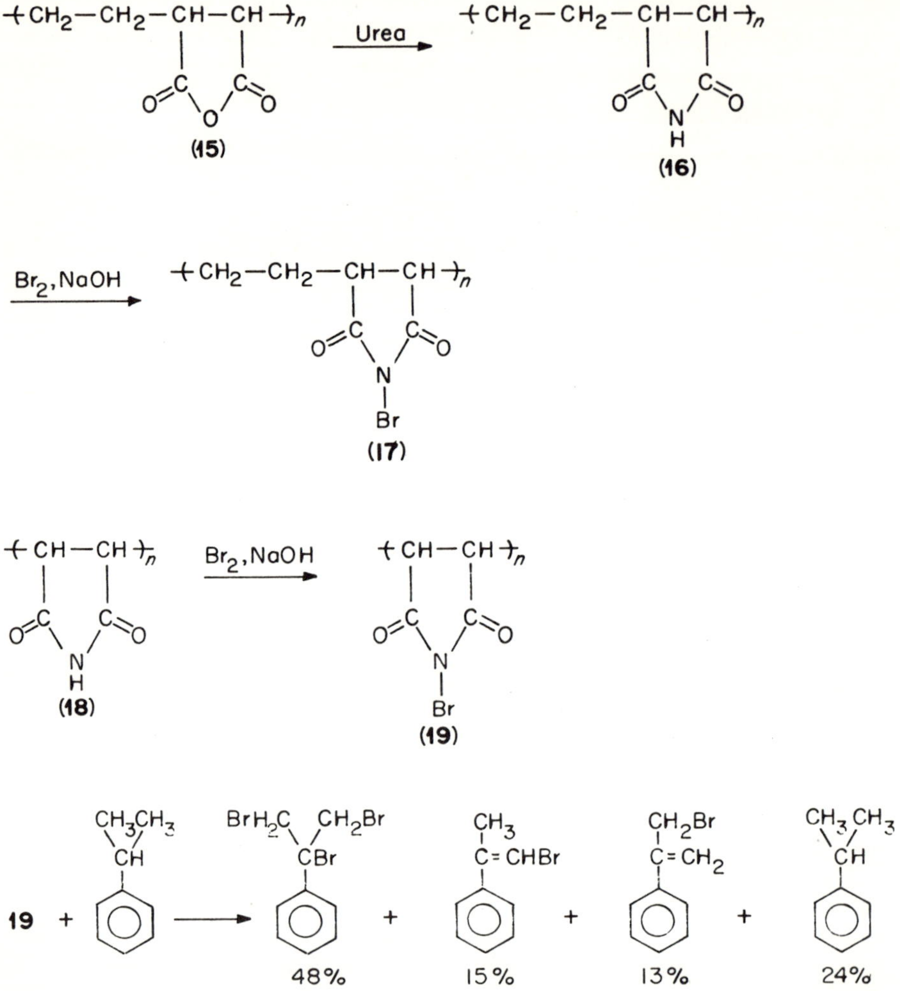

Scheme 12-10

poly(vinylpyridinium hydrobromide perbromide) resins have been reported (Fréchet et al., 1977). The resin could be regenerated without loss of activity. Graphite–bromine complexes have been investigated for their ability to brominate 2-methylcyclohexanone (Kagan, 1976). Bromine adsorbed on 5A-molecular sieves has been used to brominate a mixture of styrene and cyclohexene. Only α,β-dibromostyrene was obtained, indicating a high degree of selectivity (Risbood and Ruthven, 1978).

A BrCl complex with cross-linked polyvinylpyridine has been reported

V. Polymeric Group Transfer Reagents

to ring brominate phenol, anisole salicylanilide, and 2-(4-thiazolyl)benzimidazole (Zabicky-Zissmann et al., 1976).

B. Acylating Agents

Acylation reagents based on polymeric active esters have been discussed (Chapter 7).

Polymers incorporating mixed anhydrides of carboxylic and benzoic acids were reported by Shambhu and Digenis (1973, 1974) and Martin et al. (1978a). Yanagisawa et al. (1969a,b) reported the synthesis of a polymer containing the unsymmetrical anhydride derived from succinic and acetic acids. This polymer on reaction with cyclohexylamine gave a quantitative yield of N-acetylcyclohexylamine.

Mixed anhydrides of carboxylic acids normally react with alcohols or amines to give a mixture of esters or amides, respectively, according to Eq. (2).

$$R_1-\overset{O}{\underset{\|}{C}}-OO\overset{O}{\underset{\|}{C}}-R_2 \xrightarrow{ROH} R_1-COOR + R_2COOR \qquad (2)$$

The relative ratio of the two products will depend on two factors, either steric or electronic in origin. For example, a sterically hindered carboxylic acid will give lower yields of ester, whereas a carboxylic acid which has its carbonyl group more polarized ($>C^{\oplus}-O^{\ominus}$) will give higher yields of ester.

Steric factors are more significant in mixed anhydrides of polymeric acids and low-molecular-weight acids. Thus, the findings of Shambhu and Digenis (1973) indicate that with benzoic acid anhydride polymers (1), benzamides (2) were the only products formed [Eq. (3)].

$$\text{\textcircled{P}}-\text{\textcircled{}}-CO-O-CO-\text{\textcircled{}} \xrightarrow{RNH_2} \text{\textcircled{P}}-\text{\textcircled{}}-CONHR + \text{\textcircled{}}-CONHR \qquad (3)$$
$$(1) \qquad\qquad\qquad (2)$$

Extensive use of mixed anhydrides of carbonic acids has been made in peptide synthesis. Polymer-based mixed carbonic anhydrides prepared by Shambhu and Digenis (1974) and Martin et al. (1978a) gave exclusively the amide of benzoic acid with aliphatic amines [Eq. (4)], but with aniline, some polymeric urethane was also formed.

$$\text{\textcircled{P}}-\text{\textcircled{}}-CH_2O\overset{O}{\underset{\|}{C}}-O-\overset{O}{\underset{\|}{C}}-C_6H_5 \xrightarrow{RNH_2} \text{\textcircled{P}}-\text{\textcircled{}}-CH_2OH + CO_2 + C_6H_5-\overset{O}{\underset{\|}{C}}-NHR \qquad (4)$$

The polymer-bound mixed carbonic carboxylic anhydrides have re-

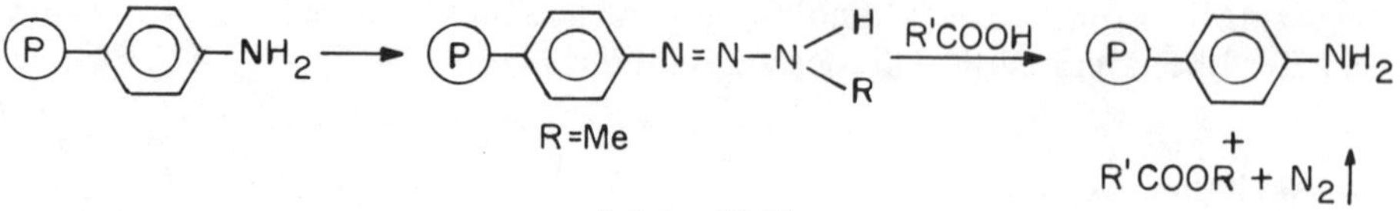

Scheme 12-11

cently been used to acylate 7-aminocephalosporanic acid (Martin et al., 1978b).

Mixed polymeric sulfonic carboxylic anhydrides prepared by Ang and Harwood (1973) were slow to react with alcohols larger than ethanol. Similarly, amines did not react quantitatively with these anhydrides. It was thought that this reaction was slowed due to the protonation of the amino groups by the free sulfonic acid groups present on the polymer. Hence the reaction, when repeated in the presence of an excess of a tertiary amine such as pyridine or triethylamine, was found to proceed smoothly (Jain, 1977). The formation of N-acylpyridinium compounds as intermediates in these reactions may also account for the enhanced reactivity and lack of steric hindrance. The reactions of acetylpolystyrene sulphonates and phosphonates with aniline have also been investigated (Laird and Spence, 1977).

C. Alkylating Agents

Alkylation of carboxylic acids with N-alkyl-N-aryltrizene groups supported on polymeric material has been reported (Adaway and Harwood, 1977) (Scheme 12-11). Homologation of alkyl iodides by a polymeric phenylthiomethyl lithium reagent was shown to give the chain-lengthened alkyl iodide in good yield (Crosby and Kato, 1977) (Scheme 12-12). Chain lengthening of diiodides was also attempted.

D. Polymeric Nucleophiles

Anion-exchange resins have been used in nucleophilic substitution reactions. Gordon et al. (1963) prepared benzyl cyanide by the reaction of

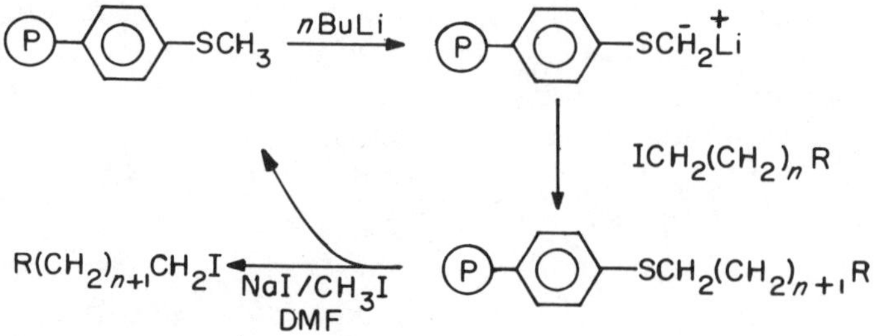

Scheme 12-12

V. Polymeric Group Transfer Reagents

$$C_6H_5CH_2Br + \text{(P)}-CH_2\overset{+}{N}Me_3CN^- \longrightarrow C_6H_5CH_2CN + \text{(P)}-CH_2\overset{+}{N}Me_3Br^-$$

Scheme 12-13

benzyl bromide with an anion-exchange resin, IRA-400 (cyanide form). When sodium or potassium cyanide is used as the source of nucleophile (CN^-), the reaction product had to be isolated from the inorganic material. With resin-bound nucleophiles, on the other hand, the reaction was carried out on a column. The product was eluted out while the by-products remained in the column (Scheme 12-13).

Benzyl ethers were similarly prepared by the reaction of benzyl bromide, with resin-bound phenoxide as the nucleophile (Rowe and Kaufmann, 1958) (Scheme 12-14).

In another report (Borders et al., 1972), arylsulfonyl fluorides were prepared by the reaction of the fluoride form of a resin with arylsulfonyl chlorides.

Further examples of the use of polymers as nucleophilic reagents will be discussed in Chapter 13, Section VI, since the polymer is used more as a catalyst than a reagent.

E. Wittig and Ylid Reagents

Alkene synthesis using a compound containing a carbonyl group and low-molecular-weight Wittig reagent are well known. Camps *et al.* (1971), McKinley and Rakshys (1972), and Heitz and Michels (1972, 1973) reported the preparation of polymers incorporating a triphenylphosphine group that could be converted into a polymeric Wittig reagent. The parent polymeric phosphines were prepared by three different routes (Scheme 12-15). In one of the preparations, a preformed monomer was copolymerized, while the styrene polymers were functionalized in the other two cases. These polymeric reagents were used to convert many low-molecular-weight carbonyl compounds to alkenes (Scheme 12-16).

Although lower yields (40%) were observed by Camps *et al.* (1971), they were improved (55–94%) by McKinley and Rakshys (1972). The separation of the insoluble polymer by-product (phosphine oxide) makes the isolation of the major alkene product very simple. The presence of lithium ion is known to lead to the formation of trans-alkenes because of the preferred complexation of the threo-form of the betaine. With the

$$C_6H_5CH_2Br + \text{(P)}-CH_2\overset{+}{N}Me_3\bar{O}C_6H_5 \longrightarrow C_6H_5CH_2OC_6H_5 + \text{(P)}-CH_2\overset{+}{N}Me_3Br^-$$

Scheme 12-14

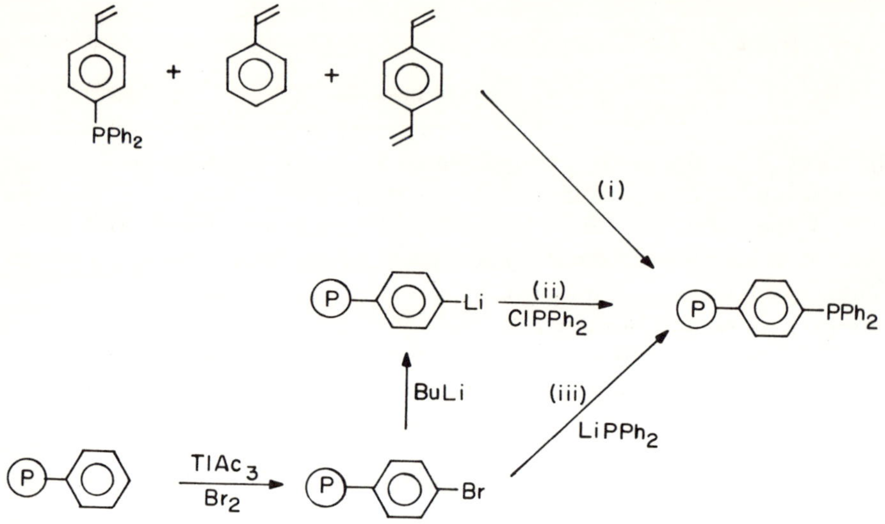

Scheme 12-15

polymeric Wittig reagent, lithium ions are washed away after betaine formation, leading to formation of pure cis-olefins. Heitz and Michels (1973) also reported a method of converting phosphine oxide into phosphine that allows the by-product resin to be regenerated and reused.

A polymeric sulfur ylid reagent containing dimethylsulfonium methylide was synthesized by Tanimoto *et al.* (1967) from a copolymer of *p*-vinylphenylmethyl thioether, styrene, and DVB, according to Scheme 12-17. The reagent was used for the synthesis of epoxides from the carbonyl compounds. It was claimed that, in addition to simplifying the procedure for isolation of the products, the polymeric reagent is nonodorous and convenient to handle, compared to the noxious, volatile, low-molecular-weight sulfides used in classical syntheses.

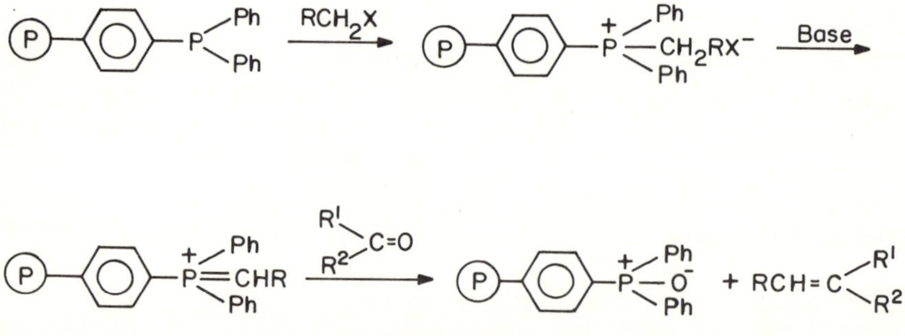

Scheme 12-16

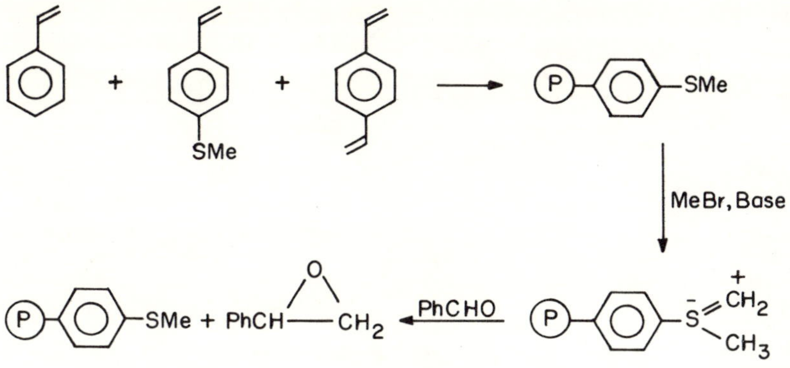

VI. Polymeric Coupling Agents

Scheme 12-17

F. Tosyl Azides as Diazo Transfer Reagents

A polymeric tosyl azide has been prepared (Roush *et al.*, 1974) and used as a diazo transfer reagent (Scheme 12-18). With its help, β-diketones can be converted into diazodiketones. The polymeric reagent, unlike its low-molecular-weight analog, is not explosive.

VI. POLYMERIC COUPLING AGENTS

A. Carbodiimides

More types of polymeric reagents have been functionalized to contain a carbodiimide group than any other functional group. Table 12-1 lists the polymeric carbodiimides and the reactions that have been carried out with their help. The carbodiimide group can be introduced into a polymer matrix according to Scheme 12-19 (Weinshenker and Shen, 1972a), or a Merrifield-type resin can be functionalized with a preformed, low-molecular-weight carbodiimide (Scheme 12-20) (Fridkin *et al.*, 1971). Determination of the carbodiimide-reactive-group concentration by reaction of the resin with oxalic acid has also been reported (Adam and Yany, 1977).

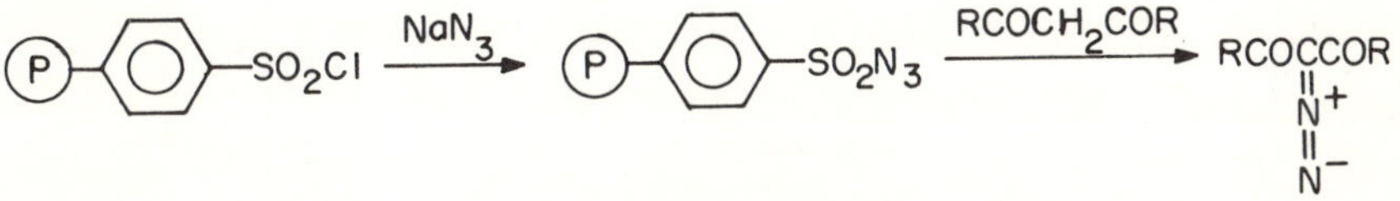

Scheme 12-18

TABLE 12-1

Polymeric Carbodiimides and Their Reactions

Carbodiimide	Reaction	Reference
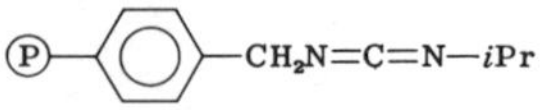 ⓅーC₆H₄ーCH₂N=C=N—iPr	(i) Anhydride synthesis (ii) Moffatt oxidation	Weinshenker and Shen, 1972a Weinshenker and Shen, 1972b
$-\!\!+\!\!(CH_2)_6-N=C=N\!+\!\!_n\!\!-$	Peptide synthesis	Wolman et al., 1967

$-\!\!+\!CH_2-CH\!+\!\!_n\!\!-$
     |
     N=C=N—R
	Peptide synthesis	Fridkin et al., 1971

ⓅーC₆H₄ーCH₂—S⁺(CH₂)₂N=C=N—R
                |
                Me
	Peptide synthesis	Fridkin et al., 1971

ⓅーC₆H₄ーCH₂—N⁺(CH₂)₃N=C=N—R
     Cl⁻ |
             Me
	Peptide synthesis	Fridkin et al., 1971

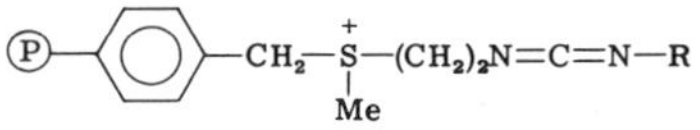

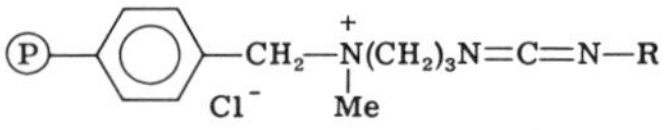

The polymeric carbodiimides have been used in peptide synthesis and for preparation of carboxylic anhydrides. The polymeric carbodiimide (Weinshenker and Shen, 1972b) has also been used for Moffat oxidation of alcohols to aldehydes. During these reactions, the carbodiimides are converted into the corresponding ureas as the end-product. The urea by-product can be reconverted to the carbodiimide (Scheme 12-19). However, the regenerated reagent is less active than the original reagent, presumably because of blocking of some diimide groups due to N-acyl formation.

B. Sulfonyl Chlorides

Fridkin et al. (1971) have employed polymeric benzenesulfonyl chloride as a coupling reagent for peptide synthesis via the polymeric mixed

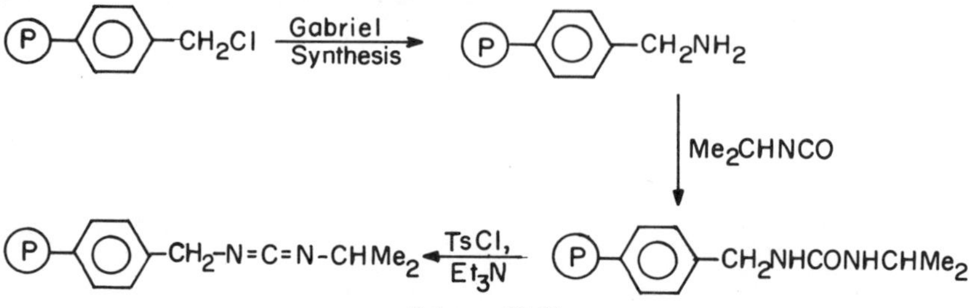

Ⓟ—C₆H₄—CH₂Cl $\xrightarrow{\text{Gabriel Synthesis}}$ Ⓟ—C₆H₄—CH₂NH₂

$\downarrow$ Me₂CHNCO

Ⓟ—C₆H₄—CH₂-N=C=N-CHMe₂ $\xleftarrow{\text{TsCl, Et}_3\text{N}}$ Ⓟ—C₆H₄—CH₂NHCONHCHMe₂

Scheme 12-19

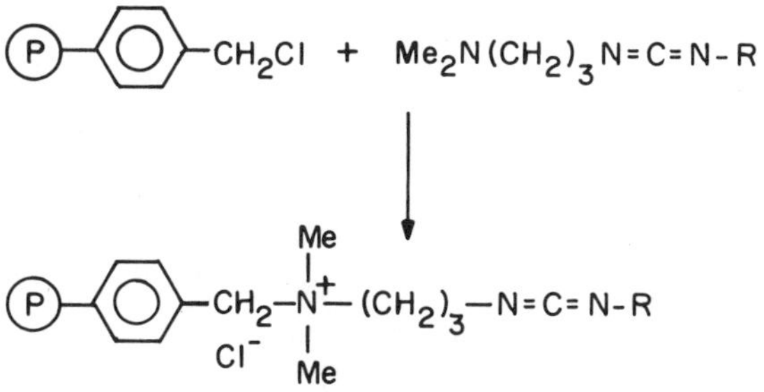

VI. Polymeric Coupling Agents

Scheme 12-20

carboxylic sulfonic anhydrides. Another hindered polymeric arylsulfonyl chloride was prepared by copolymerization of 3,5-diethylvinylbenzene, styrene, and DVB, followed by chlorosulfonation of the product. Rubinstein and Patchornik (1972, 1975) used this reagent (Fig. 12-3), the polymeric analog of mesityenesulfonyl chloride, as a coupling reagent in the synthesis of oligonucleotides. The usual side reaction of sulfonation of the alcoholic group, and the difficulty of complete removal of the sulfonic acid by-product, are avoided by the use of the polymeric reagent, without affecting the rates and yields of the coupling reaction.

C. Reagent Based on EEDQ

Brown and Williams (1971) and Williams *et al.* (1972) described the preparation of the polymeric reagent based on EEDQ (2-ethoxy-1-ethoxycarbonyl-1,2-dihydroquinoline) (Scheme 12-21) and used it to couple peptides. It was found to give clean products in high yields, and the reactions proceeded with little racemization. The reagent was also regenerable.

D. Reagents Using Polymer-Supported Triphenylphosphine

The use of polymer-supported triphenylphosphine for the coupling of peptides has been the subject of two reports. One report described the use

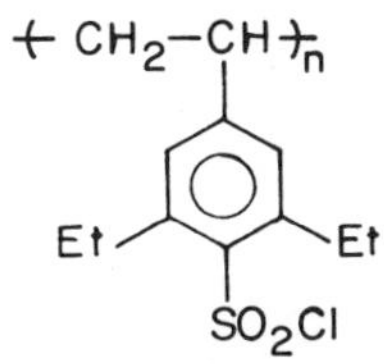

Fig. 12-3

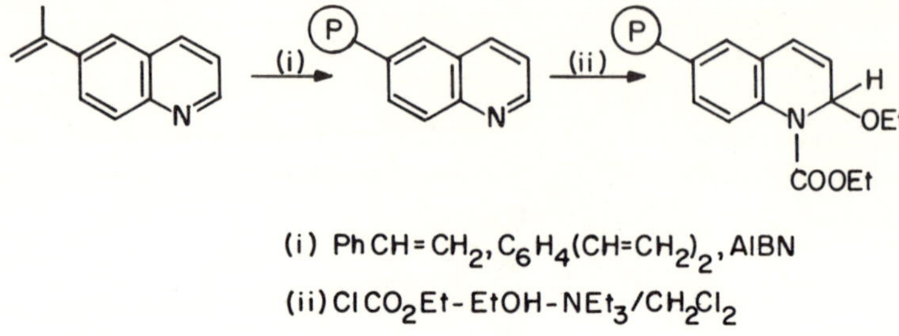
P
(i)
(ii)
N
H
OEt
COOEt
(i) PhCH=CH2, C6H4(CH=CH2)2, AIBN
(ii) ClCO2Et-EtOH-NEt3/CH2Cl2

Scheme 12-21

of polymer-supported triphenylphosphine and carbon tetrachloride to prepare simple peptides (Appel *et al.*, 1976; Appel and Willms, 1977). Partial racemization occurred, but its incidence could be reduced with 1-hydroxybenzotriazole. The other report described the use of the polymer and 2,2′-dipyridyl disulfide to prepare simple peptides (Horiki, 1976). No data on racemization were given.

VII. MISCELLANEOUS REAGENTS

A polymeric triphenylphosphine dibromide was synthesized by Michels and Heitz (1975) and was used for the cleavage of ether bonds.

The formation of dimethyl acetals by trimethyl orthoformate adsorbed on a montmorillonite clay (K-10) has been described (Taylor and Chiang, 1977).

Polymer-supported selenium reagents have been prepared to avoid some of the obnoxious and toxic properties of these reagents as monomers (Michels *et al.*, 1976). The reagents were used to convert 4-methylcyclohexanone into 4-methylcyclohex-2-enone, to hydrobrominate ethyl 2-bromopropionate, and to glycolate 2-methyl-2-heptene.

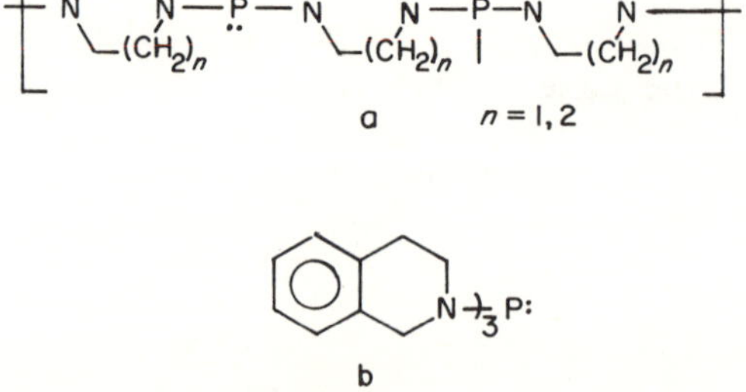
N
P
(CH2)n
a
n = 1, 2
N
P:
3
b

Fig. 12-4

Polymeric aminophosphines (Fig. 12-4) have been used to desulfurize disulfides, a thiosulfinate ester, and a thioimide (configurational inversion) (Harpp et al., 1978).

A polymeric yneamine (Ⓟ—C ≡ CNEt$_2$) has recently been prepared (Moore and Kennedy, 1978) and used to prepare a mixed anhydride, an amide and an ester of benzoic acid.

REFERENCES

Adam, W., and Yany, F. (1977). *Anal. Chem.* **49**, 676.
Adaway, T. J., and Harwood, H. J. (1977). *Polym. Prepr., Am. Chem. Soc. Div. Polym. Chem.* **18**, 661.
Ang, T. L., and Harwood, J. (1973). *Macromol. Sci. Chem.* A **7**, 1079.
Appel, R., and Willms, L. (1977). *J. Chem. Research* (S) **84**, (M) 0901.
Appel, R., Strüver, W., and Willms, L. (1976). *Tetrahedron Lett.*, 905.
Bertin, J., Luche, J. L., Kagan, H. B., and Setton, R. (1974). *Tetrahedron Lett.*, 763.
Borders, C. L., Jr., MacDonnel, D. L., and Chambers, J. L., Jr. (1972). *J. Org. Chem.* **37**, 3549.
Brown, J., and Williams, R. E. (1971). *Can. J. Chem.* **49**, 3765.
Cainelli, G., Cardillo, G., Orena, M., and Sandri, S. (1976). *J. Am. Chem. Soc.* **98**, 6737.
Cainelli, G., Manescalchi, F., and Panunzio, M. (1976). *Synthesis*, 472.
Camps, F., Castells, J., Font, J., and Vela, F. (1971). *Tetrahedron Lett.*, 1715.
Cardillo, G., Orena, M., and Sandri, S. (1976). *Tetrahedron Lett.*, 3985.
Cassidy, H. G. (1972). *J. Polym. Sci.* Part D **6**, 1.
Cassidy, H. G., and Kun, K. A. (1965). "Oxidation-Reduction Polymers." Wiley (Interscience), New York.
Christensen, L. W., and Heacock, D. J. (1978). *Synthesis*, 50.
Cook, M. M. (1978). Personal communication.
Cook, M. M., Wagner, S. E., Demko, P. R., Clements, J. G. and Mikulski, R. A. (1978). *Polym. Prepr., Am. Chem. Soc. Div. Polym. Chem.* **19**, 414.
Corey, E. J., and Jautelat, M. (1968). *Tetrahedron Lett.*, 5787.
Crosby, G. A., and Kato, M. (1977). *J. Am. Chem. Soc.* **99**, 278.
Crosby, G. A., Weinshenker, N. M., and Uh, H. S. (1975). *J. Am. Chem. Soc.* **97**, 2232.
De Klein, J. P., Seetz, J. W., Zawierko, J. F., and Van Zanten, B. (1977). *Int. J. Appl. Radiat. Isot.* **28**, 591.
Ebert, L. B., Huggins, R. A., and Brauman, J. I. (1974). *Carbon* **12**, 199.
Farrall, M. J., and Frechet, J. M. J. (1976). *J. Org. Chem.* **41**, 3877.
Fetizon, M., and Golfier, M. (1968). *Compt. rendu* **267**, 900.
Fetizon, M., Golfier, M., and Morques, P. (1972). *Tetrahedron Lett.*, 4445.
Fieser, L. F., and Fieser, M. (1977). "Reagents for Organic Synthesis," Vol. 6, p. 511. Wiley (Interscience) New York.
Flory, P. J. (1953). "Principles of Polymer Chemistry." p. 829. Cornell Univ. Press, Ithaca, New York.
Frechet, J. M. J., and Haque, K. E. (1975). *Macromolecules*, **8**, 130.
Fréchet, J. M. J., Farrall, M. J., and Nuyens, L. J. (1977). *J. Macromol. Sci. Chem.* A **11**, 507.
Frechet, J. M. J., Warnock, J., and Farrall, M. J. (1978). *J. Org. Chem.* **43**, 2618.

Fridkin, M., Patchornik, A., and Katchalski, E. (1971). *Pept. Proc. Eur. Pept. Symp., 10th, 1969,* p. 166. North-Holland, Amsterdam.
Fung, N. Y. M., de Mayo, P., Schauble, J. H., and Weedon, A. C. (1978). *J. Org. Chem.* **43,** 3977.
Gibson, H. W., and Bailey, F. C. (1977). *J. Chem. Soc. Chem. Commun.,* 815.
Gordon, M., DePamphilis, M. L., and Grifin, C. E. (1963). *J. Org. Chem.* **28,** 698.
Hallensleben, M. L. (1972). *Angew. Makromol. Chem.* **27,** 223.
Hallensleben, M. L. (1973). *Angew. Makromol. Chem.* **31,** 143.
Hallensleben, M. L. (1974). *J. Polymer Sci.,* Symposium No. 47, p. 1.
Harpp, D. N., Adams, J., Gleason, J. G., Mullins, D., and Steliou K. (1978). *Tetrahedron Lett.,* 3989.
Harrison, C. R., and Hodge, P. (1974). *J. Chem. Soc. Chem. Commun.,* 1009.
Harrison, C. R., Hodge, P., and Rogers, W. J. (1977). *Synthesis,* 41.
Heitz, W., and Michels, R. (1972). *Angew. Chem. Int. Ed.* **11,** 298.
Heitz, W., and Michels, R. (1973). *Liebigs Annalen,* 277.
Helfferich, P., and Luten, D. B., Jr. (1964). *J. Appl. Polym. Sci.* **8,** 2899.
Hodge, P., and Richardson, G. (1975). *J. Chem. Soc. Chem. Commun.,* 622.
Hodosan, F., and Servan, N. (1969). *Rev. Roum. Chim.* **14,** 121.
Horiki, K. (1976). *Tetrahedron Lett.,* 4103.
Hutchins, R. O., Natale, N. R., and Taffer, I. M. (1978). *J. Chem. Soc. Chem. Commun.,* 1088.
Jain, J. C. (1977). Ph.D. Thesis. University of Jodhpur, Jodhpur, India.
Kaczmer, R. U. (1973). *Angew. Chem. Int. Ed.* **5,** 430.
Kagan, H. B. (1976). *Pure Appl. Chem.* **46,** 177.
Kakis, F. S., Fetizon, M., Douchkine, N., Golfier, M., Mourques, P., and Prange, T. (1974). *J. Org. Chem.* **39,** 523.
Keinan, E., and Mazur, Y. (1977). *J. Org. Chem.* **42,** 844.
Lalancette, J. M., Robbin, G., and Dumas, P. (1972). *Can. J. Chem.* **50,** 3058.
Lamb, F. A., Cote, P. N., Slutsky, B., and Vittimberga, B. M. (1977). *J. Org. Chem.* **39,** 2796.
Lindsey, A. S. (1974). *In* "The Chemistry of Quininoid Compounds Part 2" (S. Patai, ed.), p. 763. Wiley, New York.
Liu, K-T., and Tong, Y-C. (1978). *J. Org. Chem.* **43,** 2717.
Manecke, G. (1974). *Pure Appl. Chem.* **38,** 181.
Martin, G. E., Shambhu, M. B., Shakhshir, S. R., and Digenis, G. A. (1978a). *J. Org. Chem.* **43,** 4571.
Martin, G. E., Shambhu, M. B., and Digenis, G. A. (1978b). *J. Pharm. Sci.* **67,** 110.
McKinley, S. V., and Rakshys, J. W., Jr. (1972). *J. Chem. Soc. Chem. Commun.,* 134.
Michels, R., and Heitz, W. (1975). *Makromol. Chem.* **176,** 245.
Michels, R., Kato, M., and Heitz, W. (1976). *Makromol. Chem.* **177,** 2311.
Moore, J. A., and Kennedy, J. J. (1978). *J. Chem. Soc. Chem. Commun.,* 1079.
Oosawa, F. (1971). "Polyelectrolytes," pp. 1–12, 113–158. Dekker, New York.
Pan, H-L., Cole, C-A., and Fletcher, T. L. (1975). *Synthesis* 716.
Patterson, J. A. (1971). "Biochemical Aspects of Reactions on Solid Supports" (G. R. Stark, ed), pp. 204–208. Academic Press, New York.
Perrier, D. M., and Benerito, R. R. (1976). *Appl. Polym. Symp.* **29,** 213.
Posner, G. H. (1978). *Angew. Chem. Int. Ed.* **17,** 487.
Posner, G. H., and Chapdelaine, M. J. (1977). *Tetrahedron Lett.,* 3227.
Posner, G. H., Perfetti, R. B., and Runquist, A. W. (1976). *Tetrahedron Lett.,* 3499.
Posner, G. H., Runquist, A. W., and Chapdelaine, M. J. (1977). *J. Org. Chem.* **42,** 1202.

References

Regen, S. L., and Koteel, C. (1977). *J. Am. Chem. Soc.* **99**, 3837.
Regen, S. L., and Lee, D. P. (1975). *J. Org. Chem.* **40**, 1669.
Relles, H. M., and Schluenz, R. W. (1974). *J. Am. Chem. Soc.* **96**, 6469.
Risbood, P. A., and Ruthven, D. M. (1978). *J. Am. Chem. Soc.* **100**, 4919.
Robinson, G. D., Jr. (1975). *J. Nucl. Med.* **16**, 561.
Robinson, G. D., Jr. (1973). *J. Nucl. Med.* **14**, 446.
Roush, W. R., Feitler, D., and Rebek, J. (1974). *Tetrahedron Lett.*, 1391.
Rowe, E. J., Kaufman, K. L., and Piantadosi, C. (1958). *J. Org. Chem.* **23**, 1622.
Rubenstein, M., and Patchornik, A. (1972). *Tetrahedron Lett.*, 2881.
Rubenstein, M., and Patchornik, A. (1975). *Tetrahedron* **31**, 1517.
Sahni, M. K. (1977). Ph.D. Thesis, University of Jodhpur, Jodhpur, India.
San Filippo Jr., J., and Chern, C. I. (1977). *J. Org. Chem.* **42**, 2182.
Sato, Y., Kunieda, N., and Kinoshita, M. (1977). *Makromol. Chem.* **178**, 683.
Schwartz, L. H., and Bathija, B. L. (1976). *J. Am. Chem. Soc.* **98**, 5344.
Sermon, P. A., Bond, G. C., and Webb, G. (1974). *J. Chem. Soc. Chem. Commun.*, 417.
Shambhu, M. B., and Digenis, G. A. (1973). *Tetrahedron Lett.*, 1627.
Shambhu, M. B., and Digenis, G. A. (1974). *J. Chem. Soc. Chem. Commun.*, 619.
Sherrington, D. C., Craig, D. J., Dagleish, J., Domin, G., Taylor, J., and Meehan, G. V. (1977). *Europ. Polym. J.* **13**, 73.
Takagi, T. (1967). *J. Polym. Sci. Polym. Lett. Ed.* **5**, Part B, 1031.
Tanimoto, S., Horikawa, J., and Oda, R. (1967). *Kogyo Kagaku Zasshi* **10**, 1269.
Taylor, E. C., and Chiang, C. S. (1977). *Synthesis*, 467.
Taylor, E. C., Chiang, C. S., McKillop, A., and White, J. F. (1976). *J. Am. Chem. Soc.* **98**, 6750.
Weinshenker, N. M., Crosby, G. A., and Wong, J. Y. (1975). *J. Org. Chem.* **40**, 1966.
Weinshenker, N. M., and Shen, C. M. (1972a). *Tetrahedron Lett.*, 3281.
Weinshenker, N. M., and Shen, C. M. (1972b). *Tetrahedron Lett.*, 3285.
Williams, R. E., Brown, J., and Lauren, D. R. (1972). *Polym. Prepr., Am. Chem. Soc. Div., Polym. Chem.* (2)**13**, 823.
Wolman, Y., Kivity, S., and Frankel, M. (1967). *J. Chem. Soc. Chem. Commun.*, 629.
Yanagisawa, Y., Akiyama, M., and Okawara, M. (1969a). *Kogyo Kagaku Zasshi* **72**, 1399.
Yanagisawa, Y., Akiyama, M., and Okawara, M. (1969b). *J. Polym. Sci. Part A-1 (Chem.).* **7**, 1905.
Yaroslavsky, C., and Katchalski, E. (1972). *Tetrahedron Lett.*, 5.
Yaroslavsky, C., Patchornik, A., and Katchalski, E. (1970a). *Isr. J. Chem.* **8**, 37.
Yaroslavsky, C., Patchornik, A., and Katchalski, E. (1970b). *Tetrahedron Lett.*, 3629.
Zabicky-Zissmann, J. Z., Oren, I., and Katzir-Katchalski, E. (1976). Israeli Patent No. 40,345 [C.A. **85**, 46681].
Zupan, M. (1977). *Collec. Czech. Chem. Commun.* **42**, 266.

13 Polymer-Bound Catalysts (I)

I.	Introduction	198
II.	Ion-Exchange Resins as General Acid–Base Catalysts	199
III.	Polystyrene–Aluminum Chloride as a Lewis Acid Catalyst	205
IV.	Polymer-Based "Super Acid" Catalysts	205
V.	Polymeric Esterolytic Catalysts	206
VI.	Polymer-Supported Phase-Transfer Catalysts	209
VII.	Polymeric Triphase Catalysts	212
VIII.	Polymer-Based Photosensitizers	213
	A. Polymer-Bound Rose-Bengal	214
	B. Other Polymeric Photosensitizers	215
	References	216

I. INTRODUCTION

Catalysts differ from other reagents normally used in a synthesis. They are not used up in the net chemical transformation and may be recovered in their original state at the end of the reaction. The use of homogeneous-phase catalysts in organic synthesis is well-known.

Catalytic species can be bound to polymeric carriers without losing their basic catalytic activity. Immobilization of a variety of conventional chemical catalysts on polymeric supports has received wide application during recent years. Ion-exchangers, esterolytic catalysts, phase-transfer catalysts, and photosensitizers have all been prepared (for a review, see Manecke and Storck, 1978). A common advantage in using a polymer-bound catalyst is found in the ease of separation at any stage of the reaction, thus offering the possibility of arresting further progress of the reaction. Besides, an immobilized catalyst is sometimes more stable to atmospheric conditions, has greater bench stability and does not cause undesirable side reactions as is the case with homogeneous catalysts. In industrial applications, they provide for the adoption of continuous flow processes in place of batch operations and, in many cases, the catalyst can be reused many times, thus lowering the cost of operation.

II. ION-EXCHANGE RESINS AS GENERAL ACID–BASE CATALYSTS

Ion-exchangers were probably the first synthetic polymeric catalysts to be used in organic synthetic reactions as substitutes for low-molecular-weight and water-soluble acid–base catalysts. The use of resin acids or resin bases as catalysts for certain reactions is so well-established that many of them were being used in industry while other polymeric catalysts were still in the process of development. As an example, isopropyl alcohol can be manufactured by direct hydration of propene using a sulfonate resin as catalyst (Neier and Woellner, 1973). The cation-exchangers generally contain sulfonic acid groups introduced by sulfonation of polystyrene (3–5% cross-linked with DVB), whereas anion-exchangers generally contain quaternary ammonium groups introduced by chloromethylation of polystyrene, followed by amination. The methods for their preparation have been discussed in detail (Helfferich, 1962) and they are commercially available in various grades. Although weakly acidic or weakly basic resins, such as those containing the carboxyl and amino groups, respectively, have found applications in analytical and separation methods, they have not been used extensively as catalysts. The major types of reactions catalyzed by polymer-bound acids and bases are listed in Table 13-1. This listing is only intended to be a representative collection since their uses are too numerous to be covered completely in this volume.

The resin acid–base catalysts, apart from their ease of removal from the reaction mixture by filtration and the fact that they are reusable, have certain other merits: (i) Their use permits a continuous flow process in manufacturing; (ii) side reactions are less significant, giving rise to purer products; and (iii) in many cases the resin acids or resin bases show increased selectivity in reactions. Examples of this latter point will be given later in this chapter.

The main limitations of ion-exchangers are a limited chemical and thermal stability. Furthermore, a well-established, quantitative theory of ion-exchangers as catalysts is lacking since it is difficult to take into consideration all the various parameters in the reactions and to work out suitable rate laws. Therefore, prediction of the outcome of the use and selectivity of a polymeric catalyst is difficult.

The selectivity of acidic or basic polymeric catalysts in certain reactions has been reported.

1. Selective mild hydrolysis of amides: A selective mild hydrolysis of amides in preference to esters has been described using the resin, IR-120

TABLE 13-1

Reactions Catalyzed by Polymeric Acid–Base Catalysts

Reaction type	Reactants	Resin form	References
(A) Acid-Catalyzed Reactions			
Ester hydrolysis	General esters, mainly aliphatic	Amberlite IR-100 [H$^+$]	Davies and Thomas, 1952
Hydrolytic decarboxylation of β-keto esters	Esters of acetoacetic acid	Amberlite IR-120 [H$^+$]	Astle and Oscar, 1961
Esterification	(i) Fatty acids including the higher ones, and alcohols or olefins	ⓟ—SO$_3$H	George and Virgil, 1971
	(ii) α-Amino acids and methanol	Amberlite IR-120 [H$^+$] or Zeo-Karb 222 [H$^+$]	Mill and Crimmin, 1957; Jain et al., 1977
	(iii) Formylphenyllactonitrile and propanol	ⓟ—SO$_3$H	
	(iv) Acetylation of n-butanol	Macroporous, Glycidyl methacrylate–ethylene methacrylate	Svec et al., 1977
Acetal formation		ⓟ—SO$_3$H	Moffat, 1963
Glycoside synthesis, from carbohydrates and alcohol	Glucose and methanol; benzaldehyde and glycerol	Dowex—SO$_3$H$^+$	Mowery, 1961
Hydrolysis of enamines		Amberlite IR-120 [H$^+$]	Hasek et al., 1963
Epoxidation of fats by H$_2$O$_2$, followed by hydrolysis to transglycol	Fats and oils	ⓟ—SO$_3$H	Kunin, 1958
Amide formation	3-(p-Nitrophenyl)propanoic acid and 2-(1-cyclohexenyl)ethylamine	Amberlite IR-120 [H$^+$]	Walter et al., 1961

Reaction	Substrate	Catalyst	Reference
Rearrangement reaction	2-Ethynyl carbinol rearranged to α,β-unsaturated ketone (HO-C₆H₁₀-C≡CH)	Dowex-50 [H⁺]	Newman, 1953
Phenol alkylation, with alkenes	Phenol and isobutene	ⓟ—SO₃H	Loev and Massengale, 1957
Coumarin synthesis (Pechman reaction)	Resorcinol, ethyl acetoacetate	Zeo-Karb or Amberlite IR-120 [H⁺]	John and Israelstam, 1961
Acylation of phenols	Pyrogallol, acetic anhydride	Amberlite IR-120 [H⁺]	Price and Israelstam, 1964
Cyclization of polyenes	Methyl *trans, trans*-farnesate	XE-100	Moriyama et al., 1968
Hydrolysis of amides	General amides	Amberlite IR-100 [H⁺]	Collins, 1957
COOH = Terminal thiohydantoin hydrolysis	Peptides for sequential degradation from the COOH-terminus	Amberlite IR-100 [H⁺]	Yamashita, 1971
Hydrolysis of hydroxy amides	COOH = terminal reduced peptides	Rexyn-100 [H⁺]	Bora et al., 1971
Complete hydrolysis of peptides and proteins	Peptides and proteins	ⓟ—SO₃H	Kunin, 1958
Hydration of isoolefins	Propene and other olefines	ⓟ—SO₃H	Neier and Woellner, 1973
Hydration of acetylenes	HO-C₆H₁₀-C≡CH	Dowex-50 [H⁺] impregnated with 1% HgSO₄, or ZeoKarb 225 [H⁺] impregnated with 1% HgSO₄	Newman, 1953; Billimoria and Maclagan, 1954
Hydrolysis of glucosylamines	*N*-(2-*O*-methyl-*O*-glycosyl)-piperidine	Amberlite IR-120 [H⁺]	Hodge and Rist, 1952
Inversion of sucrose	Sucrose	ⓟ—SO₃H	Bodamer and Kunin, 1951
Hydrolysis of carbohydrates and their derivatives	Maltose, glycogen methyl galacturonate and polymethyl galacturonate	ⓟ—SO₃H	Deuel, 1955

(Continued)

TABLE 13-1 (*Continued*)

Reaction type	Reactants	Resin form	References
Dehydration	Glycols, various sugars, cumene, etc.	Dowex 1-X10	Kunin, 1958
(B) Base-Catalyzed Reactions			
Aldehydes or ketones, condensation with active methylene group	(i) Acetone, cyclopentadiene	Dowex 1-X10	Lorette, 1957; McCain, 1958
	(ii) Furfural and aliphatic aldehydes	$\text{P}-\text{NH}_3^+-\text{OH}$	Mastagli *et al.*, 1952
Knoevenagel reaction	(iii) C_2H_5CHO, CH_3NO_2 Aldehyde, cyanoacetic acid	IRA-400(OH)$^-$ IR-4B or Dowex-3 (acetate form)	Astle and Abbott, 1956; Astle and Gergel, 1956; Hein *et al.*, 1961
Hydrolysis of esters	1-Methyl-3-carbomethoxy pyridinium iodide $\underset{\text{Me}}{\text{[pyridinium-COOMe]}}\, I^- \xrightarrow{H_2O} \underset{\text{Me}}{\text{[pyridinium-COOH]}}\, I^-$	Dowex-1 (OH)$^-$	Kosower and Patton, 1961

Hydration of nitriles to amides	Nicotinonitrile	IRA-400 (OH)⁻	Galat, 1948; Bobbitt and Scola, 1960; Bobbitt and Doolittle, 1964
Dehydrohalogenation	(i) 3-Chloro-1,1-dimethyl indolinium chloride	IRA-400 (OH)⁻	Hinman and Lang, 1964
	(ii) [structure of 3-chloro-1,1-dimethyl indolinium chloride converting to 1,1-dimethyl indolinium hydroxide]		
α- or β-Haloamide cyclization into lactams	Cyclization of α- or β-haloamides into lactams	IRA-400 (OH)⁻	Chatterjee et al; 1965
Peptide blocking group removal	Removal of fluoren-9-methoxycarbonyl (FMOC) groups	Macroreticular piperazine-containing resin	Carpino et al., 1978
O-Alkylation of carboxylate ions (ester formation)	RCOOH + R′—X → RCOOR′	IRA-904	Cainelli and Manescalchi, 1975
Acylation catalyst	Acylation of coenzyme	Polypyridine film	LeGoffic et al., 1976

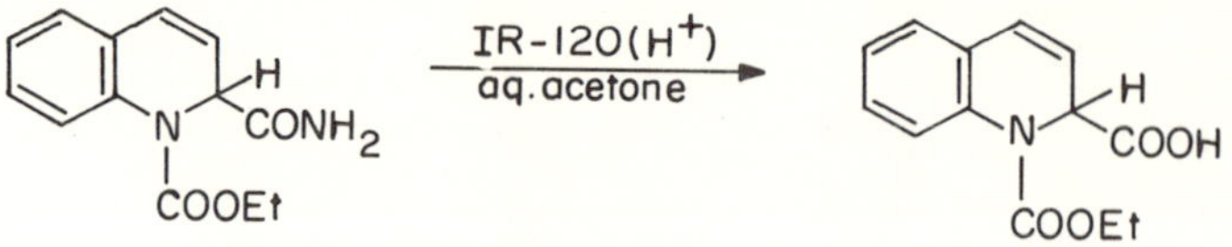

Scheme 13-1

[Structure: 1,2-dihydroquinoline with N-COOEt, 2-H, 2-CONH₂] →[IR-120(H⁺), aq. acetone]→ [Structure: 1,2-dihydroquinoline with N-COOEt, 2-H, 2-COOH]

Scheme 13-1

(H^+) (Collins, 1957) (Scheme 13-1). The same resin has been used for selective cleavage of the thiohydantoin formed from a peptide in preparation for COOH-terminal identification (Yamashita, 1971). Earlier, cold 6 N HCl was used for this cleavage reaction (Cromwell and Stark, 1969), but these conditions were considered too drastic since cleavage of the peptide bond also took place. Thus, the method using aqueous HCl was not suitable for sequential degradation from the COOH-terminus of the peptide.

2. Increased yields due to the use of resin acids: Increased yields have been reported in many cases. For example, the resin-catalyzed hydrolysis of N-(2-O-methyl-D-glucosyl)piperidine resulted in a yield of 80% as compared with the 61% obtained when the hydrolysis was conducted with 1 N sulfuric acid (Hodge and Rist, 1952). In most cases, the yields using resin acids were found to be comparable to those obtained with low-molecular-weight catalysts.

3. Resin acids have been used to catalyze the hydrolysis of β-hydroxyamides (**1**) (the reduction products of peptides). The hydrolysis proceeds via N → O acyl transfer, but an additional advantage is observed when the resin acid is used. Once the free amino group is formed, it gets bound (ionically) to the resin (Scheme 13-2). On hydrolysis of the rearranged β-amino ester (**2**), the amine component (**3**) remains on the resin while the acid component (**4**) goes into solution. Thus, the separation of the products is simplified (Bora et al., 1976).

4. During the resin-catalyzed transesterification of resin peptides, the ester of the side-chain function in protected peptides is formed whereas in the normal acidolysis of resin peptides, the benzyl-based protecting

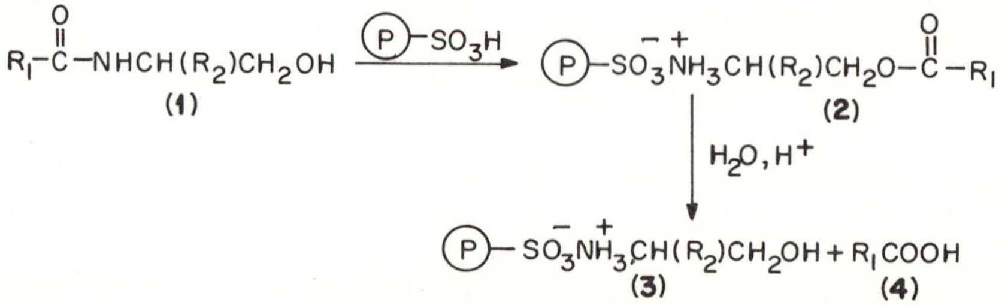

$$R_1-\overset{O}{\overset{\|}{C}}-NHCH(R_2)CH_2OH \quad \xrightarrow{\textcircled{P}-SO_3H} \quad \textcircled{P}-SO_3^-\overset{+}{N}H_3CH(R_2)CH_2O-\overset{O}{\overset{\|}{C}}-R_1$$
(**1**) (**2**)

$$\downarrow H_2O, H^+$$

$$\textcircled{P}-SO_3^-\overset{+}{N}H_3CH(R_2)CH_2OH + R_1COOH$$
(**3**) (**4**)

Scheme 13-2

groups are cleaved. The resulting protected peptide fragments can be further used for fragment condensation (Halpern *et al.*, 1968).

III. POLYSTYRENE–ALUMINUM CHLORIDE AS A LEWIS ACID CATALYST

Among a number of metal chlorides used in organic synthesis, anhydrous aluminum chloride is undoubtedly one of the most effective of Lewis acid catalysts. During the chloromethylation of polystyrene, using aluminum chloride as catalyst, it was observed that all the aluminum chloride could not be removed, even after repeated washing. This was attributed to the formation of a tightly bound polystyrene–aluminum chloride complex. Complex formation was demonstrated by the increase in color (yellow) intensity of the polymer, and by the development of a new characteristic ir band at 1650 cm^{-1}. This complex could act as a mild Lewis acid catalyst for certain organic preparations.

The catalyst has been prepared by treating co(polystyrene–DVB) (1.8% cross-linked, 50–100 mesh) with anhydrous aluminum chloride in carbon disulfide, followed by addition of cold water (Neckers *et al.*, 1972). The polymeric catalyst is filtered off and washed successively with water, ether, acetone, and hot isopropanol, and finally dried in a vacuum oven.

In solution, the catalyst–polymer swells in certain kinds of solvents, such as benzene, releasing the Lewis acid. Thus, the catalytic activity depends on the solvent. Although the catalyst was not effective in Friedel–Craft-type reactions, it was successfully used in certain dehydration reactions, e.g., in the formation of esters (Blossey *et al.*, 1973b), ethers (Neckers *et al.*, 1972), and acetals (Blossey *et al.*, 1975).

The polymer-supported catalyst has some advantages over the ordinary reagent. It has a shelf life of over a year and is not easily hydrolyzed by water. In reactions involving ether formation, the yields are generally high and, in addition, sensitive carbinols such as dicyclopropylcarbinol react more cleanly than they do with the ordinary reagent directly. The scope of the method is further demonstrated by formation of mixed ethers in high yields.

IV. POLYMER-BASED "SUPER ACID" CATALYSTS

The cracking and isomerization of alkanes is catalyzed by Lewis acids. With normal Lewis acids such reactions take place at temperatures and pressures much higher than ambient. Recently, it has been reported that certain "superacids," a combination of Lewis and proton acids, can catalyze such reactions under milder conditions.

A "superacid polymer catalyst" was obtained by binding aluminium chloride to sulfonated, macroporous co(polystyrene–DVB) (Magnotta et al., 1976; Magnotta and Gates, 1977a,b). The catalyst was active in bringing about cracking and isomerization of n-hexane at 357°C at atmospheric pressure.

These polymers have been referred to as "organometallic superacid catalysts." Electron microprobe X-ray analysis has shown that bound aluminium was uniformly distributed in the polymer beads. Such acidic catalysts may prove useful as insoluble Friedel–Craft catalysts, although many of the cracking and isomerizations reported can be conveniently done with zeolites (molecular sieves) (Meier and Uytterhoeven, 1973) albeit at higher temperatures.

Lewis acids supported on graphite as well as the perfluorinated resin sulfonic acids (Nafion-H) have been studied for their abilities to perform Friedel–Crafts alkylation of benzene and transalkylation of alkylbenzenes (Olah et al., 1977). The graphite-supported materials were less stable than the perfluorinated resin sulfonic acid. Subsequent studies investigated the use of this latter resin in the methylation of phenols, benzene, and alkylbenzenes (Kaspi and Olah, 1978; Kaspi et al., 1978), the nitration of aromatics (Olah et al., 1978a), the rearrangement of allyl alcohols to aldehydes (Olah et al., 1978b), the pinacolone rearrangement (Olah and Meidar, 1978) and isomerizations of alkylbenzenes (Olah and Kaspi, 1978).

V. POLYMERIC ESTEROLYTIC CATALYSTS

Nucleophiles such as imidazole, pyridine (Letsinger and Savereide, 1962), and hydroxylamine are known to be powerful catalysts of ester hydrolyses. An extensive study of polymeric analogs of these compounds has shown that they too act as effective esterolytic catalysts (Overberger et al., 1965; Overberger and Smith, 1975). The topic has been reviewed (Overberger and Salamone, 1969; Morawitz, 1969; Manecke and Storck, 1978; Overberger et al., 1978) and only important features of these polymeric catalysts will be discussed here.

Polymeric imidazole catalysts are the synthetic prototype of esterolytic enzymes because, like the enzymes, they have binding centers for the substrate as well as for the catalytic groups. In case of the homopolymer, polyvinylimidazole, the partially protonated groups can act as binding sites, provided the substrate is negatively charged. Thus at pH 7.5, when

V. Polymeric Esterolytic Catalysts

5% of the basic groups in polyvinylimidazole are protonated, it acts as a highly effective catalyst for the solvolysis of substrates carrying a negative charge, e.g., for p-nitrophenyl hydrogen phthalate (1) or for polymeric substrates such as a copolymer of acrylic acid with p-nitrophenyl-p-vinylbenzoate (2a) or 2,4-dinitrophenyl-p-vinylbenzoate (2b) (Letsinger and Klaus, 1964).

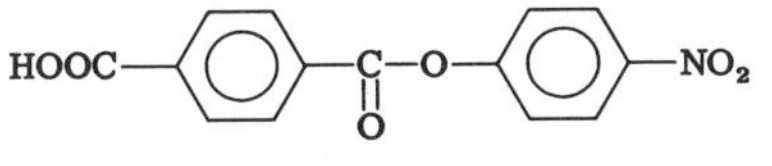

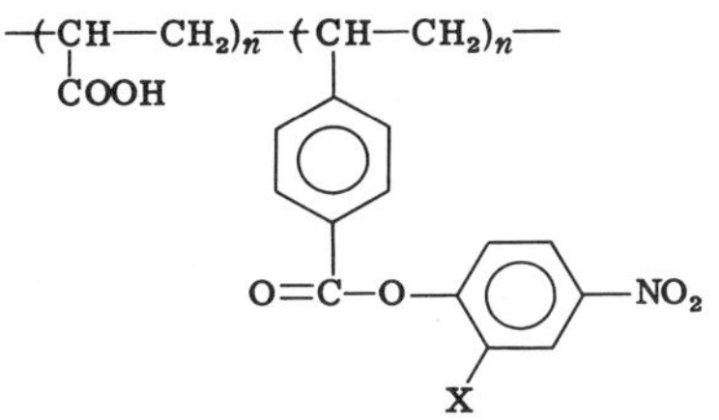

Whereas two mutually reactive compounds are rendered inert by attachment to two different polymers, Letsinger and Klaus (1964) have demonstrated that in case of soluble polymers, the compounds can react when the two polymers can associate due to the presence of oppositely charged groups (cf., Letsinger and Wagner, 1966).

The copolymer of N-vinylimidazole and acrylic acid has been found to be an effective esterolytic catalyst for the hydrolysis of p-nitrophenyl acetate (Letsinger and Klaus, 1964; Shimidzu et al., 1974). The carboxyl groups in this polymer can bind a positively charged substrate, e.g. (3). Thus, the catalytic esterolytic activity of such a polymer for the positively charged 3, the neutral 4 and the negatively charged 5 is in the order:

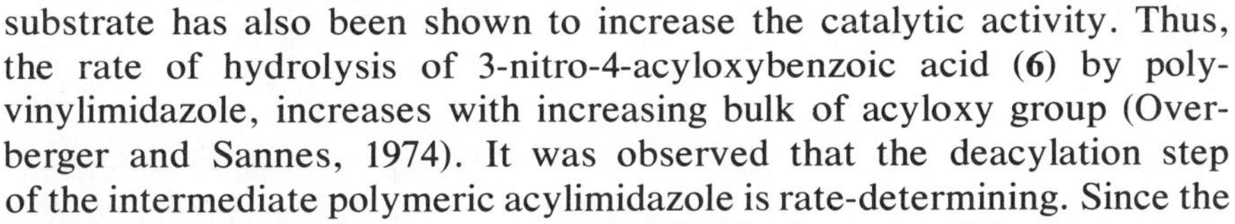

Non-ionic (hydrophobic) association of the polymeric catalyst–substrate has also been shown to increase the catalytic activity. Thus, the rate of hydrolysis of 3-nitro-4-acyloxybenzoic acid (6) by polyvinylimidazole, increases with increasing bulk of acyloxy group (Overberger and Sannes, 1974). It was observed that the deacylation step of the intermediate polymeric acylimidazole is rate-determining. Since the acylated polymer becomes more hydrophobic, because of the extra group attached to the polymer, apolar associations increase with the chain length of the acyl group, and, hence increase the catalytic activity.

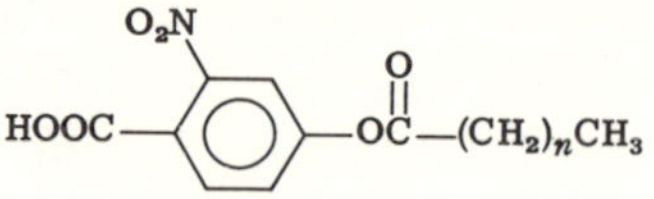

$$\text{HOOC}-\underset{(6)}{\underset{n = 1, 6, 11, 17}{\overset{O_2N}{\bigcirc}}}-O\overset{O}{\overset{\|}{C}}-(CH_2)_nCH_3$$

It was shown (Overberger and Sannes, 1974) that the polymeric imidazole was about 10^3 times as reactive as the monomeric imidazole in the hydrolysis of *p*-nitrophenyl acetate and certain 3-nitro-4-acyloxybenzoic acids. The increased catalytic activity of the polymeric imidazoles has been attributed to the cooperative interactions of the imidazole moieties anchored on the polymer chain at regular intervals (Scheme 13-3). Although practical applications of polyvinylimidazoles as catalytic nucleophiles in ester hydrolysis have not been made, it appears that they can provide quite effective and reusable solid-phase catalysts for such reactions.

Hydroxamate ion, which is also a powerful nucleophilic catalyst for ester hydrolysis (Aubort and Hudson, 1970), has been incorporated into a polymer to act as a bifunctional solid-phase catalyst (Kunitake *et al.*, 1974; Kunitake and Okahata, 1975; 1976a,b,c; Okahata and Kunitake, 1977). Two types of catalysts were tested for catalytic activity: (i) a ternary copolymer incorporating acrylamide (AAm), *N*-vinyl-2-methylimidazole (MIm), and *N*-phenylacrylohydroxamate (PHA) and (ii) a binary polymer of AAm and PHA. Acylation of PHA residues in the binary polymer by *p*-nitrophenyl acetate was rapid, but the subsequent hydrolysis of acetyl PHA was slow. On the other hand, in the case of the ternary polymer catalyst, although the PHA groups were readily acylated, they could also be deacetylated rapidly by pendate and proximate MIm groups. The ternary polymer has been suggested as an enzyme model containing two catalytic centers that react in succession.

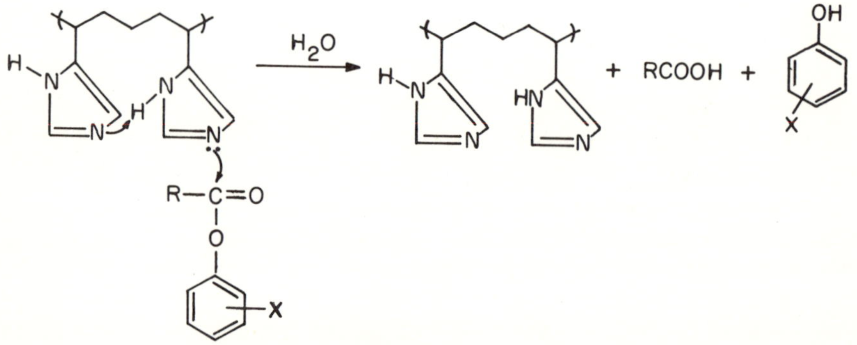

Scheme 13-3

Substituted poly(ethyleneimines) (PEIs) have been shown to be effective catalysts for esterolytic cleavages and pyrophosphate hydrolyses. PEIs containing imidazole groups and long-chain alkyl groups have high catalytic activity (Keifer et al., 1972; Klotz et al., 1969, 1971; Klotz and Stryker, 1968; Johnson and Klotz, 1973; Overberger and Dixon, 1977; Pshezhetskii et al., 1975, 1977a,b; Royer and Klotz, 1969; Spetnagel and Klotz, 1977; Turyn et al., 1974). Hollow PEI "ghosts" containing imidazole groups have been prepared and shown to be effective catalysts (Meyer and Royer, 1977).

VI. POLYMER-SUPPORTED PHASE-TRANSFER CATALYSTS

The use of phase-transfer catalysts bound to polymeric supports has been reported. The catalytic functional groups anchored to the polymer were (i) quaternary ammonium salts (Fig. 13-1a,b,c), (ii) phosphonium salts (Fig. 13-1d), (iii) Crown ethers (Fig. 13-1e), and (iv) cryptands (Fig. 13-1f). Chloromethylated, 2–4% cross-linked polystyrene and silica gel were used as the support polymers, and the catalyst groups were anchored either by the reaction with the corresponding amine or phosphine or by absorption. Spacer-arms were used for linking the crown ether and cryptand (Cinouini et al., 1976; Cinquini et al., 1975; Molinari et al., 1977; Tundo, 1977, 1978).

The catalytic activity of these anchored catalysts was lower than that of the corresponding nonimmobilized phase-transfer catalysts (Dockx, 1973; Landini et al., 1974, 1975; Cinquini et al., 1975). However, it was clearly indicated that the catalytic activity of the parent group was retained on binding to the polymer. Insertion of spacer-arms has been found to improve catalytic activity (Molinari et al., 1977; Tundo, 1978). For phase-transfer catalysts based on tetraalkyl ammonium ions, the organophilicity of the polymer-bound group was directly dependent upon the size of the alkyl group. The catalysts, recovered by filtration from the reaction mixture did not show any change in activity after several cycles.

Crown polyethers have also been synthesized (Kopolow et al., 1971, 1973) by polymerization of the corresponding vinyl monomer (4'-vinylmonobenzo-15-crown-5) (Scheme 13-4). The polymeric crown ethers were obtained as amorphous solids, softening at 122–128°C and with an average molecular weight of 11,600. The polymeric polyethers, just like the monomeric crown ethers, had the ability to bind alkali metal, particularly potassium, ions. Hence, these have been used for extraction of potassium. The crown polyethers form a complex with $KMnO_4$, which is

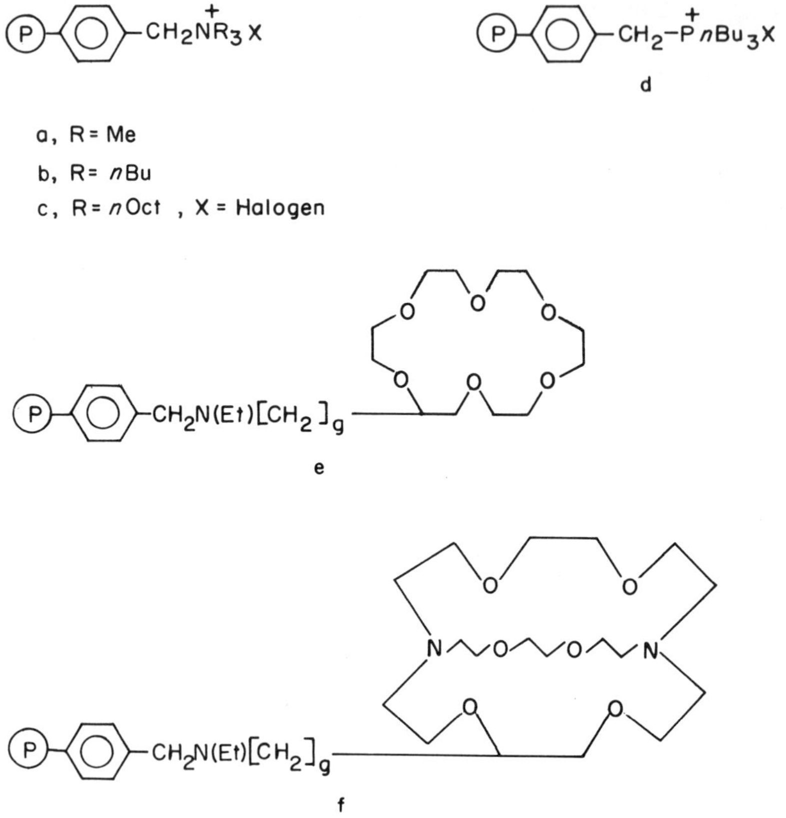

Fig. 13-1

more efficient as an oxidizing agent in nonaqueous media (Sam and Simmons, 1972).

Complexes of alkali metal ions with polymeric crown ethers have been studied by Kopolow *et al.* (1971) (cf., Pedersen and Frensdorff, 1972, and the more recent work of Smid *et al.*, 1974). Although the stoichiometric ratio of monomeric or polymeric crown ether to the alkali metal ion was the same (2:1 in case of 15C5 ether system and 1:1 in case of 18C6 ether system), the polymeric 15C5 ether formed more stable complexes with fluorenylpotassium or KCNS than did the monomeric 15C5 ethers. By contrast, the 1:1 complexes, either with 18C6 monomeric or polymeric crown ethers, had the same stability. The difference in stoichiometry of 15C5 and 18C6 ethers has been reasonably attributed to the differences in the diameters of the two ethers, but when two 15C5 crown units are held in close proximity in the polymer, the polymeric 2:1 complex can be

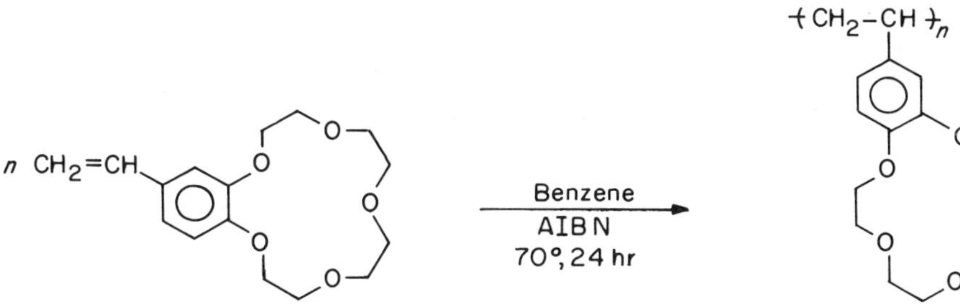

VI. Polymer-Supported Phase-Transfer Catalysts

Scheme 13-4

expected to be more stable than the three-unit complex in the 15C5 monomeric system (Scheme 13-5). Since the formation of 1:1 complexes of 18C6 ethers does not involve "a cooperative effect" of proximate groups, these would be expected to show less difference in stability between the monomeric or the polymeric system.

Polymeric quaternary ammonium salts, polyvinylpyridine dimethyl-

Crown 15C5

Crown 18C6

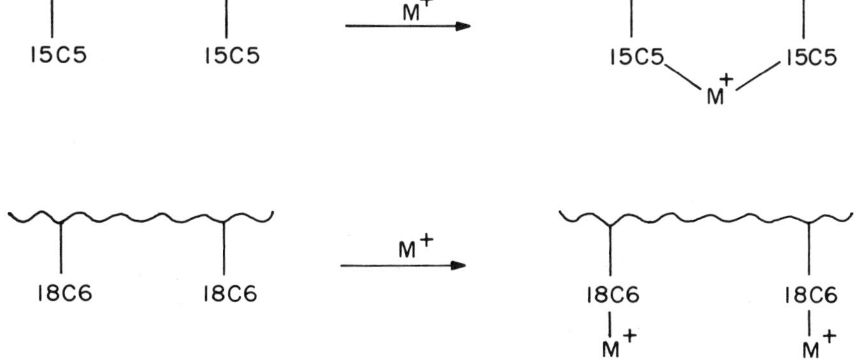

Scheme 13-5

polyethylene glycols, have also been used as phase-transfer agents in oxidations, displacement reactions, alkylations, and carbenation reactions (Chiellini and Solaro, 1977; Chiellini *et al.*, 1977; Lee and Chang, 1978; Vander Zwan and Hartner, 1978).

VII. POLYMERIC TRIPHASE CATALYSTS

The use of anion-exchangers as the source of nucleophiles in nucleophilic substitution reactions has been referred to (Chapter 12). Brown and Jenkins (1976) and Regen (1975, 1976, 1977) have developed a triphase-catalysis method for carrying out such nucleophilic substitution reactions. The resin in these reactions acts as a polymer-bound phase-transfer catalyst. Quaternary ammonium groups that were introduced into the resin contained large alkyl groups to act as the linked phase-transit catalyst.

On keeping the resin in contact with an aqueous phase, such as a solution of sodium cyanide, and an organic phase, such as an alkyl bromide, alkyl cyanides were formed by nucleophilic substitution (Scheme 13-6). The basic difference between the common heterogeneous catalytic reaction and the one being discussed is that in the latter, the catalyst and each pair of reactants are located in separate phases.

The quaternized resin acts as a phase-transfer catalyst because of the presence of both charged groups (polar, hydrophilic phase) and large alkyl groups (organophilic phase). Thus, the catalyst acts by bringing together the reactants (cyanide ion and alkyl halide) from the two phases. A kinetic study of the reaction, which showed first-order dependence of the reaction rate on the alkyl halide concentration, was made. The reaction rate also showed linear dependence on the amount of catalyst (the number of

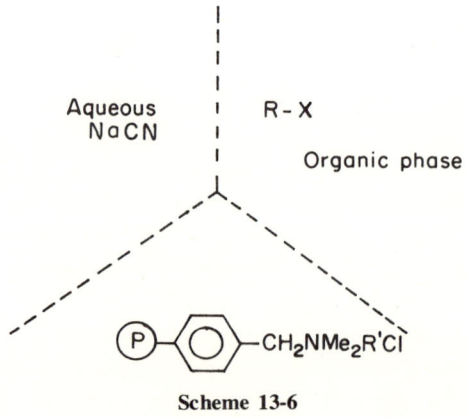

Scheme 13-6

quaternary ammonium groups on the catalyst resin). Recent results also suggest that a molecular size selectivity also exists (Regen and Nigam, 1978), and they support a cosolvent catalytic mechanism (Regen *et al.*, 1979) in which the diffusion and reaction rates may be similar (Regen and Besse, 1979). Unlike other cases where ion-exchange resins have been used to catalyze the cyanide ion displacement, no observable decomposition of quaternary groups was said to occur in this case, and the resin catalyst showed no change in activity on recycling. A check for resin decomposition (Dou *et al.*, 1977) did, however, show that dequarterization of a trimethylbenzylammonium resin occurred.

Triphase catalysts were also used for C-alkylations (Komeili-Zadeh, 1978) and have been shown to promote asymmetric addition in carbene addition reactions (Chiellini and Solaro, 1977; Colonna *et al.*, 1978). Ammonium groups have been replaced by phosphoric triamides (Tomoi *et al.*, 1978), phosphonium groups (Tundo, 1978), and polyethylene glycol (Regen and Dulak, 1977) to provide alternate phase-transfer agents.

For a more in depth treatment of this area see Regen (1979).

VIII. POLYMER-BASED PHOTOSENSITIZERS

Synthetic organic chemists have conceived of novel ways in which the immobilization of photochemically sensitive species on polymeric supports might be useful (Williams, 1974; Kamogawa, 1974; Ledwith, 1975). One such way is to use certain polymer-bound photosensitizers in organic reactions.

One of the common uses of photosensitizers in organic reactions has been to produce singlet molecular oxygen ($^1O_2$) using various dyestuffs. As early as 1967, Leermark and James reported the use of a polymer-based carbonyl sensitizer. Although there is some reason to believe that Leermark and James' photosensitizing of the reaction may have been due to the presence of the dissolved monomer, the idea still has a great deal of merit.

It has been shown that heterogeneous photosensitizers can generate singlet oxygen. Williams *et al.* (1973) reported the use of several complexes of dyes with ion-exchange resins for the photochemical generation of $^1O_2$. Similarly, Nilson and Kearns (1974) employed photosensitizers adsorbed on silica gel. In these two cases, the photosensitizer was not covalently bound to the insoluble support and could leach into the reaction mixture. Subsequently, several reports have been published regarding the use of polymer-bound photosensitizers and their applications in organic reactions (Blossey *et al.*, 1973a; Blossey and Neckers, 1974; Schaap *et al.*, 1975; Rosenthal and Archer, 1974).

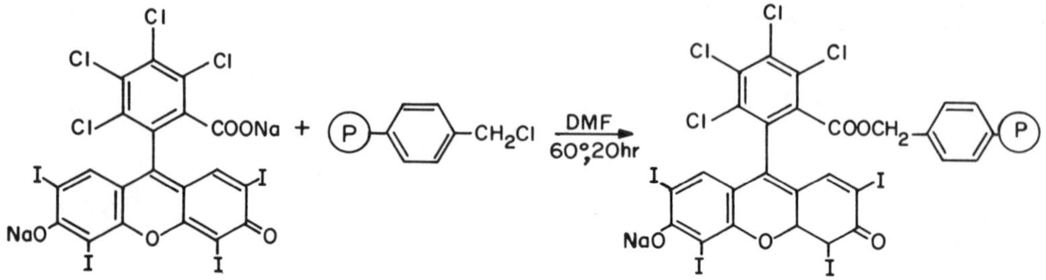

Scheme 13-7

A. Polymer-Bound Rose-Bengal

The dye Rose-Bengal, in solution or as a suspension, is a very suitable sensitizer for generating singlet oxygen (Foote, 1968; Kearns, 1971). There are several limitations to the dyes as photosensitizers: (a) When the sensitizer is used in a solution, solvents which can be used for the reaction are limited. (b) The dye bleaches if longer reaction times are used. (c) The dye can sometimes react with the reactants or with the products. (d) Separation of the products from the dye in the reaction mixture is often difficult.

In order to circumvent these limitations, Blossey et al. (1973a) and Schaap et al. (1975) reported the preparation of this photosensitizer covalently bound to Merrifield-type resins via an ester bond (Scheme 13-7).

TABLE 13-2

Photooxidation of Organic Compounds with ⓟ—Rose-Bengal

Singlet oxygen acceptor	Product	Yield (% isolated)
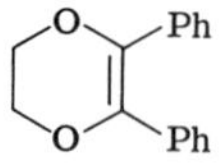 dioxene-Ph₂	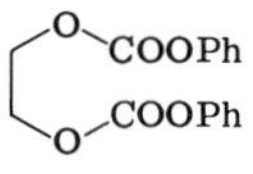 COOPh / COOPh	95
 cyclohexadiene	 endoperoxide	69
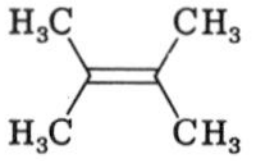 (H₃C)₂C=C(CH₃)₂	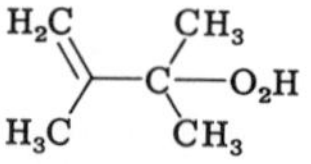 H₂C=C(CH₃)–C(CH₃)₂–O₂H	82

VIII. Polymer-Based Photosensitizers

Schaap et al. (1977) and Zaklika et al. (1978) have reported on the use of the polymer in studies of chemiluminescence.

In addition to Rose-Bengal, two other phthalein dyes (viz., eosin-Y, and fluorescein) as well as two porphyrins (chlorophyllin and hematoporphyrin) were also linked to the polymer. These proved to be considerably less efficient than polymer-bound Rose-Bengal. Three types of reactions of singlet molecular oxygen were carried out in high yield using polymer-based Rose-Bengal (Table 13-2). These were (i) 1,4-cyclo-addition with conjugated dienes to yield cyclic peroxides; (ii) an "ene"-type reaction to give allylic hydroperoxides; and (iii) 1,2-cycloaddition to give 1,2-dioxetanes, which cleave thermally to carbonyl-containing products.

A polymer-bound Rose-Bengal photosensitizer is now commercially available under the trade name, SENSITOX® (Hydron Labs, New Brunswick, New Jersey). In a technical brochure, the manufacturer claims the following advantages: (i) Being insoluble in virtually all solvents, it is readily separated from the products and is reusable. (ii) The sensitizer can be used in a variety of solvents in which the polymer can swell. (iii) Oxidation (bleaching) of the photosensitizer does not take place. (iv) The possibility of developing a continuous process to carry reactions using immobilized Rose-Bengal exists. (v) The efficiency of SENSITOX® is about 65% of that of free Rose-Bengal, but the high yields of the products and ease of isolation more than compensate for the slightly longer reaction periods.

B. Other Polymeric Photosensitizers

Anthracene (Foote and Wexler, 1964) and 9,10-diphenylanthracene (Corey and Tayer, 1964) are known to produce singlet oxygen. Rosenthal and Archer (1974) incorporated the anthracene ring into a polymer by copolymerizing 9,10-di-*p*-styrylanthracene (Scheme 13-8). Photooxidation of several compounds using this polymer was reported. Another polymer containing the benzophenone group was prepared by reaction of *p*-benzoylbenzoic acid with chloromethylated co(polystyrene–DVB) (Blossey and Neckers, 1974) (Scheme 13-9). Cycloaddition reactions of tetrachloroethylene to cyclopentadiene and the dimerization of indene and coumarin were studied with this reagent. The products formed in the

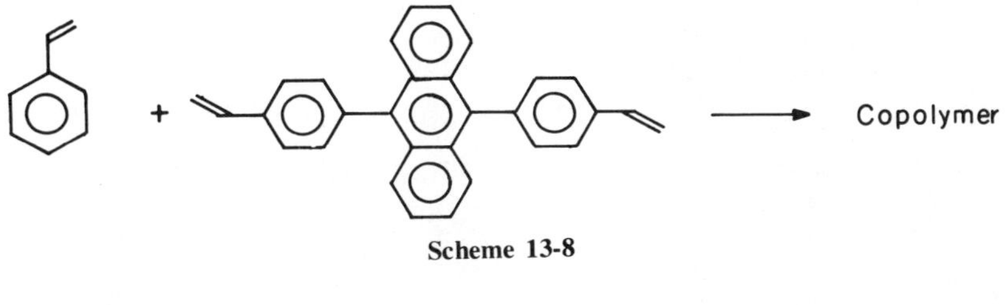

Scheme 13-8

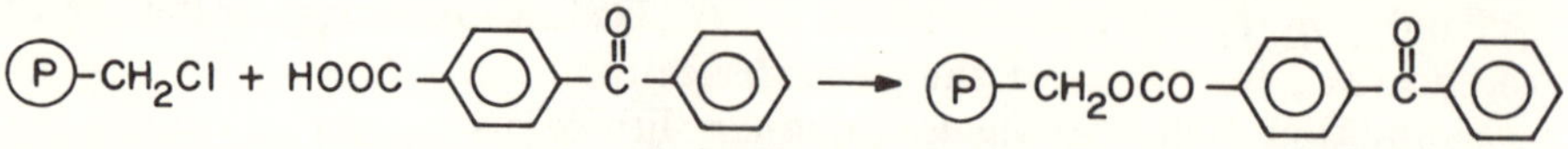

$$\text{(P)}-CH_2Cl + HOOC-\underset{}{\bigcirc}-\overset{O}{\underset{\|}{C}}-\underset{}{\bigcirc} \longrightarrow \text{(P)}-CH_2OCO-\underset{}{\bigcirc}-\overset{O}{\underset{\|}{C}}-\underset{}{\bigcirc}$$

Scheme 13-9

reaction were compared to those formed with benzophenone, and in general, purer products were obtained with the polymeric reagent. The polymeric photosensitizer offered the usual advantage of facile separation from the reaction mixture, and it was reusable. The photooxidation of methionine to methionine sulfoxide by immobilized methylene blue has been reported (Lewis and Scouten, 1976). It was shown that light accelerated the inactivation of lysozyme by the immobilized dye. The sensitization of photooxidation and reduction by copolymers of 2-(9,10-anthraquinoyl) methacrylate–methyl methacrylate has also been reported (Nakahira et al., 1978).

REFERENCES

Astle, M. J., and Abbot, F. P. (1956). J. Org. Chem. **21**, 1228.
Astle, M. J., and Gergel, W. C. (1956). J. Org. Chem. **21**, 493.
Astle, M. J., and Oscar, J. A. (1961). J. Org. Chem. **26**, 1713.
Aubort, J. D., and Hudson, R. H. (1970). J. Chem. Soc. Chem. Commun., 938.
Billimoria, J. D., and Maclagan, N. F. (1954). J. Chem. Soc. 3257.
Blossey, E. C., and Neckers, D. C. (1974). Tetrahedron Lett., 323.
Blossey, E. C., Neckers, D. C., Thayer, A. L., and Schapp, A. B. (1973a). J. Am. Chem. Soc. **95**, 5820.
Blossey, E. C., Turner, L. M., and Neckers, D. C. (1973b). Tetrahedron Lett., 1823.
Blossey, E. C., Turner, L. M., and Neckers, D. C. (1975). J. Org. Chem. **40**, 959.
Bobbitt, J. M., and Doolittle, R. E. (1964). J. Org. Chem. **29**, 2298.
Bobbitt, J. M., and Scola, D. A. (1960). J. Org. Chem. **25**, 560.
Bodamer, G., and Kunin, R. (1951). Ind. Eng. Chem. **43**, 1082.
Bora, J. M., Saund, A. K., Sharma, I. K., and Mathur, N. K. (1976). Indian J. Chem. **14B**, 722.
Brown, J. M., and Jenkins, J. A. (1976). J. Chem. Soc. Chem. Commun., 458.
Cainelli, G., and Manescalchi, F. (1975). Synthesis, 723.
Carpino, L., Williams, J. R., and Lopusinski, A. (1978). J. Chem. Soc. Chem. Commun., 450.
Chatterjee, B. G., Rao, V. V., and Mazumdar, B. N. G. (1965). J. Org. Chem. **30**, 4101.
Chiellini, E., and Solaro, R. (1977). J. Chem. Soc. Chem. Commun., 231.
Chiellini, E., Solaro, R., and D'Antone, S. (1977). Makromol. Chem. **178**, 2545, 3165.
Cinouini, M., Colonna, S., Molinari, H., Montanari, F., and Tundo, P. (1976). J. Chem. Soc. Chem. Commun., 394.
Cinquini, M., Montanari, F., and Tundo, P. (1975). J. Chem. Soc. Chem. Commun., 393.
Collins, R. F. (1957). Chem. Ind., 736.
Colonna, S., Fornasier, R., and Pfeiffer, U. (1978). J. Chem. Perkin Trans. I, 8.
Corey, E. J., and Tayer, W. C. (1964). J. Am. Chem. Soc. **86**, 3881.

References

Cromwell, L. D., and Stark, G. R. (1969). *Biochemistry* **8**, 4735.
Davies, C. W., and Thomas, G. G. (1952). *J. Chem. Soc.* 1607.
Deuel, H. (1955). *Mitt. Geb. Lebensmittelunters Hyg.* **46**, 12.
Dockx, J. (1973). *Synthesis,* 441.
Dou, H. J.-M., Gallo, R., Hassanaly, P., and Metzger, J. (1977). *J. Org. Chem.* **42**, 4275.
Foote, C. S. (1968). *Acc. Chem. Res.* **1**, 104.
Foote, C. S., and Wexler, S. (1964). *J. Am. Chem. Soc.* **80**, 3880.
Galat, A. (1948). *J. Am. Chem. Soc.* **70**, 3945.
George, F. V., and Virgil, I. S. (1971). *J. Org. Chem.* **36**, 2548.
Halpern, B., Chow, L., Close, V., and Patton, W. (1968). *Tetrahedron Lett.,* 5163.
Hasek, R. H., Gott, P. G., Meen, R. H., and Martin, J. C. (1963). *J. Org. Chem.* **28**, 2496.
Hein, R. W., Astle, M. J., and Shelton, J. R. (1961). *J. Org. Chem.* **26**, 4874.
Helfferich, F. (1962). "Ion Exchange." McGraw-Hill, New York.
Hinman, R. L., and Lang, J. (1964). *J. Org. Chem.* **29**, 1449.
Hodge, J. E., and Rist, C. E. (1952). *J. Am. Chem. Soc.* **74**, 1498.
Hydron Laboratories, Inc. (1976). Technical Bulletin on Polymer-Bound Sensitizers, "SENSITOX," New Brunswick, New Jersey.
Jain, J. C., Sharma, I. K., Sahni, M. K., Gupta, K. C., and Mathur, N. K. (1977). *Indian J. Chem.* **15B**, 766.
John, E. V. O., and Israelstam, S. S. (1961). *J. Org. Chem.* **26**, 240.
Johnson, T. W., and Klotz, I. M. (1973). *Macromolecules* **6**, 788.
Kamogawa, H. (1974). *Prog. Polym. Sci. Jpn.* **7**, 1.
Kaspi, J., and Olah, G. A. (1978). *J. Org. Chem.* **43**, 3142.
Kaspi, J., Montgomery, D. D., and Olah, G. A. (1978). *J. Org. Chem.* **43**, 3147.
Kearns, D. R. (1971). *Chem. Rev.* **71**, 395.
Keifer, H. C., Congdon, W. I., Scarpa, I. A., and Klotz, I. M. (1972). *Proc. Nat. Acad. Sci. U.S.A.* **69**, 2155.
Klotz, I. M., and Stryker, V. H. (1968). *J. Am. Chem. Soc.* **90**, 2717.
Klotz, I. M., Royer, G. P., and Sloniewsky, A. R. (1969). *Biochemistry,* **8**, 4752.
Klotz, I. M., Royer, G. P., and Scarpa, I. A. (1971). *Proc. Nat. Acad. Sci. U.S.A.* **68**, 263.
Komeili-Zadeh, H., Dou, H. J.-M., and Metzger, J. (1978). *J. Org. Chem.* **43**, 156.
Kopolow, S., Hogen, T. E., and Smid, J. (1971). *Macromolecules* **4**, 359.
Kopolow, S., Esch, T. E. H., and Smid, J. (1973). *Macromolecules* **6**, 133.
Kosower, E. M., and Phatton, J. W. (1961). *J. Org. Chem.* **26**, 1318.
Kunin, R. (1958). "Ion Exchange Resins." Wiley, New York.
Kunitake, T., and Okahata, Y. (1975). *Bioorg. Chem.* **4**, 136.
Kunitake, T., and Okahata, Y. (1976a). *J. Am. Chem. Soc.* **98**, 7793.
Kunitake, T., and Okahata, Y. (1976b). *Macromolecules* **9**, 15.
Kunitake, T., and Okahata, Y. (1976c). *Adv. Poly. Sci.* **20**, 161.
Kunitake, T., Okahata, Y., and Ando, R. (1974). *Macromolecules* **7**, 140.
Landini, D., Montanari, F., and Pirisi, F. M. (1974). *J. Chem. Soc. Chem. Commun.,* 879.
Landini, D., Montanari, F., Pirisi, F. M., and Maia, H. M. (1975). *Gazetta* **105**, 863.
Ledwith, A. (1975). *Phys. Chem. Ser. Two* **8**, 253.
Lee, D. G., and Chang, V. S. (1978). *J. Org. Chem.* **43**, 1532.
Leermark, P. A., and James, F. C. (1967). *J. Org. Chem.* **32**, 2843.
LeGoffic, F., Sicsic, S., and Vincent, C. (1976). *Tetrahedron Lett.,* 2845.
Letsinger, R. L., and Klaus, I. S. (1964). *J. Am. Chem. Soc.* **86**, 3884.
Letsinger, R. L., and Savereide, T. J. (1962). *J. Am. Chem. Soc.* **84**, 3112.
Letsinger, R. L., and Wagner, T. E. (1966). *J. Am. Chem. Soc.* **88**, 2062.
Lewis, C., and Scouten, W. H. (1976). *Biochim. Biophys. Acta* **444**, 326.

Loev, B., and Massengale, J. T. (1957). *J. Org. Chem.* **22,** 988.
Lorette, N. B. (1957). *J. Org. Chem.* **22,** 346.
McCain, G. H. (1958). *J. Org. Chem.* **23,** 632.
Magnotta, V. L., and Gates, B. C. (1977a). *J. Polym. Sci. Polym. Chem. Ed.* **15,** 1341.
Magnotta, V. L., and Gates, B. C. (1977b). *J. Catal.* **46,** 266.
Magnotta, V. L., and Gates, B. C., and Schuit, G. C. A. (1976). *J. Chem. Soc. Chem. Commun.,* 342.
Manecke, G., and Storck, W. (1978). *Angew. Chem. Int. Ed. Engl.* **17,** 657.
Mastagli, A., Floch, A., and Durr, G. (1952). *Comp. Rend.* **235,** 1402.
Meier, W. M., and Uytterhoeven, J. B. (1973). "Molecular Sieves," *Amer. Chem. Soc. Symp. No.* 121, Washington, D.C.
Meyers, W. E., and Royer, G. P. (1977). *J. Am. Chem. Soc.* **99,** 6141.
Mill, P. J., and Crimmin, R. (1957). *Biochim. Biophys. Acta.* **23,** 432.
Moffatt, J. G. (1963). *J. Am. Chem. Soc.* **85,** 1118.
Molinari, H., Montanari, F., and Tundo, P. (1977). *J. Chem. Soc. Chem. Commun.,* 639.
Morawitz, H. (1969). *Ad. Catal.* **24,** 341.
Moriyiama, H., Sugihara, Y., and Nakanishi, K. (1968). *Tetrahedron Lett.,* 2851.
Mowery, Jr. D. F. (1961). *J. Org. Chem.* **26,** 3484.
Nakahira, T., Shinomiya, E., Fukumoto, T., Iwabuchi, S., and Kojima, K. (1978). *Eur. Polym. J.* **14,** 317.
Neckers, D. C., Kooistra, D. A., and Green, G. W. (1972). *J. Am. Chem. Soc.* **94,** 9284.
Neier, W., and Woellner, J. (1973). *Chem. Technol.,* 95.
Newman, M. S. (1953). *J. Am. Chem. Soc.* **75,** 4740.
Nilson, R., and Kearns, D. R. (1974). *Photochem. Photobiol.* **19,** 181.
Olah, G. A., and Kaspi, J. (1978). *Nouveau J. Chim.* **2,** 581, 585.
Olah, G. A., and Meidar, D. (1978). *Synthesis,* 358.
Olah, G. A., Kaspi, J., and Bukala, J. (1977). *J. Org. Chem.* **42,** 4187.
Olah, G. A., Malhotra, R., and Narang, S. C. (1978a). *J. Org. Chem.* **43,** 4628.
Olah, G. A., Meidar, D., and Liang, G. (1978b). *J. Org. Chem.* **43,** 3890.
Okahata, Y., and Kunitake, T. (1977). *J. Polym. Sci. Polym. Chem. Ed.* **15,** 2571.
Overberger, C. G., and Dixon, K. W. (1977). *J. Polym. Sci. Polym. Chem. Ed.* **15,** 1863.
Overberger, C. G., and Salamone, J. C. (1969). *Macromolecules* **2,** 553.
Overberger, C. R., and Sannes, K. N. (1974). *Angew. Chem. Int. Ed. Eng.* **13,** 99.
Overberger, C. G., and Smith, T. W. (1975). *Macromolecules* **8,** 401, 407, 416.
Overberger, C. G., Sitaramaiah, R., Pierre, T. St., and Yaroslavsky, S. (1965). *J. Am. Chem. Soc.* **87,** 3270.
Overberger, C. G., Guterl, Jr. A. C., Kawakami, Y., Mathias, L. J., Meenakshi, A., and Tomono, T. (1978). *Pure Appl. Chem.* **50,** 309.
Pedersen, C. J., and Frensdorff, H. K. (1972). *Angew. Chem. Int. Ed. Eng.* **11,** 16.
Price, P., and Israelstam, S. S. (1964). *J. Org. Chem.* **29,** 2800.
Pshezhetskii, V. S., and Lukjanova, A. P. (1976). *Bioorg. Khim (U.S.S.R.)* **2,** 110.
Pshezhetskii, V. S., Lukjanova, A. P., and Kabanov, V. A. (1975). *Bioorg. Khim. (U.S.S.R.)* **1,** 1458.
Pshezhetskii, V. S., Lukjanova, A. P., and Kabanov, V. A. (1977a). *J. Mol. Catal.* **2,** 49.
Pshezhetskii, V. S., Nikolaev, G. M., and Lukjanova (1977b). *Eur. Poly. J.* **13,** 423.
Regen, S. L. (1975). *J. Am. Chem. Soc.* **97,** 5956.
Regen, S. L. (1976). *J. Am. Chem. Soc.* **98,** 6270.
Regen, S. L. (1977). *J. Org. Chem.* **42,** 875.
Regen, S. L. (1979). *Angew. Chem. Int. Ed. Engl.* **18,** 421.
Regen, S. L., and Besse, J. J. (1979). *J. Am. Chem. Soc.* **101,** 4059.

References

Regen, S. L., and Dulak, L. (1977). *J. Am. Chem. Soc.* **99,** 623.
Regen, S. L., and Nigam, A. (1978). *J. Am. Chem. Soc.* **100,** 7773.
Regen, S. L., Besse, J. J., and McLick, J. (1979). *J. Am. Chem. Soc.* **101,** 116.
Rosenthal, I., and Archer, A. J. (1974). *Israel J. Chem.* **12,** 897.
Royer, G. P., and Klotz, I. M. (1969). *J. Am. Chem. Soc.* **91,** 5885.
Sam, D. J., and Simmons, H. E. (1972). *Chem. Tech.* **2,** 450.
Schaap, A. P., Thayer, A. L., Blossey, E. C., and Neckers, D. C. (1975). *J. Am. Chem. Soc.* **97,** 3741.
Schaap, A. P., Burns, P. A., and Zaklika, K. A. (1977). *J. Am. Chem. Soc.* **99,** 1270.
Smid, J., Shah, S. C., Sintra, R., Varma, A. J., and Wong, L. (1979). *Pure Appl. Chem.* **51,** 111.
Shimidzu, T., Furuta, A., and Nakamoto, Y. (1974). *Macromolecules* **7,** 160.
Spetnagel, W. J., and Klotz, I. M. (1977). *J. Polym. Sci. Polym. Chem. Ed.* **15,** 621.
Svec, F., Bares, M., Zajic, J., and Kalal, J. (1977). *Chem. Ind. (London),* 159.
Tomoi, M., Ikeda, M., and Kakiuchi, H. (1978). *Tetrahedron Lett.,* 3757.
Tundo, P. (1977). *J. Chem. Soc. Chem. Commun.,* 641.
Tundo, P. (1978). *Synthesis,* 315.
Turyn, D., Baumgartner, E., and Fernandez-Prini, R. (1974). *Biophys. Chem.* **2,** 269.
Vander Zwan, M. C., and Hartner, F. W. (1978). *J. Org. Chem.* **43,** 2655.
Walter, M., Besendorf, H., and Schmider, O. (1961). *Helv. Chim. Acta* **44,** 1546.
Williams, J. L. R. (1974). In "Polyelectrolytes" (E. Sélégnay, M. Mandel, and U. P. Strauss, eds.), p. 507. Reidel, Dordrecht-Holland, The Netherlands.
Williams, J. R., Orton, G., and Unger, L. R. (1973). *Tetrahedron Lett.,* 4603.
Yamashita, S. (1971). *Biochem. Biophys. Acta* **229,** 301.
Zaklika, K. A., Burns, P. A., and Schaap, A. P. (1978). *J. Am. Chem. Soc.* **100,** 318.

14 Polymer-Bound Catalysts (II) Transition Metal Complexes Bound to Polymers

I.	Polymer-Supported Transition Metal Catalysts	220
II.	Principles of Homogeneous Transition Metal Complex Catalysts	221
III.	Preparation of Polymer-Bound Transition Metal Complexes	222
	A. Preparation of Polymeric Ligands	222
	B. Attachment of the Transition Metal Complex to the Polymeric Ligand	224
IV.	Structure of Polymeric Catalysts	228
V.	Types of Reactions Catalyzed by Polymer-Anchored Catalysts	229
	A. Hydrogenation	229
	B. Hydroformylation	230
	C. Hydrosilylation	232
	D. Other Reactions	232
VI.	Asymmetric Organic Synthesis via Transition Metal Catalysts Bound to Polymeric Chiral Ligands	233
	References	236

I. POLYMER-SUPPORTED TRANSITION METAL CATALYSTS

During the past decade several soluble transition metal catalysts have been developed. In contrast to the heterogeneous catalysts that are widely used in industry, the homogeneous catalysts often have higher substrate selectivity and better reproducibility in the catalyzed reactions. Even with these advantages, homogeneous catalysts have not been preferred because their recovery by separation from the low-molecular-weight reactants and products is difficult and results in a considerable loss of the expensive catalyst (metal). In addition, in many cases the material is not reusable.

More recently, this problem has been circumvented by an intermediate system that involves the preparation of a homogeneous catalyst in a heterogeneous form. This is achieved by binding the homogeneous catalyst to inorganic carrier materials or organic polymers.

The properties of the catalysts, made by chemical binding of a soluble transition metal complex to a polymer, lie between those of homogeneous and those of heterogenous catalysts. The polymer-bound transition metal complex catalysts can have the following advantages.

1. Homogeneous catalytic activity is retained by transition metal complexes on binding to the resin.
2. The economy and convenience of heterogeneous catalysts is attained.
3. The steric environment of the catalyst is altered and substrate selectivity is increased.
4. The catalytic sites (single metal atom) can be separated by binding to the rigid region of the support. By avoiding the formation of ligand-bridged complexes, greater catalytic activity is gained.
5. Polymer-bound catalysts can be employed under conditions comparable to those of conventional homogeneous catalysts, e.g., at temperatures below 100°C and at ambient pressure.

A difference of viewpoint exists about classification of polymer-bound catalysts as homogeneous or heterogeneous (Allum *et al.*, 1973; Haag and Whitehurst, 1973). The dual nature of these catalysts is well recognized (Manassen, 1971; Kohler and Dawans, 1972; Heinemann, 1971). These catalysts are best considered as "heterogenized-homogeneous" (He-Ho) catalysts.

II. PRINCIPLES OF HOMOGENEOUS TRANSITION METAL COMPLEX CATALYSTS

A transition metal complex can function as a homogeneous catalyst if it has an open coordination site to bind the reactants. The catalytic activity of such a catalyst arises from the binding of the reactants to the metal complex and their subsequent transfer to the substrate, thereby regenerating the free catalyst. Alkenes can also be bound to the metal complex via π-bonding. A loss in activity in a homogeneous catalyst takes place when the complex dimerizes during the opening of the active site.

It was, therefore, argued that if the transition metal complexes can be bound to a polymer (functionalized to have certain ligand groups), and if the polymer is rigid enough to separate the binding sites, then active catalysts, similar to the homogeneous ones, should be obtained. In addition, these polymer-bound catalysts could be easily recovered and reused. For these reasons, a large number of research groups are currently engaged in research on the preparation and application of such catalysts to industrial processes. A number of reviews (Manassen, 1971; Kohler and Dawans, 1972; Pitmann and Evans, 1973; Bailar, 1974; Michalska and Webster, 1974, 1975; Grubbs, 1976; Rylander and Greenfield, 1976; Chauvin *et al.*, 1977; Tsuchida and Nishide, 1977) and a large patent literature have appeared on this subject. It is not our aim to compile all the work published in this field, but to review and illustrate the applications of polymer-bound catalysts as one of the groups of reagents for solid-phase synthesis. Tables 1 to 14 give typical examples of polymers of different types that have been used. The table is not intended to be exhaustive. (For a fuller coverage see Chauvin *et al.*, 1977; Tsuchida and Nishide, 1977; Rylander and Greenfield, 1976; and Smith, 1977.)

Some of the polymeric ligands for attaching the homogeneous catalysts are now commercially available (Strem Chemicals Inc., Danvers, Massachusetts) and a technical bulletin in the form of a review has also been published (Grubbs, 1976).

III. PREPARATION OF POLYMER-BOUND TRANSITION METAL COMPLEXES

Some of the earlier attempts to make supported transition metal complexes were directed toward physically trapping the complexes in the pores of molecular sieves (Rony, 1969; Meir and Uytterhoeven, 1973), adsorbing them on carbon, silica or alumina (Robinson *et al.*, 1969; Acres *et al.*, 1966), and supporting them on proteins (Wilson and Whitesides, 1978). We shall exclude these from our discussion and restrict ourselves only to those cases where the metal complex is chemically bound to an organic polymer.

A. Preparation of Polymeric Ligands

Although unfunctionalized copolystyrene can bind certain transition metal complexes via π-bonding through the benzene ring (Pittman, 1971), functionalized polymers containing ligands with oxygen and nitrogen donors are usually prepared for this purpose.

Binding the metal complexes via ionic bonding to ion-exchange resins (e.g., K_2PdCl_4 on Amberlyst A27) and the use of such complexes in carbonylation, hydrogenation, and hydroformylation reactions have been reported (Haag and Whitehurst, 1968; Lazcano and Germain, 1971).

III. Preparation of Polymer-Bound Transition Metal Complexes

Functionalized polymers incorporating neutral, metal binding ligands such as phosphine were prepared as early as 1959 (Rabinowitz and Marcus, 1961; Issleib and Tzschach, 1959). After Merrifield introduced the concept of solid-phase synthesis, the basic idea of using polymer-immobilized transition metal complexes as catalysts burgeoned, and many more polymeric supports containing neutral donor ligands have been prepared.

Phosphine and various amine and cyclopentadienyl ligands are the most commonly employed groups on polymers. Polymeric phosphine ligands have been obtained from soluble, linear polystyrene (Bayer and Schurig, 1975) and 2% cross-linked polystyrene–DVB, as well as from highly cross-linked (20%) polystyrene. The highly cross-linked polymer reduces the possibility of chelation by double bonding. During the preparation, some decomposition has been noted (Abdulla et al., 1976). Polyvinyl chloride and polymethacrylate have also been functionalized to incorporate phosphine groups. In addition to polystyrene, silica has been used as a support material for binding transition metal complexes. Phosphenated silica was prepared by reaction with (2-diphenylphosphinoethyl)-triethoxysilane (Niebergall, 1962), which was itself obtained by addition of diphenylphosphine to ethenyltriethoxysilane (Scheme 14-1). Although the loading on silica was low compared to that on organic resins, the reactive sites on silica were mainly confined to the polymer surface and were more easily accessible to the reactants.

Of the amine groups used to coordinate the metal ligand, by far the most commonly used groups are those containing pyridine moieties. Linear and cross-linked polyvinylpyridines (both 2- and 4-vinyl) have been used to bind transition metals (Chauvin et al., 1977). Bipyridine, a very widely used coordinating ligand, has also been introduced into a polystyrene matrix and shown to bind various transition metals (Card and Neckers, 1977). More specific metal binding groups, such as N,N'-bis(2-pyridylmethyl)-2,2'-diaminobiphenyl (Fig 14-1), have also been incorporated in a polymer support (Melby, 1975).

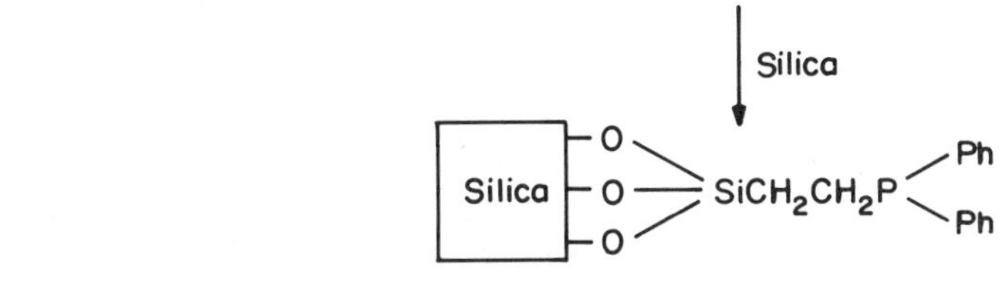

$$Ph_2PH + CH_2 = CHSi(OEt)_3 \longrightarrow Ph_2PCH_2CH_2Si(OEt)_3$$

$$\downarrow \text{Silica}$$

Silica—O, O, O—SiCH$_2$CH$_2$P(Ph)(Ph)

Scheme 14-1

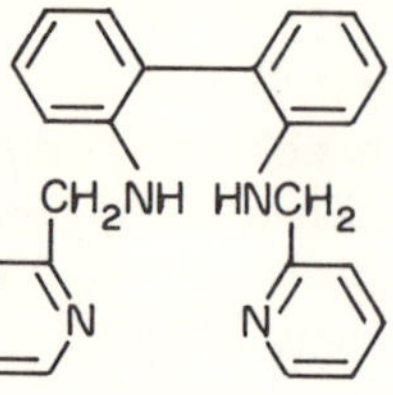

Fig. 14-1

Cyclopentadienyl (cp) ligands, another favored ligand, can be introduced into a styrene polymer by using either a chloromethylated or a lithiated polymer (Grubbs et al., 1973a).

B. Attachment of the Transition Metal Complex to the Polymeric Ligand

The desired metal complex is readily attached to the functionalized polymer by equilibration of the polymer ligand with a metal complex having similar or weaker ligands. The attachment may involve either the displacement of a low-molecular-weight ligand or the addition of the polymeric ligand. The mode of attachment depends upon the nature of the transition metal complex and the amount of cross-linking in the polymer.

For example, with low-cross-linked (2%) polymers, generally two ligands are displaced from complexes such as $RhCl_3(PPh_3)_3$, giving rise to the polymer-bound chelated complex (Scheme 14-2). It appears that the polymer chains are sufficiently mobile to bring non-adjacent sites together (Grubbs et al., 1973b). Even with a 20% cross-linked polystyrene polymer, two ligands are liberated from $RhCl(PPh_3)_3$. However, with this support, ligands attached to the polymer are much less mobile (Bonds et al., 1975).

For the reaction of $[RhCl(COD)_2]_2$ (COD = cyclooctadiene) with a polymeric phosphine ligand, varying amounts of COD displacement have been reported (Grubbs et al., 1973b; Collman et al., 1972). Since 2 and 1.4 moles of COD were displaced from $[RhCl(COD)_2]_2$ using 2 and 20% cross-linked polystyrene, respectively, it was concluded that 20% cross-linked polystyrene was more rigid than 2% cross-linked.

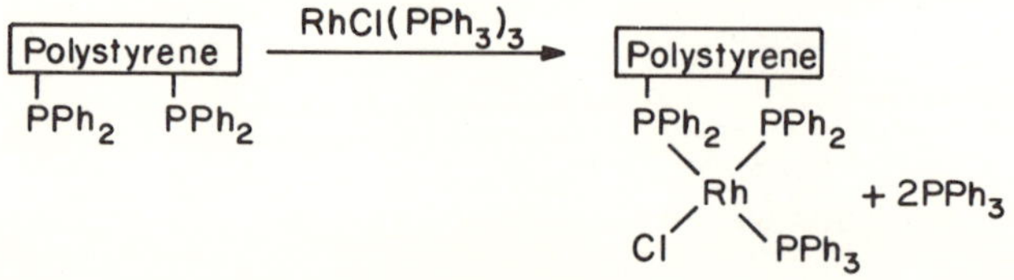

Scheme 14-2

III. Preparation of Polymer-Bound Transition Metal Complexes

According to Strathdee and Given (1974), it is difficult to get metal species that are exact analogs of active homogeneous catalysts such as $RhCl(PPh_3)_3$ (Wilkinson's catalyst) or $RuCl_2(PPh_3)_3$ (Osborn et al., 1966) attached to resins. Steric requirements do not permit more than two adjacent or alternate —PPh_3 units to coordinate to the same metal. In rare cases a third —PPh_3, in close proximity but on a neighboring chain, may participate to provide the equivalent of a tridentate coordination. Hence, the supported catalysts are invariably mixed complexes of polymeric and monomeric ligands.

Capka et al. (1971) observed that while $RhCl(PPH_3)_3$ reacted with polymeric phosphine to give polymers in which one —PPh_3 was displaced by polymeric ligands, reaction of $RhCl_3$ followed by reaction of PPh_3 with polymeric phosphine gave polymers of varying compositions (Scheme 14-3). Pittman and Evans (1973) reported the preparation of some mixed polymeric phosphine–carbonyl complexes. In some cases, the ligand displacement was facilitated by ultraviolet irradiation (Scheme 14-4).

One of the advantages of incorporating metals into polymeric supports is exemplified by a set of titanocene complexes. Soluble titanocene complexes have poor hydrogenation catalytic activity because of the formation of dimers. Attachment of titanocene to a 2% cross-linked resin (Grubbs et al., 1973a) produced a resin complex (**1**) (Scheme 14-5) that showed only 15% catalytic activity in comparison with the homogeneous complex. In contrast to this, attachment of the same complex to a 20% cross-linked polymeric ligand produced a catalyst 60 times more active than the homogeneous catalysts. In each case it is necessary to produce a coordinated unsaturated species (**2**) by reductive elimination of a ligand from the polymeric complex (Scheme 14-5). The results suggest that rigid matrices can isolate these highly reactive species and prevent dimerization.

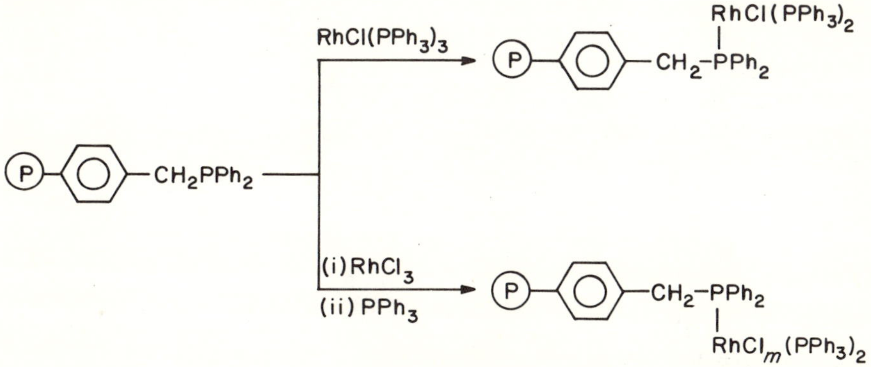

Scheme 14-3

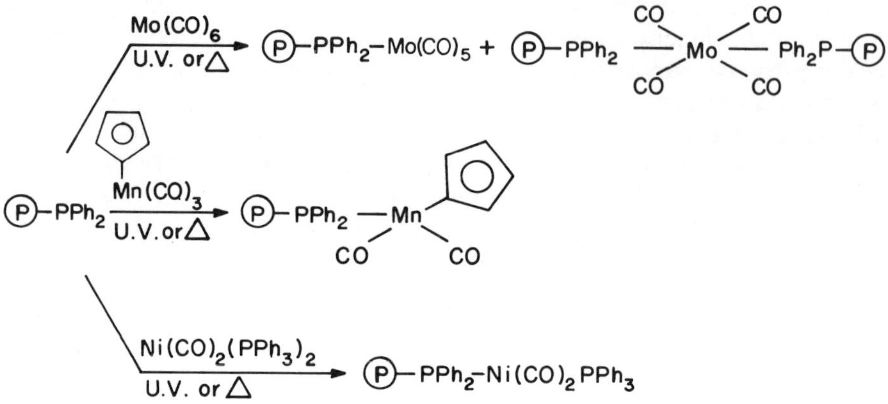

226   14. Polymer-Bound Catalysts (II)

Scheme 14-4

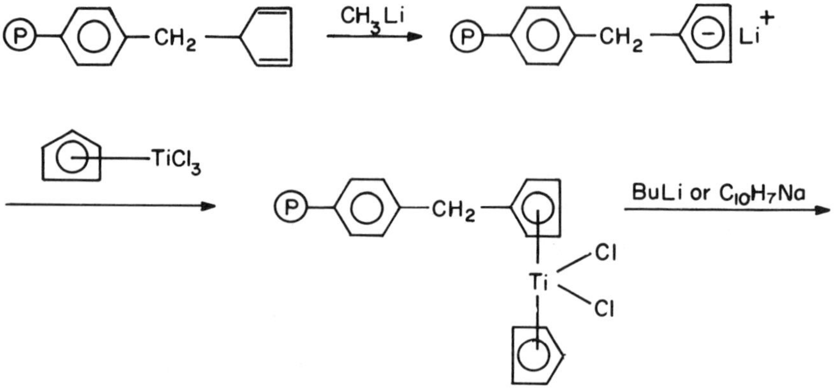

(1) Salmon-pink inactive complex

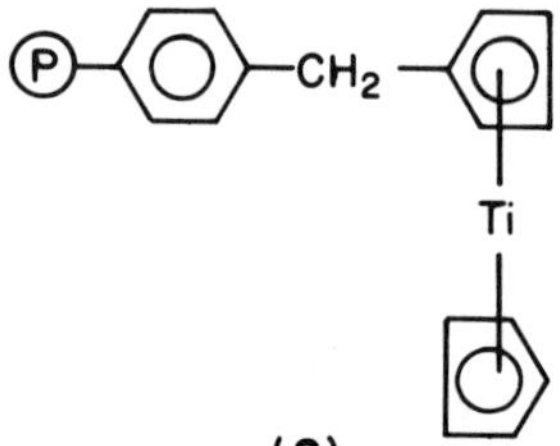

(2) Grey reactive polymer

Scheme 14-5

III. Preparation of Polymer-Bound Transition Metal Complexes

A list of some of the polymer-bound transition metal complexes and their catalytic functions are given in Tables 14-1 to 14-4. Although these lists are by no means comprehensive, it can be seen that only a few ligands other than phosphine and cyclopentadienyl have been used. Among the less frequently used polymeric ligands are —SH, —CH_2NMe_2, —$P(OMe)_2$ (Haag and Whitehurst, 1973), —CH_2NMe_2, —CH_2CN (Capka et al., 1972), —$N(CH_2COOH)_2$ (Nakamura and Hirai, 1975), and the recently reported polymer-supported carborane (Sosinsky et al., 1977).

TABLE 14-1

Hydrogenation Catalysts Based on Polymeric Transition Metal Complexes

Polymer	Metal complex	References
Phosphenated 2% cross-linked polystyrene	$[RhCl(C_2H_4)_2]_2$; $RhCl(PPh_3)_3$; $RhCl_3$; $RhCl_3 + PPh_3$; $RhCl_3 + PHPh_2$; $RhCl_3 + C_2H_4$; $RhCl(PPh_3)_3$; $RhCl(PHPh_2)_3$	Manassen, 1970; Grubbs and Kroll, 1971
Phosphenated Amberlite XAD-2	$RhCl(COD)_2$	Allum et al., 1973
Phosphenated co(polystyrene–DVB)	$PtCl_2$; $PdCl_2$	Bruner and Bailer, 1972
Phosphenated polystyrene (soluble)	$[Rh(CO)Cl]_2$; $Rh(PPh_3)_3Cl$; $PdCl_2(PhCN)_2$ $PtCl_2(PhCN)_2$	Bayer and Schurig, 1975
Phosphenated silica or silica + $(EtO)_3Si(CH_2)_2PPh_2$	$IrCl(COD)_2$; $RhH(CO)(PPh_3)_3$; $RhCl_3$	British Petroleum Co. Ltd., 1973; Michalska and Webster, 1975; Allum et al., 1973
Phosphenated PVC	$NiCl_2$, followed by $NaBH_4$; $Rh(acac)(CO)_2$; $RhCl(COD)_2$; $RhCl_3$	British Petroleum Co. Ltd., 1972
Amerlyst A27	K_2PdCl_4	Lazcano and Germain, 1971
Cross-linked polystyrene–DVB incorporating cyclopentadienyl group	Titanocene	Grubbs et al., 1973b; Bonds et al., 1975
Co(polystyrene–DVB) incorporating iminodiacetic acid group (Chelox 100)	$RhCl_3$	Nakamura and Hirai, 1975
Phosphenated cellulose	Rhodium complex	Bursian and Pracejus, 1972
Polystyrene incorporating chiral phosphine	$RhCl(C_2H_4)_2$; $NaPPh_2$	Dumont et al., 1973
Copolymer of styrene and hydroxyethyl methacrylate incorporating chiral phosphine	$RhCl(C_2H_4)_2$; $NaPPh_2$	Takaishi, 1976

Among the metals which have been bound to polymers as complexes are Rh, Ru, Pd, Ni, Ir, Pt, Fe, Co, Mn, Mo, and Cr.

IV. STRUCTURE OF POLYMERIC CATALYSTS

It is difficult to obtain detailed information about the structure of a metal complex bound to a polymer. Elemental analysis of the polymer only reflects the mole ratio of the polymeric ligand and the metal present. The exact mode of linking of the metal atom to the polymeric and nonpolymeric ligands is very difficult to deduce.

In general, it has been observed that all catalysts that have been attached to a polymer support retain their basic catalytic activity, hence a correlation between the structures of the low-molecular-weight catalyst complex and the corresponding polymeric complex can be made. Since the linking of a metal complex to a polymer generally proceeds by substitution, the number of polymeric ligands coordinating to the metal can be taken to be equal to the number of low-molecular-weight ligands displaced, unless there has been chelation due to bridging.

A basic requirement for metal complexes to function as catalysts is the availability of an open coordination site and the isolation of the metal complex on the polymeric matrix. This is only achieved when the polymer is rigid. Thus, when (mesotetraphenylporphyrin) iron [$Fe^{(II)}TPP$] was reacted with 2% cross-linked polystyrene incorporating imidazole, a six-coordinated diamagnetic complex with two resin-bound imidazoles (Fig. 14-2a) was formed (Collman and Reed, 1973). In subsequent experiments, a paramagnetic, five-coordinated $Co^{(II)}TPP$ (Fig. 14-2b) was also prepared (Collman et al., 1974) by using a 20% cross-linked resin.

Further characterization of the polymer can be done using some of the methods discussed in Chapter 3.

Wherever a low-molecular-weight analog of the polymeric complex is

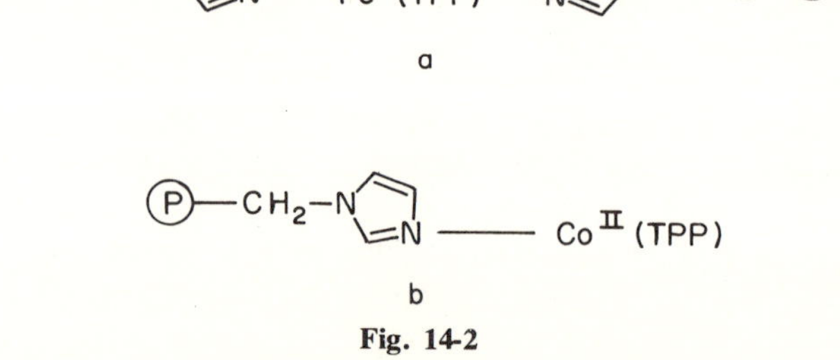

Fig. 14-2

available, comparison of ir spectra has been helpful for deducing the structure, e.g., the ir spectrum of (benzyl-Cp)CpTiCl (Cp = cyclopentadienyl) is comparable to that of the corresponding polymeric complex shown in Scheme 14-5.

Similarly, Collman *et al.* (1972) compared the ir spectra of the complex $MCl(CO)(PPh_3)_2$ (M = Rh or Ir) and the corresponding polymeric complexes, $MCl(CO)L_2$ (L = phosphenated polystyrene) and concluded that the complexes were identical except that the two phosphorus atoms on polystyrene form a chelate ring. A carbonyl stretching frequency for the CO group was detected in both cases.

Diffuse reflectance electronic spectra of the metals have also been studied in polymeric complexes (Allum *et al.*, 1973).

Magnetic measurements, including those based on esr, have been helpful in evaluating the paramagnetism of resin–titanocene. Stepwise reduction of polymer-bound titanocene complexes can be observed by two distinct esr signals of comparable intensities near g = 2 (Grubbs *et al.*, 1973a). Finally, X-ray back scattering from an ion microprobe has been used to scan the distribution of metal complexes in polymer beads (Grubbs and Sweet, 1975; Magnotta *et al.*, 1976).

V. TYPES OF REACTIONS CATALYZED BY POLYMER-ANCHORED CATALYSTS

A. Hydrogenation

Tris(triphenylphosphine)chlororhodium (I) (Wilkinson's catalyst) has been the most widely studied homogeneous hydrogenation catalyst. Analogous polymeric catalysts have been prepared for use in hydrogenation. A variety of alkenes, cycloalkenes, dienes, and alkynes have been reduced by polymer-bound catalysts, indicating the wide scope of their use (Grubbs *et al.*, 1977).

In addition, greater substrate selectivity and sensitivity to the size of alkene was observed in the case of polymeric catalysts. As an example, the rate of hydrogenation with a polymeric catalyst has been found to be 6.25 times greater for cyclohexene than that for cyclooctene (Grubbs and Kroll, 1971). The substrate selectivity has been attributed to the restriction of the substrate entering the pores of the ligand beads and indicates that the reaction mainly takes place inside the resin particles.

Platinum and palladium have also been linked to organic polymer supports. Polymers containing phosphine ligands (Kaneda *et al.*, 1975), and acid groups [e.g., polyacrylic acid (Nakamura and Hirai, 1976) and anthranilic acid (Holy, 1978)] have been prepared and shown to be catalytically active.

Silica-supported hydrogenation catalysts show less substrate selectivity than polystyrene-supported catalysts (Michalska and Webster, 1975; Ichikawa, 1976; Sinfelt, 1977). The rates of hydrogenation with silica-supported catalysts are high, although they are still lower than the equivalent homogeneous catalysts. It also appears that, unlike polystyrene, the reactive groups are confined only to the surface of silica. Ruthenium has also been immobilized in a zeolite matrix (Coughlan et al., 1977).

The efficacy of polystyrene-supported rhodium complexes decreases as the following complexes are used for equilibration of polystyrene–phosphine ligand: $RhCl_3$, $RhCl_3 + PPh_3$, $RhCl_3 + PHPh_2$, $RhCl_3 + C_2H_4$, $RhCl(PPh_3)_3$, $RhCl(PHPh_2)_3$ (Capka et al., 1971). It has also been possible to carry out selective partial hydrogenation of polyunsaturated compounds (Bruner and Bailar, 1972).

Since many organometallic complexes used as catalysts are very expensive, losses during their use are minimized by binding them to a polymer support. It may be expected that in the near future, they may replace such currently used metal catalysts as Pd and Pt.

Since the transition metal complexes can also be bound to a polymer via a chiral ligand, they can also be used to bring about asymmetric hydrogenation of alkenes, producing optically active hydrocarbons. Such catalysts will be discussed later in this chapter.

B. Hydroformylation

This is a reaction for which polymer-supported catalysts have been used extensively. The reaction sequence (shown in Scheme 14-6) involves the addition of an aldehyde group to the terminal or the internal carbon atom of an alkene. The ratio of the two aldehydes formed is dependent on the catalyst used, e.g., in the case of the two homogeneous catalysts, $Rh(acac)(CO)_2$ and $Rh(acac)(CO)PPh_3$, the ratios of normal to branched chain aldehydes were found to be 1.2:1 and 2.9:1, respectively (Allum et

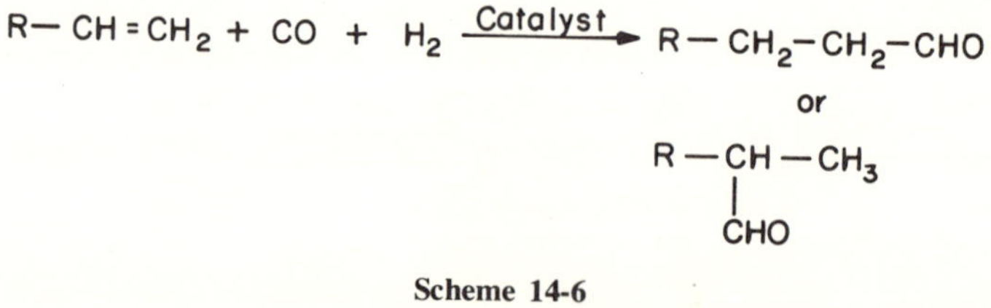

$$R-CH=CH_2 + CO + H_2 \xrightarrow{Catalyst} R-CH_2-CH_2-CHO$$
$$\text{or}$$
$$R-CH(CHO)-CH_3$$

Scheme 14-6

V. Types of Reactions Catalyzed by Polymer-Anchored Catalysts

al., 1973). When these catalysts are bound to the polymer, the ratio is further changed. Thus, when $Rh(acac)(CO)_2$ is equilibrated with a phosphenated polymer (polystyrene, polyvinyl chloride, or silica), normal and branched aldehydes are formed in the ratio 2.0–2.5:1. The increase in the ratio is an indication of displacement of CO by a polymeric phosphine ligand. Changing the amount of cis- and trans-bound catalyst also influences the ratio of normal to branched product (Pittman and Hiro, 1978). Another hydroformylation catalyst has been prepared from the precursor complex $RhCl(CO)(Ph_3P)_2Cl$ (Haag and Whitehurst, 1968). Many of the hydrogenation catalysts are also active for the hydroformylation reaction (Table 14-2) (Capka et al., 1971). In addition to the nature of the catalyst, the reaction conditions (e.g., temperature and pressure) also determine the n-alkanal–isoalkanal ratio (Pittman and Evans, 1973).

A side reaction in catalytic hydroformylation is aldehyde hydrogenation. In fact, for the same catalyst, the experimental conditions can be so modified that aldehyde hydrogenation takes place, producing the corresponding alcohols. As an example, binulcear cobalt polymeric complex brings about the conversion of aldehydes to alcohols when higher temperatures are used (Pittman and Evans, 1973).

TABLE 14-2

Hydroformylation Catalysts Based on Polymer-Bound Transition Metal Complexes

Polymer	Metal complex	References
Phosphenated polystyrene (soluble)	$Rh(CO_2(acac)$; $RhH(CO)(PPh_3)_3$	Bayer and Schurig, 1975
Phosphenated co(polystyrene–DVB)	$Co_2(CO)_8$; $RhCl(CO)(PPh_3)_2$	Pittman, Jr., 1971; Pittman, Jr. and Evans, 1973; Evans et al., 1974
Phosphenated 20% cross-linked polystyrene–DVB	$RhCl_3$, then C_2H_4	Capka et al., 1971
Phosphenated Amberlite XAD-2, and phosphenated PVC	$Rh(acac)(CO)_2$	Allum et al., 1973; British Petroleum Co. Ltd., 1970
Co(polystyrene–DVB) incorporating-PPh_2;$P(Bu)_2$; -SH; -CH_2NMe_2 or -$P(OMe)_2$	$Rh(CO)_2Cl_2$	Haag and Whitehurst, 1968; 1973
Polyvinylpyridine	$Co_2(CO)_8$	Collman et al., 1972; Collman and Reed, 1973
Silica + $(EtO)_3Si(CH_2)_2PPh_2$	$Rh(acac)(CO)_2$	Allum et al., 1972; British Petroleum Co. Ltd., 1970

C. Hydrosilylation

The reaction of an organosilicon hydride with an alkene in the presence of a catalyst resulting in the formation of a C—Si bond is referred to as hydrosilylation [Eq. (1)].

$$\text{R—CH} = \text{CH}_2 + \text{HSi(OEt)}_3 \xrightarrow{\text{catalyst}} \text{R—CH}_2\text{—CH}_2\text{Si(OEt)}_3 \quad (1)$$

Many homogeneous and some polymer-bound transition metal complexes are known to catalyze this reaction (Table 14-3). The reaction is of considerable interest and potential to the organosilicon industry.

Polymer-supported Rh and Pt complexes catalyze the hydrosilylation reaction (Capka *et al.*, 1971, 1972; Svoboda *et al.*, 1972). The rate of hydrosilylation decreases as electron-withdrawing substituent groups are substituted in the alkene molecule. Although hydrosilylation with triethoxysilane proceeds uniformly with all supported catalysts, complexes that are good catalysts for hydrosilylation with Et_3SiH are poor catalysts for the reaction with Cl_3SiH, and vice versa. The reaction conditions for hydrosilylation using a polymeric catalyst are mild (ambient temperature and near normal pressure) and comparable to those with homogeneous catalysts such as chloroplatinic acid.

D. Other Reactions

Among other reactions that can be catalyzed by polymer-bound transition metal complexes (or the corresponding low-molecular-weight

TABLE 14-3

Hydrosilylation Catalysts Based on Polymeric Transition Metal Complexes

Polymer	Metal complex	References
Phosphenated 20% cross-linked polystyrene–DVB	$RhCl_3$; $RhCl_3$, then C_2H_4	Capka *et al.*, 1971; Svoboda *et al.*, 1972
Phosphenated polystyrene	$RhCl(C_2H_4)_2$; $NaPPh_2$	Dumont *et al.*, 1973
Cross-linked co(polystyrene–DVB incorporating —CH_2PPh_2, CH_2NMe_2, or CH_2CN	H_2PtCl_6 or $RhCl_3$	Capka *et al.*, 1972
Polymethacrylate incorporating -$OC_6H_4PPh_2$, -$O(CH_2)_3PPh_2$, $O(Ch_2)_2NMe_2$, or -$O(CH_2)_2CN$	H_2PtCl_6 or $RhCl_3$	Capka *et al.*, 1972
Amberlyst A21, Co(polyallyl chloride–DVB) incorporating -CH_2PPh_2 group	H_2PtCl_6 or $RhCl_3$	Capka *et al.*, 1972

homogeneous catalysts) are oxidative polymerization of phenols (Tsuchida *et al.*, 1973), acetoxylation (British Petroleum Co. Ltd., 1973), hydrogen-deuterium exchange in alcohols (Strathdee and Given, 1974), polymerization and oligomerization, acetylene cyclotrimerization and tetramerization (Table 14-4), ketone synthesis (Pittman and Hanes, 1977), allylic addition of nucleophiles (Trost and Keiman, 1978), and olefin metathesis (Warwel and Buschmeyer, 1978; Tamagaki *et al.*, 1978).

Linear dimerization of dienes is of commercial interest, but the reaction proceeds with simultaneous formation of cyclic oligomers. After examining a number of transition metal complexes of Pt, Pd, and Ni it was found that the nickel (O) complex formed *in situ* by reduction of $(PPh_3)_2NiBr_2$ with $NaBH_4$ gave a high yield of the acyclic dimer of butadiene. The corresponding polymer-bound catalyst (Pittman and Smith, 1975) was also effective in bringing about the linear dimerization of butadiene.

This catalytic system has the unique advantage of giving a high yield of one specific linear dimer. At the same time, the purification problem is drastically reduced and continuous operation is made possible. The anchored catalyst can also be exposed to air for a short time without losing activity.

VI. ASYMMETRIC ORGANIC SYNTHESIS VIA TRANSITION METAL CATALYSTS BOUND TO POLYMERIC CHIRAL LIGANDS

A polymer containing chiral centers, poly(isobutylethyleneimine), was found to catalyze the asymmetric addition of HCN to benzaldehyde (Tsuboyama, 1962). Asymmetric synthesis using soluble chiral complexes of transition metal ions has been one of the most recent and important advances (Kagan and Dang, 1972). Dumont *et al.* (1973) conceived of the idea of preparing polymers having bound chiral ligands, so that the corresponding chiral metal complexes could function as asymmetric catalysts. The preparation of polymer-bound transition metal complexes having a chiral group can be achieved by two different methods. An achiral phosphine complex can either be attached to a chiral natural polymer or a chiral phosphine complex can be linked covalently to a synthetic achiral resin. The latter approach has been found more convenient and is, therefore, preferred.

The first report of an asymmetric hydrogenation catalyst refers to a German Patent (Bursian and Pracejus, 1972) where a rhodium complex with phosphinated cellulose was used. In two recent reports, an elaborate synthesis of a rhodium complex of a polymer having a chiral phosphine

TABLE 14-4

Polymerization Reactions (Including Cyclooligomerization) of Alkenes and Alkynes Using Polymeric Transition Metal Catalysts

Polymer	Metal complex	Reaction type	References
Phosphenated polystyrene	$NiCl_2$ $PdCl_2$–$AgBF_4$	Cyclotri- and tetramerization of ethyl propiolate and cyclotrimerization of phenylacetylene and various olefins	British Petroleum Co. Ltd., 1972; Kaneda et al., 1977
Phosphophenated co(polystyrene–DVB)	$Ni(CO)_2(PPh_3)_2$	Same as above	Evans et al., 1974; Pittman and Smith, 1975
Phosphenated PVC	$NiCl_2$ then $NaBH_4$	Cyclotrimerization of phenyl acetylene	Allum et al., 1973
Phosphinated polystyrene	$NiBr_2$ + $NaBH_4$	Cyclotri- and tetramerization of ethyl propiolate	Pittman and Smith, 1975
Phosphenated silica	$Ni(COD)_2$	Dimerization of cyclobutadiene	British Petroleum Co. Ltd., 1973

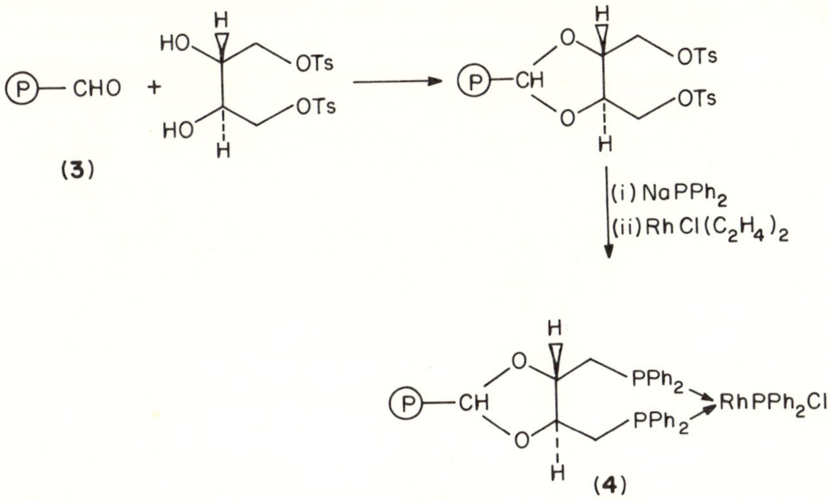

VI. Asymmetric Organic Synthesis

Scheme 14-7

ligand has been achieved. In addition, the polymer chiral ligand complex of Rh (4) was prepared by Dummont *et al.* (1973) from the aldehyde resin (3) (Frechet and Schuerch, 1971) (Scheme 14-7). In a slightly different approach, Takaishi *et al.* (1976) have prepared the polymer having the same chiral ligand by using a preformed chiral styryl monomer, 2-*p*-styryl-4,5-bis(tosyloxymethyl)1-4-dioxolane (5) (Scheme 14-8). This was copolymerized with hydroxyethyl methacrylate (6) to give a polymer (7) which was phosphinated and finally converted into the rhodium complex (8). This second polymer, which incorporated a hydrophilic group, swelled well in nonpolar solvents and was found to be more effective as a catalyst. Similar polymers have been used in the synthesis of optically active amino acids (Takaishi *et al.*, 1978; Masuda and Stille, 1978). In both cases, the phosphinated polymer was converted into the active catalyst by attaching the rhodium complex by a simple ligand exchange reaction.

These two examples provide a very good illustration of modifying the properties of a polymer by changing the polymer backbone. They also demonstrate that using preformed monomer should give a more uniform distribution of the reactive functional groups in the polymer molecule.

The insoluble, polymer-supported, optically-active rhodium complex prepared by Dumont *et al.* (1973) was used for asymmetric reduction of alkenes, leading to optically active hydrocarbons. Thus, 2-phenylbutene produced (*R*)-2-phenyl butane in 1.5% optical purity. The catalyst also hydrogenated methyl atropate to (*S*)-(+) methyl hydroatropate with an optical yield 2.5%. With α-acetamidocinnamic acid there was no reduc-

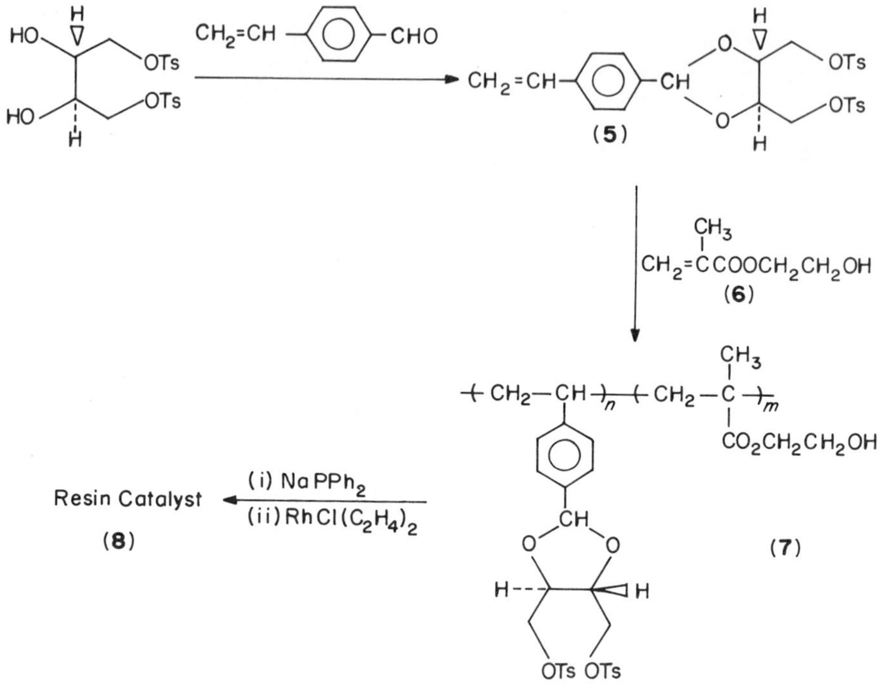

Scheme 14-8

tion at all, presumably because of complete lack of solubility of the substrate in benzene.

The catalyst prepared by Takaishi et al. (1976), with the modified polar polymer backbone, swelled even in polar solvents and reduced α-acetamidoalkenonic acids to the corresponding (R)-N-acetylamino acids with optical yields of 52–86%.

The catalytic reduction of aldehydes and ketones when carried out via hydrosilylation, gave high optical yield (60%) (Dumont et al., 1973). The mechanism of this reduction is thought to involve the intermediate formation of an asymmetric siloxane.

These asymmetric hydrogenation catalysts can be recovered by filtration and reused. However, some decrease in stereoselectivity has been observed in the case of recovered catalysts.

REFERENCES

Abdulla, K. A., Allen, N. P., Badran, A. H., Burns, R. P., Dwyer, J., McAuliffe, C. A., and Toma, N. D. A. (1976). *Chem. Ind. (London),* 273.

Acres, G. J. K., Bond, G. C., Cooper, B. J., and Dawson, J. A. (1966). *J. Catal.* **6,** 139.

References

Allum, K. G., Hancock, R. D., McKenzie, S., and Pitkethly, R. G. (1973). *Proc. Int. Cong. Catal., 5th, 1972*, p. 477. North-Holland Publ., Amsterdam.
Bailar Jr., J. C. (1974). *Catal. Rev.* **10**, 17.
Bayer, E., and Schurig, V. (1975). *Angew. Chem. Int. Ed. Engl.* **14**, 493.
Bonds, Jr., W. D., Brubaker, Jr., C. H., Chandrasekran, E. S., Gibbons, C., Grubbs, R. H., and Kroll, L. C. (1975). *J. Am. Chem. Soc.* **97**, 2128.
British Petroleum Co. Ltd. (1970). Dutch Patent 70, 06740.
British Petroleum Co. Ltd. (1972). Brit. Patent 1, 295675.
British Petroleum Co. Ltd. (1973). U.S. Patent 3,726,809.
Bruner, H. S., and Bailar, J. C. (1972). *J. Am. Chem. Soc.* **49**, 533.
Bursian, M., and Pracejus, H. (1972). East German Patent 92,031 [*C.A.* 1973, **78**, 72591].
Capka, M., Svoboda, P., Cerny, M., and Hetflejs, J. (1971). *Tetrahedron Lett.*, 4787.
Capka, M., Svoboda, P., Kraus, M., and Hetflejs, J. (1972). *Chem. Ind. (London)*, 650.
Card, R. J., and Neckers, D. C. (1977). *J. Am. Chem. Soc.* **99**, 7733.
Chauvin, Y., Commereuc, D., and Dawans, F. (1977). *Prog. Polym. Sci.* **5**, 95.
Collman, J. P., and Reed, C. A. (1973). *J. Am. Chem. Soc.* **95**, 2048.
Collman, J. P., Hegedus, L. S., Cooke, M. P., Norton, J. R., Dolcetti, G., and Marquardt, D. N. (1972). *J. Am. Chem. Soc.* **94**, 1789.
Collman, J. P., Gagne, R. R., Kouba, J., Ljustberg-Wahren, D. D. (1974). *J. Am. Chem. Soc.* **96**, 6800.
Coughlan, B., Narayanan, S., McCann, W. A., and Carroll, W. M. (1977). *Chem. Ind. (London)*, 125.
Dumont, W., Poulin, J. C., Dang, T. P., and Kagan, H. B. (1973). *Chem. Ind. (London)* **95**, 8295.
Evans, G. O., Pittman, Jr., C. U., McMillan, R., Beach, R. J., and Jones, R. (1974). *J. Organomet. Chem.* **7**, 296.
Frechet, J. M. J., and Schuerch, C. (1971). *J. Am. Chem. Soc.* **93**, 492.
Grubbs, R. H. (1976). "The Strem Chemiker," Polymer-Attached Homogeneous Catalysts, No. 4, Vol. 3, Strem Chemicals Inc., Danvers, Massachusetts.
Grubbs, R. H., and Kroll, L. C. (1971). *J. Am. Chem. Soc.* **93**, 3062.
Grubbs, R. H., and Sweet, E. M. (1975). *Macromolecules*, **8**, 241.
Grubbs, R. H., Gibbons, G., Kroll, L. C., Bonds, Jr., W. D., and Brubaker, Jr., C. H. (1973a). *J. Am. Chem. Soc.* **95**, 2373.
Grubbs, R. H., Kroll, L. C., and Sweet, E. M. (1973b). *J. Macromol. Sci. Chem.* **A7**(5), 1047.
Grubbs, R. H., Sweet, E. M., and Phisanbut, S. (1977). *In* "Catalysis in Organic Synthesis, 1976" (P. N. Rylander and H. Greenfield, eds.), Academic Press, New York.
Haag, W. O., and Whitehurst, D. D. (1968). Belgian Patent 721, 686.
Haag, W. O., and Whitehurst, D. D. (1973). *Proc. Int. Cong. Catal., 5th, 1972*. North-Holland Publ., Amsterdam.
Heinemann, H. (1971). *Chem. Technol.* 286.
Holy, N. L. (1978). *J. Org. Chem.* **43**, 4686.
Ichikawa, M. (1976). *J. Chem. Soc. Chem. Commun.*, 26.
Issleib, K., and Tzschach, A. (1959). *Chem. Ber.* **92**, 1118.
Kagan, H. B., and Dang, T. P. (1972). *J. Am. Chem. Soc.* **94**, 6429.
Kaneda, K., Terasawa, M., Imanaka, T., and Teranishi, S. (1975). *Chem. Lett.*, 1005.
Kaneda, K., Terasawa, M., Imanaka, T., and Teranishi, S. (1977). *Tetrahedron Lett.*, 2957.
Kohler, N., and Dawans, F. (1972). *Rev. Inst. Fr. Pet.* **27**, 105.
Lazcano, R. L., and Germain, J. E. (1971). *Bull. Soc. Chim. Fr.*, 1869.
Magnotta, V. L., Gates, B. C., and Schuit, G. C. A. (1976). *J. Chem. Soc. Chem. Commun.*, 342.

Manassen, J. (1970). *Isr. J. Chem.* **8**.
Manassen, J. (1971). *Platinum Met. Rev.* **15**, 142.
Masuda, T., and Stille, J. K. (1978). *J. Am. Chem. Soc.* **100**, 268.
Meir, W. M., and Uytterhoeven, J. B. (1973). *Adv. Chem. Ser.* **121**.
Melby, L. R. (1975). *J. Am. Chem. Soc.* **97**, 4044.
Michalska, Z. M., and Webster, D. E. (1974). *Platinum Met. Rev.* **18**, 65.
Michalska, Z. M., and Webster, D. E. (1975). *Chem. Technol.* 117.
Nakamura, Y., and Hirai, H. (1975). *Tetrahedron Lett.*, 823.
Nakamura, Y., and Hirai, H. (1976). *Chem. Lett.*, 1197.
Niebergall, H. (1962). *Macromol. Chem.* **52**, 218.
Osborn, J. A., Jardine, F. H., Young, J. F., and Wilkinson, G. (1966). *J. Chem. Soc.* A 1711.
Pittman, Jr., C. U. (1971). *Chem. Tech.* 416.
Pittman, Jr., C. U., and Evans, G. O. (1973). *Chem Technol.* 560.
Pittman, Jr., C. U., and Hanes, R. M. (1977). *J. Org. Chem.* **42**, 1194.
Pittman, Jr., C. U., and Hiro, A. (1978). *J. Org. Chem.* **43**, 640.
Pittman, Jr., C. U., and Smith, L. R. (1975). *J. Am. Chem. Soc.* **97**, 341.
Rabinowitz, R., and Marcus, R. W. (1961). *J. Org. Chem.* **26**, 4157.
Robinson, K. K., Paulik, P. E., Hershman, A., and Roth, J. F. (1969). *J. Catal.* **15**, 245.
Rony, P. R. (1969). *J. Catal.* **14**, 142.
Rylander, P. N., and Greenfield, H. (1976). "Catalysis in Organic Synthesis, 1976." Academic Press, New York.
Sinfelt, J. H. (1977). *Science* **195**, 641.
Smith, G. V. (1977). "Catalysis in Organic Synthesis, 1977." Academic Press, New York.
Sosinsky, B. A., Kalb, W. C., Grey, R. A., Uski, V. A., and Hawthorne, M. F. (1977). *J. Am. Chem. Soc.* **99**, 6768.
Strathdee, G., and Given, R. (1974). *Can. J. Chem.* **52**, 3000.
Svoboda, P., Capka, M., Chvalovsky, V., Bazant, V., Hetflejs, J., Jahr, H., and Pracejus, H. (1972). *Angew. Chem. Int. Ed. Engl.* **12**, 153.
Takaishi, N., Imai, H., Bertelo, C. A., and Stille, J. K. (1976). *J. Am. Chem. Soc.* **98**, 5400.
Takaishi, N., Imai, H., Bertelo, C. A., and Stille, J. K. (1978). *J. Am. Chem. Soc.* **100**, 264.
Tamagaki, S., Card, R. J., and Neckers, D. C. (1978). *J. Am. Chem. Soc.* **100**, 6635.
Trost, B. M., and Keinam, E. (1978). *J. Am. Chem. Soc.* **100**, 7779.
Tsuboyama, S. (1962). *Bull. Chem. Soc. Jpn.* **35**, 1004.
Tsuchida, E., and Nishide, H. (1977). *Adv. Poly. Sci.* **24**, 1.
Tsuchida, E., Kaneko, M., and Nishide, H. (1973). *Die Makromolekulare Chemie.* **164**, 203.
Warwel, S., and Buschmeyer, P. (1978). *Angew. Chem. Int. Ed. Engl.* **17**, 131.
Wilson, M. E., and Whitesides, G. M. (1978). *J. Am. Chem. Soc.* **100**, 306.

15 Polymers as Aids in Related Areas of Chemistry

I.	Introduction	239
II.	Applications in Analytical Chemistry	239
	A. Solid-Phase Reusable pH Indicators	240
	B. Analytical Electrochemical Applications	241
	C. Ion-Complexing Polymeric Materials	241
III.	Polymer-Bound Agriculturally and Pharmacologically Active Agents	242
IV.	Applications in Biochemistry	242
	A. Enzyme and Whole Cell Immobilization	242
	B. Applications of Immobilized Enzyme and Whole Cell Systems	244
	C. Principles of Affinity Chromatography	246
	D. Applications of Affinity Chromatography	248
	E. Support Materials for Immobilization of Enzymes and Affinity Chromatography	248
	F. Biologically Related Reducing Reagents Including Dihydrolipoic Acid and Related Thiopolymers	249
	References	252

I. INTRODUCTION

The field of organic chemistry has seen the most extensive use of polymeric materials as aids in effecting chemical transformation and product isolations. Polymers have been used in other, related areas of chemistry. Applications have been made in analytical chemistry (pH indicators and electrode modifiers), pharmaceutical and agricultural chemistry (controlled-release drugs, pesticides, herbicides, and fertilizers), and biochemistry (enzyme immobilization and affinity chromatography). Applications of polymers to solid-phase enzymo- and radioimmune assays (Landon, 1977; Chard, 1978) will not be discussed since they are mainly analytical in scope.

II. APPLICATIONS IN ANALYTICAL CHEMISTRY

Aside from using polymeric materials as chromatographic and enzyme supports (see later), applications in this area include the following categories.

A. Solid-Phase Reusable pH Indicators

Binding of reagents to an insoluble polymeric support has found some unusual applications. One such application has been the development of pH indicators bound to solid supports.

Conventional pH indicators must be dissolved in a liquid to be tested. Their presence may be undesirable in some applications, e.g., in food and pharmaceuticals. Even when pH papers are used to indicate the pH of a solution, lengthy immersion of the paper can result in bleeding of the indicator dye into solution, as well as promoting chromatographic effects that cause the dye to creep and form uneven color zones.

In order to overcome these difficulties, E. Merck Co. has marketed nonbleeding pH paper, which has the indicator dye covalently bound to the paper (cellulose) via its 2-sulfoxyethylsulfonyl derivative. The paper-bound indicators, although better than the conventional pH papers, were subject to bacterial attack and were less stable to acids and bases.

To overcome this difficulty, vinyl polymers containing various pH sensitive groups have been prepared (Hatanaka et al., 1974). Harper (1975) also conceived of making glass-bound, reusable pH indicators. The idea of the covalent binding of indicator dyes was taken on the one hand from the binding of enzymes to highly porous glass and, on the other hand, from the use of reactive dyes in textile industries. The reactive dyes have basically two types of reactive groups by which they can be bound chemically to nucleophilic groups on the fabric (Venkataraman, 1972). These are mono- or dichloro S-triazinyl and allied groups and β-sulfatoethylsulfonyl-type groups.

The introduction of dyes into the vinyl polymers was via their acrylic acid esters. The color of the bound dyes was shown to be dependent on the pH (Hatanaka et al., 1974). In order to introduce nucleophilic groups onto the more water-compatible surface, the porous glass was silylated with 3-aminopropyltriethoxysilane. These compounds, known as alkylamine glasses, can also be converted into arylamine glasses. The arylamine glass can be diazotized and coupled to N,N-dimethylaniline (Scheme 15-1) and a number of phthalein or sulfothalein dyes (indicators) to give glass-bound indicators corresponding to well-known soluble indicators such as phenolphthalein, phenol red, methyl red, and bromocresol.

A conventional form of the glass-bound indicator is one made on glass filter sticks. These filter sticks can be dipped into the solution and the pH can be read from the color of fritted disk. After washing, the indicator glass disk can be reused. It has been claimed that such an indicator can be used over a period of at least one year. They are quite stable at room temperature and resistant to most organic and aqueous solutions. Compared to indicator paper, they are completely resistant to microbial at-

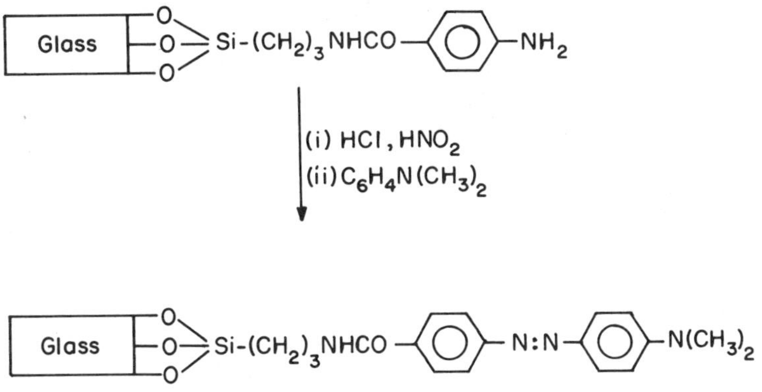

II. Applications in Analytical Chemistry

Scheme 15-1

Glass—Si-(CH$_2$)$_3$NHCO—C$_6$H$_4$—NH$_2$

(i) HCl, HNO$_2$
(ii) C$_6$H$_4$N(CH$_3$)$_2$

↓

Glass—Si-(CH$_2$)$_3$NHCO—C$_6$H$_4$—N:N—C$_6$H$_4$—N(CH$_3$)$_2$

tack. Harper (1975) has discussed the possibility of developing other types of glass-bound indicators, including chelating (ion detector) indicators.

One may hope that in the not-to-distant future, these reusable solid-supported indicators will be commercially available and may find a useful place in analytical procedures.

B. Analytical Electrochemical Applications

Polymeric substances have been used in analytical electrochemical applications to impart ion selectivity to electrodes (Buck, 1974, 1976, 1978). They have been used either to form solid electrodes [see also (SN)$_n$, Nowak *et al.*, 1977; Voulgaropoulos *et al.*, 1978] or to form liquid membrane systems and neutral carrier systems exhibiting ion selectivity.

Chemically modified electrodes have been used in place of the conventional materials (e.g., platinum) to modify the electron-transfer rates (Heineman and Kissinger, 1978; Kissinger, 1974, 1976). Polymers have also recently been used in this application (Miller and Van de Mark, 1978). It is expected that future applications of these electrodes will include applications in synthetic-scale organic chemistry.

C. Ion-Complexing Polymeric Materials

These are used to remove and concentrate various ions from solution and have been alluded to in Chapter 14. Analytical applications of such polymers are obvious and may be exemplified by the polymer–PDT(PDT = 3-(2-pyridyl)-5,6-diphenyl-1,2,4-triazine) combination that has been used to extract and concentrate Fe^{2+} and other divalent ions from solution (Lundgren and Schilt, 1977). In this case, PDT was physically adsorbed on the polystyrene (Amberlite XAD-2) rather than being covalently linked.

III. POLYMER-BOUND AGRICULTURALLY AND PHARMACOLOGICALLY ACTIVE AGENTS

The indiscriminate use of such agricultural chemicals as pesticides, herbicides, and fertilizers is an important source of environmental pollution. A novel application of polymer-bound materials has been made in the controlled release of agricultural chemicals (Allan et al., 1973; Beasley and Collins, 1970; Shambu et al., 1976; Schacht et al., 1977, 1978; reviews of Neogi and Allan, 1974; Scher, 1977). When these chemicals are covalently bound to a polymer from which they can be slowly released into the environment, they not only check pollution but their duration of action is prolonged. The same effect can be obtained by encapsulation of the chemicals in polymeric beads from which they can be released slowly, e.g., 2,4-dichlorophenoxyacetic acid and 4-amino-3,5,6-trichloropicolinic acid have been used in polymer-bound form.

Recently, interest has also been shown in binding pharmacologically active compounds to polymers from which they can be released slowly (Campbell, 1963; Batz et al., 1973; Netter et al., 1976; Schindler et al., 1977; Chytry et al., 1978; reviews of Batz, 1977; Donaruma and Vogl, 1978).

IV. APPLICATIONS IN BIOCHEMISTRY

By far the largest number of applications of polymeric materials in the field of biochemistry have been associated with enzyme and whole cell immobilization, affinity and covalent chromatography, and the immobilization of biologically-related reducing agents.

A. Enzyme and Whole Cell Immobilization

The chemical transformations and product isolations discussed in the previous chapters have been effected with reagents and catalysts generally accepted as within the realm of organic chemistry. Chemical transformations and product isolations employing techniques and principles which originate within the realm of biological chemistry have not tended to be adopted by organic chemists. Recent work in this area, however,

IV. Applications in Biochemistry

has emphasized the utility to which enzymes and various biochemical techniques may be put in organic chemistry (Jones et al., 1976).

In Chapters 13 and 14 of this book the applications of conventional chemical catalysts were described. The use of enzymes or whole cells as catalysts for chemical transformations is well known. They can bring about various reactions at ambient temperature and pressure and afford high reaction velocities. In fact, enzymatic reaction sequences may be designed to give the ideal efficiency embodied in the second law of thermodynamics. Thus, hundreds of compounds that are very difficult to prepare by purely chemical methods may be obtained quite readily and economically with the help of enzymes. Until recently, most laboratory investigations and manufacturing processes employed soluble enzymes in dilute aqueous solutions. Before use, the required enzyme must be obtained from biological sources as a concentrated extract. It is not uncommon for a particular type of cell to contain many proteins in addition to the one desired. Therefore, the purification and concentration of enzymes in preparation for use is a very cumbersome process. When used in solution, enzyme catalysts are invariably lost after each batch operation. The use of immobilized enzymes and whole cells has been proposed as a means that could eliminate such losses and preserve hard won stocks of specialized enzymes.

The idea of immobilizing enzymes on polymeric supports is older than the use of polymer-immobilized chemical reagents (Grubhofer and Schleith, 1954). A number of reviews have been published on enzyme immobilization (Manecke, 1962; Silman and Katchalski, 1966; Goldstein and Katchalski, 1968; Goldstein, 1969, 1970; Mosbach, 1971; Vieth and Venkatsubramanian, 1973; Goldman et al., 1974; Mosbach, 1976; Manecke and Storck, 1978). Several books have also appeared on the subject (Stark, 1969; Zaborsky, 1973; Messing, 1975; Weetall, 1975; Pierce, 1977; Chibata, 1978). Microbial and animal whole cell immobilization has also been reviewed (Vandamme, 1976; van Wezel, 1972; Chibata and Tosa, 1977; van Wezel and van der Velder-de Groot, 1978). Since a full treatment of the area is outside the scope of this monograph, we shall limit ourselves to a discussion of the basic principles and will present only illustrative examples of the use of immobilized enzymes as catalysts.

Most enzymes in their natural environments do not behave exactly as they do *in vitro*. Cellular enzymes, for example, function either in an environment resembling a gel or while absorbed at interfaces or in solid-state arrays, such as those existing in mitochondria. It follows that if an enzyme could be immobilized on a polymeric support by covalent binding through a group other than the active site of the enzyme, it could be used in heterogeneous phase. Employing enzymes in the laboratory in an environment similar to that existing in a cell can have many advantages.

1. The enzymes can be recovered and reused.
2. A continuous process instead of a batch operation can be used.
3. Immobilized enzymes have been found to show improved stability, e.g., proteolytic enzymes do not undergo autolysis when bound to polymeric supports.
4. Immobilized enzymes also show greater environmental stability toward changes in pH or temperature. They also show improved storage properties.
5. The isolation and purification of the products is facilitated.
6. In some cases, immobilized enzymes show enhanced activity toward a substrate as compared with the soluble enzymes.
7. The reaction involving an immobilized enzyme can be stopped at any desired stage by simple filtration of the insolubilized enzyme.
8. In certain cases, two successive enzymatic reactions can be carried out on a substrate solution by reacting it with two different immobilized enzymes. Thus, the isolation of the intermediate product may not be necessary. This is equivalent to events in a cell where successive chemical transformations of a substrate take place via a cascade of different enzymes present in the cell system.

B. Applications of Immobilized Enzyme and Whole Cell Systems

Immobilized enzymes and whole cells have found well-documented applications in industry, medicine, and analytical chemistry. Theoretically, it should be possible to carry out any enzymatic reaction with the help of the respective immobilized enzyme or whole cell containing the enzyme. The technique of using an immobilized enzyme for a chemical transformation is not basically different from using the soluble enzymes. In commercial applications, the immobilized enzymes can be used in a continuous-flow reactor. However, the optimum conditions for a specific reaction will have to be redetermined before maximum turn-over can be achieved. Thus, proteolytic enzymes such as trypsin, when immobilized on an anionic matrix such as co(polyethylene–maleic anhydride), require a much lower pH for reaction than in solution. Some typical applications of immobilized enzymes that are currently being made, or are in the process of development, are mentioned in Table 15-1.

IV. Applications in Biochemistry

TABLE 15-1

Typical Examples of Processes Using Immobilized Enzymes and Whole Cells

Enzyme	Substrate	Reaction type
Amino acid acylase	DL-Amino acids	Acylation resolution
α-Amylase	Starch	Hydrolysis
Invertase	Sucrose	Hydrolysis
Glucose isomerase	Glucose	Isomerization to fructose
Pectinase	Pectin	Hydrolysis to clarify fruit juice and wine
Pepsin, rennin	Casein	Cheese making
Proteases	Proteins	Hydrolysis
Glucose oxidase	Glucose	Oxidation to gluconic acid
Penicillin amidase	Penicillin	Deamidation
β-Galactosidase	Lactose	Hydrolysis
Steroidal modifying enzymes	Steroids	Steroidal transformations
Nucleases	Nucleic acids	Hydrolysis
Urease	Urea	Hydrolysis
Radioactive labeling of proteins	Tyrosine	Iodination with ^{125}I

A typical application of immobilized enzymes has been made in the preparation of pharmaceutically important compounds (Abbott, 1976). Prednisolone (a potent steroidal drug for suppressing rheumatoid arthritis) can be made from cortisol (Mosbach, 1971). This transformation can be brought about by a multistage organic synthesis. By the use of immobilized enzymes, the synthesis is achieved much more easily. Thus the enzyme, 11-β-hydroxylase has been used for the introduction of 11-β-OH groups into the steroid nucleus, while another enzyme, $\Delta^{1\text{-}2}$-dehydrogenase, brings about the dehydrogenation of the intermediate hydroxylated product to the potent drug (Scheme 15-2). In contrast to the previous example where two different enzymes were bound to two different sets of supports, there is a possibility of immobilizing two enzymes (which react in succession on a substrate) on the same support to carry out the transformation. This would produce a high concentration of the substrate for the second enzyme (product from the reaction of first enzyme) near the immobilized enzymes. This type of behavior has been observed in the case of the successive reactions of hexokinase and 6-phosphoglucoisomerase, which were bound to the same support. Here, transformation rates with immobilized enzymes were much higher than when the enzymes were present in solution.

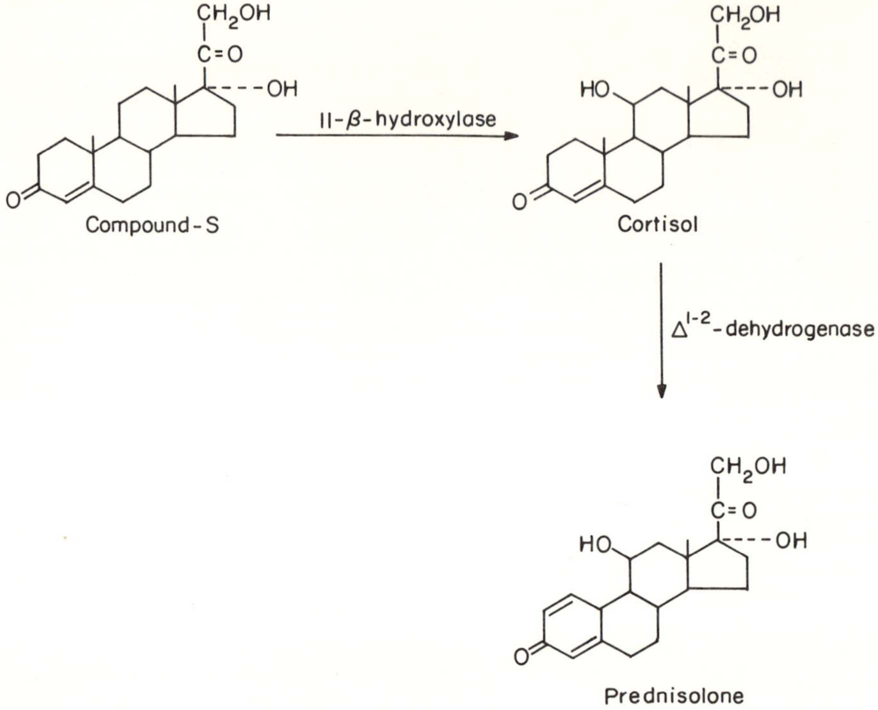

Scheme 15-2

C. Principles of Affinity Chromatography

Another well-known application of immobilized enzymes and certain other biochemicals (including nonpolymeric materials) is the affinity chromatographic method of separation and purification of biochemicals. Many of the supports used are the same as those employed in the preparation of immobilized enzymes, and some of the separations employ a certain differential reactivity among groups to achieve the separation. A large number of reviews and books have been published on the subject (Silman and Katchalski, 1966; Anonymous, 1971; Anonymous, 1972; Wingard, 1972; Wilchek and Givol, 1973; Weetall, 1973; Corning Glass Works, 1972, 1973, 1974; Olson and Richardson, 1974; Jakoby and Wilchek, 1974; Whitesides and Nishikawa, 1976). The method is based on the fact that certain substrates have selective and strong affinity for specific ligands, e.g., there is a specific interaction between an enzyme and its inhibitor, or between an antigen and its antibodies. If such a ligand is immobilized on an insoluble polymer column and an impure substrate is passed through the column, the specific interaction results in the separa-

IV. Applications in Biochemistry

tion of the substrate. Campbell *et al.* (1951) were the first to apply this principle to the separation of anti-hapten antibodies.

The interaction between the immobilized ligand and the substrate may result in covalent interaction in certain cases (e.g., an affinity medium having hydrazide groups can specifically bind steroidal ketones by the formation of a hydrazone) or in some type of noncovalent (molecule–molecule) interaction including hydrogen bonding. Such interactions are far more specific than the physical methods based on distribution of a component between two phases as in a conventional chromatographic method. The ligand–substrate interactions are even more specific than those involved in ion-exchange processes. Such methods are particularly suitable for the separation and purification of biopolymers. Where chemical interaction between the ligand and the substrate can occur, the affinity material can be designed so that it can have selective reaction with the functional groups of certain small molecules. Brocklehurst *et al.* (1974) have named such processes "covalent chromatography." The technique has been used for the separation of peptides (synthetic or natural) and proteins having specific amino acids such as thiol groups (Neuman *et al.*, 1967; Egorov *et al.*, 1975), tryptophan (Rubinstein *et al.*, 1976), COOH-terminal arginines (Yokosawa and Ishii, 1976), and methionine (Schechter *et al.*, 1977). Schemes 15-3 and 15-4 illustrate the general principles of

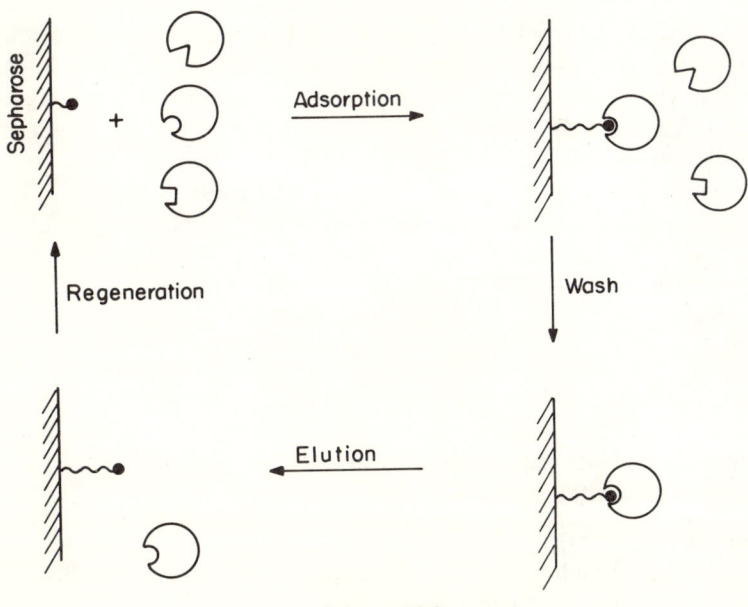

Scheme 15-3

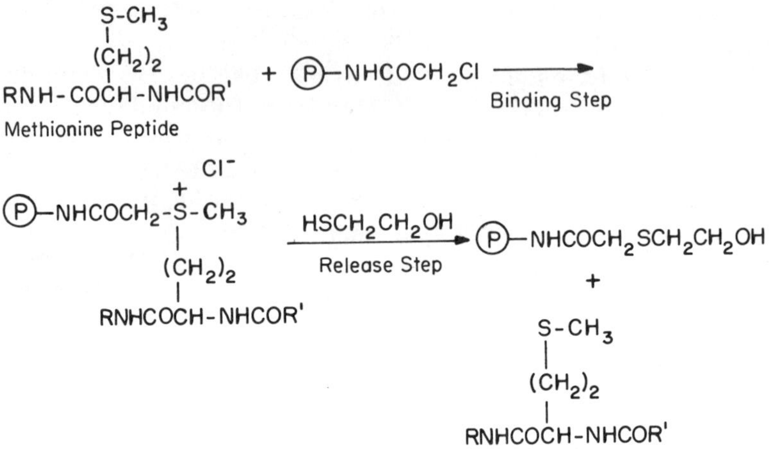

$$\begin{array}{c}
\text{S-CH}_3 \\
| \\
(\text{CH}_2)_2 \\
| \\
\text{RNH-COCH-NHCOR'} \\
\text{Methionine Peptide}
\end{array} + \text{(P)}-\text{NHCOCH}_2\text{Cl} \xrightarrow{\text{Binding Step}}$$

$$\begin{array}{c}
\text{Cl}^- \\
+ \\
\text{(P)}-\text{NHCOCH}_2-\overset{|}{\text{S}}-\text{CH}_3 \\
| \\
(\text{CH}_2)_2 \\
| \\
\text{RNHCOCH-NHCOR'}
\end{array} \xrightarrow[\text{Release Step}]{\text{HSCH}_2\text{CH}_2\text{OH}} \begin{array}{c} \text{(P)}-\text{NHCOCH}_2\text{SCH}_2\text{CH}_2\text{OH} \\ + \\ \text{S-CH}_3 \\ | \\ (\text{CH}_2)_2 \\ | \\ \text{RNHCOCH-NHCOR'} \end{array}$$

Scheme 15-4

affinity chromatographic operations involving molecule–molecule interaction and covalent interaction respectively.

D. Applications of Affinity Chromatography

As early as 1951, affinity chromatography was used for the separation of anti-hapten antibodies (Campbell *et al.*, 1951). Shortly afterwards, Lerman (1953) developed a similar method for purification of tyrosinase. The recent applications of affinity chromatography are too numerous to be detailed here, but broadly speaking, the method has been used for separation/purification of proteins, sugars and their derivatives, nucleic acids, nucleotides and their derivatives, amino acids and peptides, and various other systems that include thiols and disulfides (ligands-thiols or disulfides or organomercurial supports), steroidal hormones, coenzymes, vitamins, morphine and related drugs, antibiotics, protein receptors, and antibodies.

E. Support Materials for Immobilization of Enzymes and Affinity Chromatography

A variety of support materials have been used for immobilizing enzymes and other biochemicals. In general, the polymeric supports should have the following properties.

IV. Applications in Biochemistry

1. It should be completely insoluble in common solvents employed in the process.
2. It should be chemically stable to the reagents, solvents and environmental conditions (i.e., pH and temperature) employed.
3. The insoluble matrix should form a loose, porous network that permits the easy passage of macromolecules and retains a good flow rate.
4. Since most reactions of immobilized enzymes and affinity chromatographic processes are carried in aqueous medium. The support should be compatible with the aqueous solutions of the substrate.
5. In the case of affinity chromatography, the main support material should have minimal interaction with the substrates both before and after coupling of the specific ligand.
6. The support material should be capable of undergoing functionalization reactions, such as attachment of a ligand either directly to the resin or via an "extension arm."
7. The polymeric support should not be susceptible to bacterial attack and degradation.

In one of the earliest reports (Grubhofer and Schleith, 1954), polyaminostyrene was used for binding pepsin. These workers were basically interested in the technological potential of their new technique. Noncompatibility of styrene with relatively polar biomaterials and aqueous solvents is the most important limitation of such supports. Since then, stress has been on more compatible solid supports such as cross-linked dextrans (Sephadex and Sepharose), Agarose, and starch. Cellulose is not generally preferred because its crystalline regions make it less permeable to macromolecules. Other polymer supports used include polyacrylamide, polyamino acids, co(polyethylene–maleic anhydride), and acrylic polymers. Modified natural carbohydrate polymers such as periodate-oxidized carbohydrate polymers have also been used. Functionalized porous glass has also been extensively employed for immobilization of enzymes and other ligands. Extensive compilations of the polymeric supports that have been used can be found in Jakoby and Wilchek's review (1974). The most commonly used polymeric supports, and their suitability for specific ligands are listed in Table 15-2. Most of the polymeric supports listed are commercially available.

F. Biologically Related Reducing Reagents Including Dihydrolipoic Acid and Related Thiopolymers

The disulfide bond occurs extensively in proteins and peptides. The disulfide bridge in a protein is readily reduced by mercaptoethanol or

TABLE 15-2

Polymeric Supports Used for Immobilization of Enzymes and Whole Cells and for Affinity Chromatography

Polymeric carrier	Ligand	Remarks
Agarose[a]	Enzymes/proteins, carbohydrates, nucleic acids and other low-molecular-weight ligands	Almost a universal support for macromolecular and small molecule ligands
Cellulose,[a] including carboxymethyl- and aminoethylcellulose	Proteins, nucleic acids	Less permeable to macromolecules and unsuitable for affinity chromatography
Sephadex and Sepharose[b] including carboxymethyl and aminoethyl derivatives	Proteins, nucleic acids and carbohydrates, whole cells	
Functionalized polyacrylamides[a]	Proteins, nucleic acids and low-molecular-weight ligands	Less prone to bacterial degradation than the carbohydrate polymers
Co(polyethylene–maleic anhydride)	Enzymes	Capable of creating a microenvironment of low pH at the reactive site
Functionalized controlled pore Glass[c]	Enzymes, carbohydrates and a variety of low-molecular-weight ligands	Low capacity but varied type of functionalization; high mechanical strength and stability towards variety of chemicals

[a] Available from Bio-Rad Laboratories, Richmond, California.
[b] Available from Pharmacia, Uppsala, Sweden.
[c] Available from Pierce Chemical, Rockford, Illinois.

thioglycolic acid, but such a disulfide–thiol interchange reaction being reversible eventually attains an equilibrium [Eq. (1)].

$$R-S-S-R + 2\,R'-SH \rightleftharpoons R'-S-S-R' + 2\,RSH \qquad (1)$$

It has been possible to push the reduction of a disulfide to completion by using a bisthiol, e.g., dithioerythritol (Cleland, 1964). The bisthiols owe their strong reducing power (lower redox potential) to the formation of a stable and sterically favorable dithiolane ring system upon oxidation [Eq. (2)].

$$R-S-S-R + R'\!\!\begin{array}{c}\diagup SH\\ \diagdown SH\end{array} \longrightarrow 2\,RSH + R'\!\!\begin{array}{c}\diagup S\\ |\\ \diagdown S\end{array} \qquad (2)$$

In spite of the strong reducing properties of bisthiols (redox potential, -0.325 V vs -0.2 V for cysteine at pH 7.0) (Fruton and Clarke, 1934), the

IV. Applications in Biochemistry

separation of the by-product, dithiolane, from the reaction mixture is sometimes very difficult.

Gorecki and Patchornik (1973) made a polymer in which dihydrolipoic acid was covalently bound to the polymer via its carboxyl group. Since the reagent polymer was ultimately designed to be used for the reduction of proteins and peptides, hydrophilic support polymers (e.g., aminoethyl Sephadex G-25, aminoethyl polyacrylamide and aminoethyl cellulose) were preferred. In a more recent application, the compound was linked to controlled-pore glass (Scouten and Firestone, 1976).

Lipoic acid was joined to the polymeric amino group via an amide bond by coupling with the lipoic acid N-hydroxysuccinimide ester (Scheme 15-5). Alternatively 2-(lipoamido)ethylamine was linked to CNBr-activated Sephadex or Sepharose. These coupling reactions have been used to immobilize enzymes on polymer supports (Porath et al., 1967; Wilchek and Gorecki, 1969). The resulting polymers swelled in water and other polar solvents such as DMF.

The polymer-bound lipoic acid can be reduced by using $NaBH_4$ (see Scheme 15-5). The reduced polymers are stable between pH 2–10 and can be reused after reduction. The resulting reduced polymer bisthiol is a potential reducing agent for disulfides. The degree of lipoyl substitution in the polymer can be found by reduction of 5,5'-dithiobis-(2-nitrobenzoic acid) (DTNB) (Ellman, 1959). The thiol content of the polymer can also be determined by reaction of the polymer with iodo[^{14}C]acetic acid and by measurement of the uptake of radioactivity by liquid scintillation counting.

The reduction of cystine glutathione and proteins with polymer-bound lipoic acid can be carried out either by polymer suspended in the reaction mixture or in a column.

Earlier, another thiolated polymer (Sephadex) (Eldjarn and Jellum, 1963) containing the homocysteinyl group was prepared. This polymer

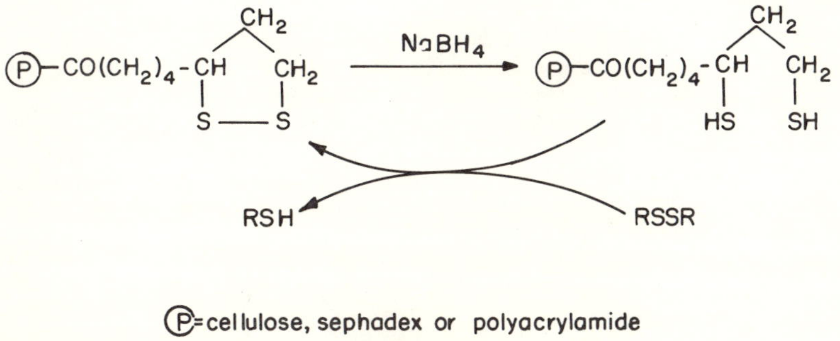

(P)=cellulose, sephadex or polyacrylamide

Scheme 15-5

was used as an antioxidant for thiol-containing proteins and for maintaining a high level of thiol concentration in solution, but the dihydrolipoic aminoethyl polymers are more effective due to the proximity of SH-groups, favoring the formation of a stable dithiolane ring.

REFERENCES

Abbott, B. J. (1976). *Adv. Appl. Microbiol.* **20,** 203.
Allan, G. G., Chopra, C. S., Friedhoff, J. F., Gara, R. I., Maggi, M. W., Neogi, A. N., Roberts, S. C., and Wilkines, R. M. (1973). *Chem. Tech.* **171,** and references therein.
Anonymous (1971). *Chem. Eng. News* **49,** 86.
Anonymous (1972). "Enzyme Manual." Worthington Chem. Corp., Freehold, New Jersey.
Axen, R., Porath, J., and Arnbach, S. (1967). *Nature (London),* **214,** 1302.
Batz, H.-G. (1977). *Adv. Polymer Sci.* **23,** 25.
Batz, H. G., Franzmann, G., and Ringsdorf, H. (1973). *Makromol. Chem.* **172,** 27.
Beasley, M. L., and Collins, R. L. (1970). *Science* **169,** 769.
Brocklehurst, K., Carlsson, J., Kierstan, M. P. J., and Grook, E. M. (1974). *Methods Enzymol.* **34,** 531.
Buck, R. P. (1974). *Anal. Chem.* **48,** 28R.
Buck, R. P. (1976). *Anal. Chem.* **48,** 23R.
Buck, R. P. (1978). *Anal. Chem.* **50,** 17R.
Campbell, D. H., Leuscher, E., and Lerman, L. S. (1951). *Proc. Nat. Acad. Sci. U.S.A.* **37,** 575.
Campbell, J. A. (1963). *J. Pharm. Sci.* **52,** 76.
Chytry, V., Vrana, A., and Kopecek, J. (1978). *Makromol. Chem.* **179,** 329.
Cleland, W. W. (1964). *Biochemistry* **3,** 480.
Corning Glass Works (1972, 1973, 1974). "Immobilized Enzymes," Vols. I, II, and III, Corning, New York.
Donaruma, L. G., and Vogl, O. (1978). "Polymeric Drugs." Academic Press, New York.
Chard, T. (1978). "An Introduction to Radioimmunoassay and Related Techniques." North-Holland Publ., Amsterdam.
Chibata, I. (1978). "Immobilized Enzymes." Halsted Press, New York.
Chibata, I., and Tosa, T. (1977). *Adv. Appl. Microbiol.* **22,** 1.
Egorov, T. A., Svenson, A., Ryden, L., and Carlsson, J. (1975). *Proc. Nat. Acad. Sci. U.S.A.* **72,** 3029.
Eldjarn, L., and Jellum, E. (1963). *Acta Chem. Scand.* **17,** 2610.
Ellman, G. L. (1959). *Arch. Biochem. Biophys.* **82,** 70.
Fruton, J. S., and Clarke, H. R. (1934). *J. Biol. Chem.* **106,** 667.
Goldman, R., Goldstein, L., and Katchalski, E. (1974). *In* "Biochemical Aspects of Reactions on Solid Supports" (G. R. Stark, ed.), p. 1. Academic Press, New York.
Goldstein, L. (1969). *In* "Fermentation Advances" (D. Perlmann, ed.), p. 391. Academic Press, New York.
Goldstein, L. (1970). *Methods Enzymol.* **19,** 935.
Goldstein, L., and Katchalski, E. (1968). *Z. Anal. Chem.* **243,** 375.
Gorecki, M., and Patchornik, A. (1973). *Biochim. Biophys. Acta* **303,** 36.
Grubhofer, N., and Schleith, L. Z. (1954). *Z. Physiol. Chem.* **297,** 108.
Harper, G. B. (1975). *Anal. Chem.* **47,** 348.

References

Hatanaka, H., Sugiyama, K., Nakaya, T., and Imoto, M. (1974). *Makromol. Chem.* **175,** 1855.
Heineman, W. R., and Kissinger, P. T. (1978). *Anal. Chem.* **50,** 166R.
Jakoby, W. B., and Wilchek, M. (Eds.) (1974). *Methods Enzymol.* **34,** 3.
Jones, J. B., Sih, C. J., and Perlman, D. (1976). "Applications of Biochemical Systems to Organic Chemistry," Parts 1 and 2. Wiley, New York.
Kissinger, P. T. (1974). *Anal. Chem.* **46,** 15R.
Kissinger, P. T. (1976). *Anal. Chem.* **48,** 17R.
Laird, R. M., and Spence, M. J. (1977). *J. Appl. Chem Biotechnol.* **27,** 214, 357.
Landon, J. (1977). *Nature (London)* **268,** 483.
Lerman, L. S. (1953). *Proc. Nat. Acad. Sci. U.S.A.* **39,** 232.
Lundgren, J. L., and Schilt, A. A. (1977). *Anal. Chem.* **49,** 974.
Manecke, G. (1962). *Pure Appl. Chem.* **4,** 507.
Manecke, G., and Storck, W. (1978). *Angew Chem. Int. Ed. Engl.* **17,** 657.
Messing, R. A. (1975). "Immobilized Enzymes for Industrial Reactors." Academic Press, New York.
Miller, L. L., and Van de Mark, M. R. (1978). *J. Am. Chem. Soc.* **100,** 639.
Mosbach, K. (1971). *Sci. Amer.* **224,** 26.
Mosbach, K. *In* "Applications of Biochemical Systems to Organic Chemistry" (J. B. Jones, C. J. Sih, and D. Perlman, eds.), p. 969. Wiley, New York.
Neogi, A. N., and Allan, G. G. (1974). *In* "Controlled Release of Biologically Active Agents" (A. C. Tanguary and R. E. Lacey, eds.), Vol. 47, p. 195. Plenum, New York.
Netter, K. J., Ringsdorf, H., and Wilk, H.-C. (1976). *Makromol. Chem.* **177,** 3527.
Neuman, H., Steinberg, Y. Z., Brown, J. R., Goldberger, F. F., and Sela, M. (1967). *Eur. J. Biochem.* **3,** 171.
Nowak, R. J., Mark, Jr. H. B., MacDiarmid, A. G., and Weber, D. (1977). *J. Chem. Soc. Chem. Commun.,* 9.
Olson, N. F., and Richardson, T. (1974). *J. Food Sci.* **30,** 653.
Pierce Chem. Co. (1977). "Handbook and General Catalog," p. 273. Corning Biomaterial Supports, Rockford, Illinois.
Porath, J., Axen, R., and Ernback, S. C. (1967). *Nature (London),* **215,** 1491.
Rubinstein, M., Schechter, Y., and Patchornik, A. (1976). *Biochem. Biophys. Res. Commun.* **70,** 1257.
Scher, H. B. (1977). "Controlled Release Pesticides," *Amer. Chem. Soc. Symp. No. 53,* American Chemical Society, Washington, D.C.
Schacht, E., Desmarets, G., and St. Pierre, T. (1977). *Makromol. Chem.* **179,** 543.
Schacht, E., Desmarets, G., and Bogaert, Y. (1978). *Makromol. Chem.* **179,** 837.
Schindler, A., Jelfcoat, R., Kimmel, G. L., Pitt, C. G., Wall, M. E., and Zweidinger, R. (1977). *In* "Contemporary Topics in Polymer Science" (E. M. Pearce and J. R. Schaefgen, eds.), Vol. 2. Plenum, New York.
Scouten, W. H., and Firestone, G. L. (1976). *Biochim. Biophys. Acta* **453,** 277.
Shambhu, M. B., Digenis, G. A., Gulati, D. K., Bowman, K., and Sabharwal, P. S. (1976). *J. Agric. Food Chem.* **24,** 666.
Shechter, Y., Rubinstein, M., and Patchornik, A. (1977). *Biochemistry,* **16,** 1424.
Silman, H., and Katchalski, E. (1966). *Annu. Rev. Biochem.* **35,** 873.
Stark, G. R. (1969). "Immobilized Enzymes." Academic Press, New York.
Vandamme, E. J. (1976). *Chem. Ind. (London),* 1070.
van Wezel, A. L. (1972). *In* "Methods and Applications of Tissue Culture" (P. R. Kruse, Jr., and M. K. Patterson, Jr., eds.), p. 372. Academic Press, New York.
van Wezel, A. L., and van der Velder-de Groot, C. A. M. (1978). *Process Biochem.* **13,** 6.

Venkataraman, K. (1972). "The Chemistry of Synthetic Dyes." Vol. 6. Academic Press, New York.
Vieth, W. R., and Venkatsubramanian, K. (1973). *Chem Technol.* 677.
Voulgaropoulos, A. N., Nowak, R. J., Kutner, W., and Mark, Jr. H. B. (1978). *J. Chem. Soc. Chem. Commun.*, 244.
Weetal, H. H. (1973). *Food Prod. Dev.* 7(3), 46 and 7(4), 94.
Weetal, H. H. (1975). "Immobilized Enzymes, Antigens, Antibodies and Peptides." Dekker, New York.
Whitesides, G. M., and Nishikawa, H. H. *In* "Applications of Biochemical Systems to Organic Chemistry" (J. B. Jones, C. J. Sih, and D. Perlman, eds.), p. 929. Wiley, New York.
Wilchek, M., and Givol, D. (1973). *Pept. Proc. Eur. Pept. Symp., 11th, 1971,* p. 203. North-Holland Publ., Amsterdam.
Wilchek, M., and Gorecki, M. (1969). *Eur. J. Biochem.* **11**, 491.
Wingard, L. B. (1972). "Enzyme Engineering," *Biotechnol. Bioeng. Symp. No. 3,* Wiley, New York.
Yokosawa, H., and Ishii, S.-H. (1976). *Biochem. Biophys. Res. Commun.* **72**, 1443.
Zaborsky, O. (1973). "Immobilized Enzymes." CRC Press, Cleveland, Ohio.

Index

A

Active esters, 29, 30, 65, 117
Acylation, 139, 170, 187
Affinity chromatography, 246
Agricultural applications, 242
Alginic acid, 159
Alkylation, 139, 188, 213
Alumina, 32, 176, 180, 181, 222
Aluminum chloride, 22, 205
Amberlite, 141, 175, 183, 242
Amberlyst, 141, 175, 183, 222
Amide, coupling, 49, 64
 hydrolysis, 199
 linkage, 62, 127
Amino acid, protecting groups, 63
 resolution, 158
Aminophosphines, 195
Analytical applications, 239
Anhydride, 65, 187
Anion exchange, 188, 199
Antibodies, 246, 247, 248
Arginine, 247
Asymmetric synthesis, 114, 156, 233
Automation, 69, 133

B

Benzophenone, 215
Benzyne, 49, 167
Bidirectional peptide extension, 73
Biochemical applications, 242
Bipyridine, 194, 223
Borohydride, 183, 251
Brominating reagents, 185

C

Carbodiimide, 55, 64, 65, 82, 101, 129, 191
Cannizzaro reduction, 181
Carbene, 213
Carborane, 227
Catalysts, 198, 220
 enzymatic, 243
 esterolytic, 206
 heterogeneous, 221
 homogeneous, 198, 220, 221
 hydrogenation, 225, 229
 phase-transfer, 19, 209, 212
 solid-phase, 198, 220
 transition metal, 221
 triphase, 166, 212
Catenanes, 152
Celite, 32, 176
Cell, immobilization, 242
Cellulose, 27, 100, 158, 240, 249
Chiral, 114, 156, 233
Chloromethyl groups, 18, 42, 55, 56, 63, 113, 184, 205, 209
Chlorinating reagents, 184
Chromatography, affinity, 246
 covalent, 247
 high pressure liquid, 100, 129
 ion exchange, 100, 158
 racemate resolution, 158
Chromium, reagents, 175
Claisen condensation, 142, 145
Clay, 32, 180, 194
Cleavage, nucleotide, 98
 oligosaccharide, 110
 peptide, 65
Colloids, 2
Conant-Swan reaction, 172

255

Condensation polymers, 4
Controlled-release materials, 242
Coupling reagents, nucleotide, 82, 98, 192, 193
  peptide, 64, 191
  saccharide, 112
Crown ethers, 62, 209
Cryptands, 161
C-terminal peptide sequencing, 133
Cyanoborohydride, 183
Cyclobutadiene, 166
Cyclopentadiene, 224, 229
Cyclooctadiene, 224
Cysteine, 250
Cystine, 251

D

Determination of reactive groups, 37
Degradation, Edman, 125
Desulfurizing agent, 195
Diacids, 149
Dialdehydes, 149
Dialkylation, 141
Diamines, 151
Diazotransfer, 191
Dicyclohexylcarbodiimide, 55, 64, 65, 82
Dieckmann cyclization, 49, 50, 143
Diels Alder reaction, 166
Diffuse reflectance electronic spectra, 229
Dihydrolipoic acid, 249
Dihydroxy compounds, 113, 146, 151
Dimerization, dienes, 233
  diketoesters, 145
Diols, 146
Dipyridine, see bipyridine
Disulfides, 195, 250
Dithiols, 151

E

Edman degradation, 125
EEDQ, 65, 193
Electrochemical applications, 241
Electrode, ion specific, 241
Electron-nuclear double resonance (ENDOR), 44
Electron-spin resonance (ESR), 45, 47, 229
Enzyme, immobilization, 242

Eosin-Y, 215
Error peptides, 67, 72
Esterolytic catalysts, 206

F

Failure sequences, 72, 100
Fetizon's reagent, 176
Fluorescamine, 68
Fluorescein, 215
Fluoride ion, 184
Fragment condensation, 74, 205
Friedel-Crafts reaction, 24, 205
Fries rearrangement, 153, 172
Functionalization, 18
  determination, 37

G

Glass, 27, 58, 70, 129, 132, 206, 240
Graphimets, see Graphite
Graphite, 32, 141, 176, 181, 184, 186
Grignard reaction/reagent, 151
Group transfer reagents, 183

H

Halogenating reagents, 178, 184, 188
Histidine-containing polymers, 160
Hooplanes, 152
Host–guest molecules, 162
Hydroformylation, 230
Hydrogenation, 229, 233
Hydrogen–deuterium exchange, 233
Hydroxamate ion, 208
Hydrosilylation, 232
Hyperentropic factor, 49, 144, 148

I

Imidazole, catalysts, 167, 206
Immobilization, enzymes, 242
  whole cells, 242
Indicators, pH, 32, 240
Infrared spectroscopy, 41, 49, 229
Insect pheromone, 148
Intersite reactions, 48, 72, 144, 225
Ion complexing materials, 222, 241
Ion-exchange resin, 189, 199
Isotactic polystyrene, 18, 85

Index

K

Kaolin, 159
Ketone, reduction, 181
　synthesis, 140, 176, 233
Kinetics, solid-phase, 9
Koenigs-Knorr, 112, 113, 177

L

Lalancette's reagent, see Graphite
Lanthanides, 220
Lewis acids, 205
Ligand-exchange chromatography, 159
Ligands, metal binding, 222, 241
Light-sensitive solid support, 66, 110
Lipoic acid, 249
Loading, determination, 38
　peptide sequencing, 129
　peptide synthesis, 62, 67
　reporting results, 90

M

Macroreticular resin, 17, 56, 94, 108, 129, 175
Mass spectrometry, 67
Meerwein-Pondorf-Verley reaction, 181
Merrifield, resin, 18, 56, 170, 214
　synthesis, 55, 117, 129
Methionine, 216, 247
　oxidation, 216
Michael reaction, 158
Microgels, 94
Micronet polymers, 160
Modification reactions, 18
Moffat oxidation, 192
Molecular sieves, see Zeolites
Monitoring, nucleotide synthesis, 100
　oligosaccharide synthesis, 114
　peptide synthesis, 67
Monofunctionalization, 146

N

Nafion-H, 206
N-bromosuccinimide, 185
N-carboxyanhydride, 27, 73
N-halopolyamides, 179, 185
Nuclear magnetic resonance (nmr), 44

Nucleophile, 168, 188, 206, 212, 233
Nucleoside, protection, 96, 100
Nucleotide, protection, 96

O

Olefin metathesis, 233
Oligonucleotide, synthesis, 81, 192
Oligopeptide, sequencing, 125
　synthesis, 53, 117, 191
Oligosaccharide synthesis, 105
Optical activity, 42, 155, 233
Oxidation-reduction, potential, 250
　reagents, 180, 249
Oxidizing reagents, 175, 213
Oxygen, singlet, 213

P

Palladium catalysts, 229
Pectic acid, 159
Peptide, sequencing,
　C-terminal, 133
　N-terminal, 125
　synthesis, 53, 117, 191
Peracids, 27, 175
pH indicators, 240
Pharmacological applications, 54, 245
Phase transfer catalysts, 19, 209, 212
Phenoplast resins, 1, 15
Phenylthiohydantoins, 125
Pheromones, insect, 148
Phosphine, ligands, 223, 225
Phosphine oxide, 183, 184
Phosphodiester, synthesis, 82
Phosphoric triamides, 213
Phosphorylation, 101, 169, 172
Phosphotriester, synthesis, 82
Photochemical, applications, 153, 213
Phthalein dyes, 215
Pinacolene rearrangement, 206
Platinum, 229, 241
Polyacrylamide, 26, 159, 249
Polyamide, 27, 62, 85, 179
Polyethylene glycol, 26, 85, 214
Polyethyleneimine (PEI), 27, 209
Polylysine, 27
Polymer, classification, 2
　functionization, 6, 18
　kinetics; polymer analogous reactions, 9

preparation, 4
properties, 5
synthesis, 4
Polynucleotide synthesis, 81, 192
Polypeptide synthesis, 53, 117, 191
Polysaccharide, synthesis, 105
Polysiloxane, 159
Polystyrene, 15, 45, 47, 56, 85, 108, 127, 129, 199, 222
Polystyrene–aluminium chloride, 205
Polystyryl boronic acid, 113
Polyvinyl alcohol, 95
Polyvinyl pyridine, 26, 176, 186
Polyvinylpyridinium chlorochromate, 176
Potassium permanganate, 32, 209
Prednisolene, 245
Proline, 160
Porphyrins, 215
Purification, nucleotide, 100
  peptide, 71

R

Racemates, resolution, 158
Racemization, peptide, 70
Radioactivity, 45, 50, 145
Redox reagents, 180
Reducing reagents, 181, 249
Resolution, amino acid, 159
Rhodium complexes, 229, 233, 244
Ribonucleotides, 100
Rose Bengal, 214
Ruthenium, 230

S

Saccharide synthesis, 105
Safety-catch principle, 66, 99
Selenium, 194
Sensitizer, photochemical, 213
Sensitox, 215
Sephadex, 27, 85, 159, 249
Sepharose, 27, 151, 249
Sequencing, peptides, 125
Silica, 32, 153, 159, 176, 180, 213, 223, 231
Silver carbonate-Celite, 176

Singlet oxygen, 213
Site separation, 48, 72, 144, 225
Sodium borohydride, 183, 251
Solvent effects, on polymer structure, 46
Spectroscopy, ultraviolet, 42
Starch, 159, 249
Sulfonyl chloride, 101
Sulfur-containing polymers, 249
Super acid catalysts, 205

T

Templates, asymmetric, 161
Thioanisole, 177
Thiol group, 227, 249
Tin dihydride, 181
Titanocene complexes, 225
Tosyl azides, 191
Transacylation, 167
Transesterification, 63, 65, 66, 204
Transition metals, 220
Triphase catalysts, 87, 212
Triphenylphosphine, 193
Tryptophan, 247

V

Valine polymer, 160

W

Wilkinson's catalyst, 225, 229
Wittig reagent, 151, 189

X

X-ray backscattering, 45, 229

Y

Ylide reagents, 189
Yneamine reagent, 195

Z

Zeolite, 180, 206